Managing Cover Crops Profitably

THIRD EDITION

Handbook Series Book 9

Published by the Sustainable Agriculture Research and Education (SARE) program, with funding from the National Institute of Food and Agriculture, U.S. Department of Agriculture

Second Printing - Jan. 2010

This book was published in 2007 by the Sustainable Agriculture Research and Education (SARE) program under cooperative agreements with the National Institute of Food and Agriculture, USDA, the University of Maryland and the University of Vermont.

SARE works to increase knowledge about—and help farmers and ranchers adopt—practices that are profitable, environmentally sound and good for communities. For more information about SARE grant opportunities and informational resources, go to www.sare.org. SARE Outreach is the national outreach arm of SARE. For more information, contact:

SARE Outreach
1122 Patapsco Building
University of Maryland
College Park, MD 20742-6715
(301) 405-8020
(301) 405-7711 (fax)
info@sare.org
www.sare.org

To order copies of this book, ($19.00 plus $5.95 s/h) contact (301) 374-9696, sarepubs@sare.org, or order online at www.sare.org/WebStore.

Project manager and editor: Andy Clark
Graphic design and layout: Diane Buric
Interior illustrations: Marianne Sarrantonio and Elayne Sears
Copy editing: Andy Clark
Proofreading: Aneeqa Chowdhury
Indexing: Claire Brannen
Printing: United Book Press, Inc.

Library of Congress Cataloging-in-Publication Data

Managing cover crops profitably / project manager and editor, Andy Clark.—3rd ed.
p. cm. – (Sustainable Agriculture Research and Education (SARE) program handbook series ; bk. 9)
Includes bibliographical references and index.
ISBN 978-1-888626-12-4 (pbk.)
1. Cover crops—United States—Handbooks, manuals, etc. I. Clark, Andy. II. Sustainable Agriculture Research & Education (SARE) program

SB284.3.U6M36 2007
631.5'82—dc22

2007024273

Cover photos (clockwise from top left):

Jeff Moyer, farm manager for The Rodale Institute, kills a hairy vetch cover crop with a newly designed, front-mounted roller while a no-till planter drops seed corn behind the tractor. Photo by Matthew Ryan for the Rodale Institute.

Annual ryegrass overseeded into kale is already providing cover crop benefits before cash crop harvest. Photo by Vern Grubinger, Univ. of VT.

Guihua Chen, a Univ. of MD graduate student, studies the ability of forage radish to alleviate soil compaction. Photo by Ray Weil, Univ. of MD.

A winter smother crop of yellow mustard minimizes weed growth in a vineyard. Photo by Jack Kelly Clark, Univ. of CA.

"Purple Bounty" hairy vetch, an early-maturing, winter hardy variety for the Northeast, was developed by Dr. Tom Devine, USDA-ARS in collaboration with The Rodale Institute, Pennsylvania State University and Cornell University Agricultural Experiment Stations. Photo by Greg Bowman, NewFarm.org.

Red clover, frostseeded into winter wheat, is well established just prior to wheat harvest. Photo by Steve Deming, MSU Kellogg Biological Station.

Back cover photo: Sorghum-sudangrass increased irrigated potato yield and tuber quality in Colorado, whether it was harvested for hay or incorporated prior to potato planting. Photo by Jorge A. Delgado, USDA-ARS.

Printed in the United States of America on recycled paper.

FOREWORD

Cover crops slow erosion, improve soil, smother weeds, enhance nutrient and moisture availability, help control many pests and bring a host of other benefits to your farm. At the same time, they can reduce costs, increase profits and even create new sources of income. You'll reap dividends on your cover crop investments for years, because their benefits accumulate over the long term.

Increasing energy costs will have a profound effect on farm economics in coming years. As we go to press, it is impossible to predict how fast energy costs will increase, but since cover crop economics are rooted in nitrogen dynamics (how much N you save or produce with cover crops), fuel costs (the cost of N and trips across the field) and commodity prices, energy prices will certainly impact the economics of cover crop use.

Economic comparisons in the 2nd edition were based on the old economy of two-dollar corn, twenty-cent nitrogen and cheap gas. Some studies showed that cover crops become more profitable as the price of nitrogen increases. We retained some of these excellent studies because data from new studies is not yet available. What we do know is that cover crops can help you to increase yields, save on nitrogen costs, reduce trips across the field and also reap many additional agronomic benefits.

There is a cover crop to fit just about every farming situation. The purpose of this book is to help you find which ones are right for you.

Farmers around the country are increasingly looking at the long-term contributions of cover crops to their whole farm system. Some of the most successful are those who have seen the benefits and are committed to making cover crops work for them. They are re-tooling their cropping systems to better fit cover crop growth patterns, rather than squeezing cover crops into their existing system, time permitting.

This 3rd edition of *Managing Cover Crops Profitably* aims to capture farmer and other research results from the past ten years. We verified the information from the 2nd edition, added new results and updated farmer profiles and research data throughout. We also added two new chapters.

Brassicas and Mustards (p. 81) lays out the current theory and management of cover crops in the BRASSICACEAE family. Brassica cover crops are thought to play a role in management of nematodes, weeds and disease by releasing chemical compounds from decomposing residue. Results are promising but inconsistent. Try brassicas on small plots and consult local expertise for additional information.

Managing Cover Crops in Conservation Tillage Systems (p. 44) addresses the management complexities of reduced tillage systems. If you are already using cover crops, the chapter will help you reduce tillage. If you are already using conservation tillage, it shows you how to add or better manage cover crops. Cover crops and conservation tillage team up to reduce energy use on your farm and that means more profits.

We have tried to include enough information for you to select and use cover crops appropriate to your operation. We recommend that you define your reasons for growing a cover crop—the section, *Selecting the Best Cover Crops for Your Farm* (p. 12) can help with this—and take as much care in selecting and managing cover crops as you would a cash crop.

Regional and site-specific factors can complicate cover crop management. No book can adequately address all the variables that make up a crop production system. Before planting a cover crop, learn as much as you can from this book and talk to others who are experienced with that cover crop.

We hope that this updated and expanded edition of *Managing Cover Crops Profitably* will lead to the successful use of cover crops on a wider scale as we continue to increase the sustainability of our farming systems.

Andy Clark, Outreach Coordinator
Sustainable Agriculture Research and Education (SARE)
June, 2007

MANAGING COVER CROPS PROFITABLY

THIRD EDITION

APPENDICES

ACKNOWLEDGMENTS

This 3rd edition could not have been written without the help of many cover crop experts. It is based in large part on the content of the 2nd edition, researched and written by Greg Bowman, Craig Cramer and Christopher Shirley. The following people reviewed the 2nd edition, suggested revisions and updates and contributed new content.

Aref Abdul-Baki, retired, USDA-ARS
Wesley Adams, Ladonia, TX
Kenneth A. Albrecht, Univ. of Wisconsin
Jess Alger, Stanford, MT
Robert G. Bailey, USDA Forest Service
Kipling Balkcom, USDA-ARS
Ronnie Barentine, Univ. of Georgia
Phil Bauer, USDA-ARS
R. Louis Baumhardt, USDA-ARS
Rich and Nancy Bennett, Napoleon, OH
Valerie Berton, SARE
Robert Blackshaw, Agriculture and Agri-Food Canada
Greg Bowman, NewFarm
Rick Boydston, USDA-ARS
Lois Braun, Univ. of Minnesota
Eric B. Brennan, USDA-ARS
Pat Carr, North Dakota State Univ.
Max Carter, Douglas, GA
Guihua Chen, Univ. of Maryland
Aneeqa Chowdhury, SARE
Hal Collins, USDA-ARS
Craig Cramer, Cornell Univ.
Nancy Creamer, North Carolina State Univ.
William S. Curran, The Pennsylvania State Univ.
Seth Dabney, USDA-ARS
Bryan Davis, Grinnell, IA
Jorge Delgado, USDA-ARS
Juan Carlos Diaz-Perez, Univ. of Georgia
Richard Dick, Ohio State Univ.
Sjoerd W. Duiker, The Pennsylvania State Univ.
Gerald W. Evers, Texas A&M Univ.
Rick Exner, Iowa State Univ. Extension
Richard Fasching, NRCS
Jim French, Partridge, KS
Eric Gallandt, Univ. of Maine
Helen Garst, SARE

Dale Gies, Moses Lake, WA
Bill Granzow, Herington, KS
Stephen Green, Arkansas State Univ.
Tim Griffin, USDA-ARS
Steve Groff, Holtwood, PA
Gary Guthrie, Nevada, IA
Matthew Harbur, Univ. of Minnesota
Timothy M. Harrigan, Michigan State Univ.
Andy Hart, Elgin, MN
Zane Helsel, Rutgers Univ.
Paul Hepperly, The Rodale Institute
Michelle Infante-Casella, Rutgers Univ.
Chuck Ingels, Univ. of California
Louise E. Jackson, Univ. of California
Peter Jeranyama, South Dakota State Univ.
Nan Johnson, Univ. of Mississippi
Hans Kandel, Univ. of Minnesota Extension
Tom Kaspar, USDA-ARS
Alina Kelman, SARE
Rose Koenig, Gainesville, FL
James Krall, Univ. of Wyoming
Amy Kremen, Univ. of Maryland
Roger Lansink, Odebolt, IA
Yvonne Lawley, Univ. of Maryland
Frank Lessiter, No-Till Farmer
John Luna, Oregon State Univ.
Barry Martin, Hawkinsville, GA
Todd Martin, MSU Kellogg Biological Station
Milt McGiffen, Univ. of California
Andy McGuire, Washington State Univ.
George McManus, Benton Harbor, MI
John J. Meisinger, USDA/ARS
Henry Miller, Constantin, MI
Jeffrey Mitchell, Univ. of California
Hassan Mojtahedi, USDA-ARS
Gaylon Morgan, Texas A&M Univ.
Matthew J. Morra, Univ. of Idaho
Vicki Morrone, Michigan State Univ.
Jeff Moyer, The Rodale Institute
Paul Mugge, Sutherland, IA
Dale Mutch, MSU Kellogg Biological Station
Rob Myers, Jefferson Institute
Lloyd Nelson, Texas Agric. Experiment Station
Mathieu Ngouajio, Michigan State Univ.
Eric and Anne Nordell, Trout Run, PA
Sharad Phatak, Univ. of Georgia
David Podoll, Fullerton, ND
Paul Porter, Univ. of Minnesota
Andrew Price, USDA-ARS
Ed Quigley, Spruce Creek, PA
RJ Rant, Grand Haven, MI
Bob Rawlins, Rebecca, GA
Wayne Reeves, USDA-ARS
Ekaterini Riga, Washington State Univ.
Lee Rinehart, ATTRA
Amanda Rodrigues, SARE
Ron Ross, No-Till Farmer
Marianne Sarrantonio, Univ. of Maine
Harry H. Schomberg, USDA-ARS
Pat Sheridan, Fairgrove, Mich.
Jeremy Singer, USDA-ARS
Richard Smith, Univ. of California
Sieglinde Snapp, Kellogg Biological Station
Lisa Stocking, Univ. of Maryland
James Stute, Univ. of Wisconsin Extension
Alan Sundermeier, Ohio State Univ. Extension
John Teasdale, USDA-ARS
Lee and Noreen Thomas, Moorhead, MN
Dick and Sharon Thompson, Boone, IA
Edzard van Santen, Auburn Univ.
Ray Weil, Univ. of Maryland
Charlie White, Univ. of Maryland
Dave Wilson, The Rodale Institute
David Wolfe, Cornell Univ.

HOW TO USE THIS BOOK

Think of this book as a tool chest, not a cookbook. You won't find the one simple recipe to meet your farming goals. You will find the tools to select and manage the best cover crops for the unique needs of your farm.

In this tool chest you will find helpful maps and charts, detailed narratives about individual cover crop species, chapters about specific aspects of cover cropping and extensive appendices that will lead you to even more information.

1. Start with *Top Regional Cover Crop Species* (p. 66). This chart will help you narrow your search by listing the benefits you can expect from the top cover crops adapted to your region. You'll discover which are the best nitrogen (N) sources, soil builders, erosion fighters, subsoil looseners, weed fighters and pest fighters.

2. Next, find out more about the performance and management of the cover crops that look like good candidates for your farm. You'll find two streams of information:

- **Charts** quickly provide you with details to help you compare cover crops. Performance and Roles (p. 67) lists ranges for N and dry matter production and ranks each cover crop's potential for providing 11 benefits. *Cultural Traits* (p. 69) and *Planting* (p. 70) explains the growth, environmental tolerances, seeding preferences and establishment costs for each crop.
- **Narratives.** The *Table of Contents* (p. 4) and the page numbers accompanying each species in Charts 2, 3 and 4 direct you to the heart of the book, the chapters on each cover crop. The chapters offer even more practical descriptions of how to plant, manage, kill and make the best use of each species. Don't overlook *Up-and-Coming Cover Crops* (p. 191) that briefly describes promising but lesser known cover crops. One of them may be right for your farm.

3. With some particular cover crops in mind, step back and look at the big picture of how you can fit cover crops into your farming operations. Sit down with a highlighter and explore these chapters:

- *Benefits of Cover Crops* (p. 9) explains important cover crop roles such as reducing costs, improving soil and managing pests.
- *Selecting the Best Cover Crops* (p. 12) helps you evaluate your operation's needs and niches (seasonal, cash-crop related, and profit potential). Several examples show how to fit crops to detailed situations.
- *Building Soil Fertility and Tilth* (p. 16) shows how cover crops add organic matter and greater productivity to the biological, chemical and physical components of soil.

SORGHUM-SUDANGRASS *is a tall, warm-season grass that stifles weeds and decomposes to build soil organic matter.*

• *Managing Pests with Cover Crops* (p. 25) explores how cover crops change field environments to protect cash crops from insects, disease, weeds and nematodes.

• **New this edition:** *Managing Cover Crops in Conservation Tillage Systems* (p. 44) provides management details for cover crops in reduced tillage systems.

• *Crop Rotations* (p. 34) explains how to integrate cover crops and cash crops in sequence from year to year for optimum productivity from on-farm resources.

• *Citations Bibliography* (p. 208) lists many of the publications and specialists cited in the book. Citations within the book are numbered in parentheses. Refer to the numbered citation in the bibliography if you want to dig deeper into a topic.

• *Climatic Zone Maps* inside the front and back covers help you understand differences in cover crop performance from location to location. You may find that some cover crops have performed well in tests far from where you farm, but under comparable climatic conditions.

The USDA **Plant Hardiness Zone Map** (inside front cover) shows whether a crop will survive the average winter in your area.We refer to the USDA hardiness zones throughout the book. **Readers' note:** A new version of this map is expected in 2008. See the online version of this book at www.sare.org.

The U.S. Forest Service map, **Ecoregions of the United States** (inside back cover), served in part as the basis for the adaptation maps included at the beginning of each cover crop chapter.This ecosystem map, while designed to classify forest growth, shows localized climate differences, such as rainfall and elevation, within a region. See Bailey (citation #17 in Appendix F, p. 209) for more information about ecoregions.

Cultivars of SUBTERRANEAN CLOVER, *a low-growing, reseeding annual legume, are adapted to many climates.*

4. Now that you've tried out most of the tools, revisit the charts and narratives to zero in on the cover crops you want to try. The *Appendices* include information to help you run reliable on-farm cover crop comparison trials.You'll also find contact information for cover crop experts in your region, seed and inoculant suppliers, references to books and academic papers cited in this book and websites with more cover crop information.

5. Finally, share your cover crop plans with farmers in your area who have experience with cover crops.Your local Extension staff, regional IPM specialist or a sustainable farming group in your area may be able to provide contacts. Be sure to tap local wisdom. You can find out the cover crop practices that have worked traditionally, and the new wrinkles or crops that innovative practitioners have discovered.

Abbreviations used in this book

A = acre or acres
bu. = bushel or bushels
DM = dry matter, or dry weight of plant material
F = (degrees) Fahrenheit in. = inch or inches
K = potassium lb. = pound or pounds
N = nitrogen
OM = organic matter
P = phosphorus
p. = page
pp. = pages
T = ton or tons
> = progression to another crop
/ = a mixture of crops growing together 

BENEFITS OF COVER CROPS

Cover crops can boost your profits the first year you plant them. They can improve your bottom line even more over the years as their soil-improving effects accumulate. Other benefits—reducing pollution, erosion and weed and insect pressure—may be difficult to quantify or may not appear in your financial statements. Identifying these benefits, however, can help you make sound, long-term decisions for your whole farm.

What follows are some important ways to evaluate the economic and ecological aspects of cover crops. These significant benefits (detailed below) vary by location and season, but at least two or three usually occur with any cover crop. Consult local farming groups and agencies with cover crop experience to figure more precise crop budgets.

- Cut fertilizer costs
- Reduce the need for herbicides and other pesticides
- Improve yields by enhancing soil health
- Prevent soil erosion
- Conserve soil moisture
- Protect water quality
- Help safeguard personal health

Evaluate a cover crop's impact as you would any other crop, balancing costs against returns in all forms. Don't limit your calculations, however, to the target cover crop benefit. A cover often has several benefits. Many cover crops offer harvest possibilities as forage, grazing or seed that work well in systems with multiple crop enterprises and livestock.

SPELLING IT OUT

Here's a quick overview of benefits you can grow on your farm. Cover crops can:

Cut fertilizer costs by contributing N to cash crops and by scavenging and mining soil nutrients.

RED CLOVER *is an annual or multi-year legume that improves topsoil. It is easily overseeded into standing crops or frostseeded into grains in early spring.*

Legume cover crops convert nitrogen gas in the atmosphere into soil nitrogen that plants can use. See *Nodulation: Match Inoculant to Maximize N* (p. 122). Crops grown in fields after legumes can take up at least 30 to 60 percent of the N that the legume produced. You can reduce N fertilizer applications accordingly. For more information on nitrogen dynamics and how to calculate fertilizer reductions, see *Building Soil Fertility and Tilth with Cover Crops* (p. 16). The N value of legumes is the easiest cover crop benefit to evaluate, both agronomically and economically. This natural fertility input alone can justify cover crop use.

- **Hairy vetch** boosted yield for no-till corn more than enough to cover its establishment costs, a three-year study in Maryland showed. Further, the vetch can reduce economic risk and usually will be more profitable than no-till corn after a winter wheat cover crop (1993 data). The result held true even if corn were priced as low as $1.80 per bushel, or N fertilizer ($0.30/lb.) was applied at the rate of 180 lb. N/A (173).
- **Medium red clover** companion seeded with oats and hairy vetch had estimated fertilizer replacement value of 65 to 103 lb. N/A in a four-year study in Wisconsin, based on a two year rotation of oats/legume > corn. Mean corn grain yield following these legumes was 163 bu./A for red

clover and 167 bu./A for hairy vetch, compared with a no legume/no N fertilizer yield of 134 bu./A (400).

• **Austrian winter peas, hairy vetch** and NITRO alfalfa can provide 80 to 100 percent of a subsequent potato crop's nitrogen requirement, a study in the Pacific Northwest showed (394).

• **Fibrous-rooted cereal grains** or grasses are particularly good at scavenging excess nutrients—especially N—left in the soil after cash crop harvest. Much of the N is held within the plants until they decompose. Fall-seeded grains or grasses can absorb up to 71 lb. N/A within three months of planting, a Maryland study showed (46). Addition of cover crops to corn>soybean and corn>peanut>cotton rotations and appropriate timing of fertilizer application usually reduce total N losses, without causing yield losses in subsequent crops, a USDA-ARS computer modeling study confirms (354).

Reduce the Need for Herbicides

Cover crops suppress weeds and reduce damage by diseases, insects and nematodes. Many cover crops effectively suppress weeds as:

- A smother crop that outcompetes weeds for water and nutrients
- Residue or growing leaf canopy that blocks light, alters the frequency of light waves and changes soil surface temperature
- A source of root exudates or compounds that provide natural herbicidal effects

Managing Pests with Cover Crops (p. 25) describes how cover crops can:

- Host beneficial microbial life that discourages disease
- Create an inhospitable soil environment for many soilborne diseases
- Encourage beneficial insect predators and parasitoids that can reduce insect damage below economic thresholds
- Produce compounds that reduce nematode pest populations
- Encourage beneficial nematode species

Using a rotation of malting barley>cover crop radish>sugar beets has successfully reduced sugar beet cyst nematodes to increase yield of sugar beets in a Wyoming test. Using this brassica cover crop after malting barley or silage corn substituted profitably for chemical nematicides when nematode levels were moderate (231). A corn>rye>soybeans> wheat>hairy vetch rotation that has reduced pesticide costs is at least as profitable as conventional grain rotations without cover crops, a study in southeastern Pennsylvania shows (174). Fall-planted brassica cover crops coupled with mechanical cultivation help potato growers with a long growing season maintain marketable yield and reduce herbicide applications by 25 percent or more, a study in the inland Pacific Northwest showed (394).

To estimate your potential N fertilizer savings from a cover crop, see the sidebar, *How Much N?* (p. 22).

Improve Yields by Enhancing Soil Health

Cover crops improve soil by:

- Speeding infiltration of excess surface water
- Relieving compaction and improving structure of overtilled soil
- Adding organic matter that encourages beneficial soil microbial life
- Enhancing nutrient cycling

Building Soil Fertility and Tilth with Cover Crops (p. 16) details the biological and chemical processes of how cover crops improve soil health and nutrient cycling. Leading soil-building crops include rye (residue adds organic matter and conserves moisture); sorghum-sudangrass (deep penetrating roots can break compaction); and ryegrass (stabilizes field roads, inter-row areas and borders when soil is wet).

Prevent Soil Erosion

Quick-growing cover crops hold soil in place, reduce crusting and protect against erosion due to wind and rain.The aboveground portion of covers also helps protect soil from the impact of raindrops. Long-term use of cover crops increases water infiltration and reduces runoff that can carry away soil.The key is to have enough stalk and leaf growth to guard against soil loss. Succulent legumes decompose quickly, especially in warm weather. Winter cereals and many brassicas have a better chance of overwintering in colder climates. These late-summer or fall-planted crops often put on significant growth even when temperatures drop into the 50s, and often are more winter-hardy than legumes (361). In a no-till cotton system, use of cover crops such as winter wheat, crimson clover and hairy vetch can reduce soil erosion while maintaining high cotton yields, a Mississippi study shows (35).

Conserve Soil Moisture

Residue from killed cover crops increases water infiltration and reduces evaporation, resulting in less moisture stress during drought. Lightly incorporated cover crops serve dual roles. They trap surface water and add organic matter to increase infiltration to the root zone. Especially effective at covering the soil surface are grass-type cover crops such as rye, wheat, and sorghum-sudangrass hybrid. Some water-efficient legumes such as medic and INDIANHEAD lentils provide cover crop benefits in dryland areas while conserving more moisture than conventional bare fallow (383). Timely spring termination of a cover crop avoids the negative impact of opposite water conditions: excess residue holding in too much moisture for planting in wet years, or living plants drawing too much moisture from the soil in dry years.

Protect Water Quality

By slowing erosion and runoff, cover crops reduce nonpoint source pollution caused by sediments, nutrients and agricultural chemicals. By taking up excess soil nitrogen, cover crops prevent N leaching to groundwater. Cover crops also provide habitat for wildlife. A rye cover crop scavenged from 25 to 100 percent of residual N from conventional and no-till Georgia corn fields, one study showed. Up to 180 lb. N/A had been applied. A barley cover crop removed 64 percent of soil nitrogen when applied N averaged 107 lb./A (220).

Help Safeguard Personal Health

By reducing reliance on agrichemicals for cash crop production, cover crops help protect the health of your family, neighbors and farm workers. They also help address community health and ecological concerns arising from nonpoint source pollution attributed to farming activities.

Cumulative Benefits

You can increase the range of benefits by increasing the diversity of cover crops grown, the frequency of use between cash crops and the length of time that cover crops are growing in the field.

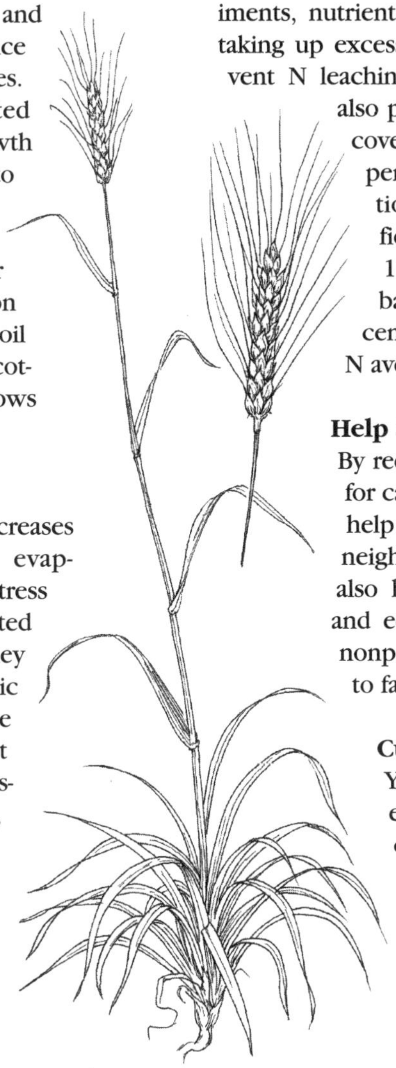

WINTER WHEAT *grows well in fall, then provides forage and protects soil over winter.*

SELECTING THE BEST COVER CROPS FOR YOUR FARM

by Marianne Sarrantonio

Cover crops provide many benefits, but they're not do-it-all "wonder crops." To find a suitable cover crop or mix of covers:

- **Clarify** your primary **needs**
- **Identify** the best **time** and **place** for a cover crop in your system
- **Test** a few options

This book makes selection of cover crops a little easier by focusing on some proven ones. Thousands of species and varieties exist, however. The steps that follow can help you find crops that will work best with a minimum of risk and expense.

1. Identify Your Problem or Use

Review *Benefits of Cover Crops* (p. 9) to decide what you want most from a cover crop. Narrowing your goals to one or two primary and perhaps a few secondary goals will greatly simplify your search for the best cover species. Some common goals for cover crops are to:

- Provide nitrogen
- Add organic matter
- Improve soil structure
- Reduce soil erosion
- Provide weed control
- Manage nutrients
- Furnish moisture-conserving mulch

You might also want the cover crops to provide habitat for beneficial organisms, better traction during harvest, faster drainage or another benefit.

2. Identify the Best Place and Time

Sometimes it's obvious where and when to use a cover crop. You might want some nitrogen before a corn crop, or a perennial ground cover in a vineyard or orchard to reduce erosion or improve weed control. For some goals, such as building soil, it may be hard to decide where and when to schedule cover crops.

To plan how and where to use cover crops, try the following exercise:

Look at your rotation. Make a timeline of 18 to 36 monthly increments across a piece of paper. For each field, pencil in current or probable rotations, showing when you typically seed crops and when you harvest them.

If possible, add other key information, such as rainfall, frost-free periods and times of heavy labor or equipment demand.

Look for open periods in each field that correspond to good conditions for cover crop establishment, underutilized spaces on your farm, as well as opportunities in your seasonal work schedule. Also consider ways to extend or overlap cropping windows.

Here are examples of common niches in some systems, and some tips:

Winter fallow niche. In many regions, seed winter covers at least six weeks before a hard frost. Winter cereals, especially rye, are an exception and can be planted a little later. If ground cover and N recycling needs are minimal, rye can be planted as late as the frost period for successful overwintering.

You might seed a cover right after harvesting a summer crop, when the weather is still mild. In cooler climates, consider extending the window by **overseeding** (some call this **undersowing**) a shade-tolerant cover before cash crop harvest. White clover, annual ryegrass, rye, hairy vetch, crimson clover, red clover and sweetclover tolerate some shading.

If overseeding, irrigate afterwards if possible, or seed just before a soaking rain is forecast. Species with small seeds, such as clovers, don't need a lot of moisture to germinate and can work their way through tiny gaps in residue, but larger-seeded species need several days of moist conditions to germinate.

When overseeding into cash crops early in the season, vigorous growth of the cover crop may cause water stress, increase disease risks due to lower air circulation or create new insect pest risks. Changing cover crop seeding rate, seeding time, or the rotation sequence may lessen this risk. To ensure adequate sunlight for the cover crop, overseed before full canopy closure of the primary crop (at last cultivation of field corn, for example) or just before the canopy starts to open again as the cash crop starts to die (as soybean leaves turn yellow, for example).

Expect excessive field traffic around harvest time? Choose tough, low-growing covers such as grasses or clovers. Limit foot traffic to alternate rows, or delay a field operation to allow for cover crop establishment.

Another option could be to use a reseeding winter annual that dies back and drops seed each summer but reestablishes in fall. Subclovers reseed well in regions south of Hardiness Zone 6. Shorter-season crimson clovers—especially varieties with a high hard-seed percentage that germinate over an extended period—work well in the Southeast where moisture is sufficient. Even rye and vetch can reseed if managed properly.

Summer fallow niche. Many vegetable rotations present cover crop opportunities—and challenges. When double cropping, you might have fields with a three- to eight-week summer fallow period between early planted and late planted crops. Quick-growing summer annuals provide erosion control, weed management, organic matter and perhaps some N.

Consider overseeding a spring crop with a quick-growing summer grain such as buckwheat, millet or sorghum-sudangrass, or a warm-season legume such as cowpeas. Or, you might till out strips in the cover crop for planting a fall vegetable crop and control the remaining cover between the crop rows with mowing or light cultivation.

Small grain rotation niche. Companion seed a winter annual cover crop with a spring grain, or frost seed (broadcasting seed onto frozen ground) a cover into winter grains. Soil freezing and thawing pulls seed into the soil and helps germination. Another option if soil moisture isn't a limiting factor in your region: broadcast a cover before the grain enters boot stage (when seedheads start elongating) later in spring or plant after harvest.

Look for open periods in each field or open spaces on your farm.

Full-year improved fallow niche. To rebuild fertility or organic matter over a longer period, perennials or biennials—or mixtures—require the least amount of maintenance. Spring-seeded yellow blossom sweetclover flowers the following summer, has a deep taproot and gives plenty of aboveground biomass. Also consider perennial forages recommended for your area. The belowground benefit of a tap rooted perennial can have tremendous soil improving benefits when allowed to grow for several years.

Another option is sequential cover cropping. Plant hairy vetch or a grass-legume mixture in fall, terminate it the following spring at flowering, and plant sorghum-sudangrass. The winter cover crop provides weed suppression and ground cover, but also nitrogen for the high-N sorghum-sudangrass, which can produce tons of biomass to build soil organic matter.

Properly managed, **living mulches** give many growers year-round erosion protection, weed control, nutrient cycling and even some nitrogen if they include a legume. Some tillage, mowing or herbicides can help manage the mulch (to keep it from using too much soil moisture, for example) before crops are strip-tilled into the cover or residue. White clover could be a good choice for sweet corn and tomatoes. Perennial ryegrass or some less aggressive turfgrasses such as sheep fescue may work for beans, tomatoes and other vegetables.

Create new opportunities. Have you honed a rotation that seems to have few open time slots? Plant a cover in strips the width of a bed or wider, alternating with your annual vegetable, herb or field crop. Switch the strips the next year. Mow the strips periodically and blow the topgrowth onto adjoining cash crops as mulch. In a bed system, rotate out every third or fourth bed for a soil-building cover crop.

Another option: Band a cover or some insect-attracting shrubs around fields or along hedgerows to suppress weeds or provide beneficial habitat where you can't grow cash crops. These hedgerows could also be used to produce marketable products such as nuts, berries or even craft materials.

3. Describe the Niche

Refer to your timeline chart and ask questions such as:

- How will I seed the cover?
- What's the weather likely to be then?
- What will soil temperature and moisture conditions be like?
- How vigorous will other crops (or pests) be?
- Should the cover be low-growing and spreading, or tall and vigorous?
- What weather extremes and field traffic must it tolerate?
- Will it winterkill in my area?
- Should it winterkill, to meet my goals?
- What kind of regrowth can I expect?
- How do I kill it and plant into it?
- Will I have the time to make this work?
- What's my contingency plan—and risks—if the crop doesn't establish or doesn't die on schedule?
- Do I have the needed equipment and labor?

4. Select the Best Cover Crop

You have identified a goal, a time and a place, now specify the traits a cover crop would need to work well.

Example 1. A sloping **orchard** needs a ground cover to **reduce erosion**. You'd like it to **contribute N** and **organic matter** and **attract beneficial organisms** but not rodents, nematodes or other pests. The cover **can't use too much water** or **tie up nutrients** at key periods. Too much N might stimulate excessive tree leaf growth or prevent hardening off before winter. Finally you want a cover crop that is easy to maintain. It should:

- be a perennial or reseeding annual
- be low-growing, needing minimal management
- use water efficiently
- have a soil-improving root system
- release some nutrients during the year, but not too much N
- not harbor or attract pests

For this orchard scenario, white clover is probably the best option north of Zone 8. A mixture of low-growing legumes or a legume and grass mix could also work. In warm regions, low-growing clovers such as strawberry clover and white clover work well together, although these species may attract pocket gophers. BLANDO brome and annual ryegrass are two quick-growing, reseeding grasses often suitable for orchard floors, but they will probably need some control with mowing. Or, try a reseeding winter annual legume such as crimson clover, rose clover, subclover, an annual vetch or an annual medic, depending on your climate.

Example 2. A **dairy** lacks adequate storage in fall and winter for the **manure** it generates, which **exceeds the nutrient needs** for its silage corn and grass/legume hay rotation. The cover crop needs to:

- establish effectively after (or tolerate) silage corn harvest
- take up a lot of N and P from fall-applied manure and hold it until spring

For this dairy scenario, rye is usually the best choice. Other cereal grains or brassicas could work if planted early enough.

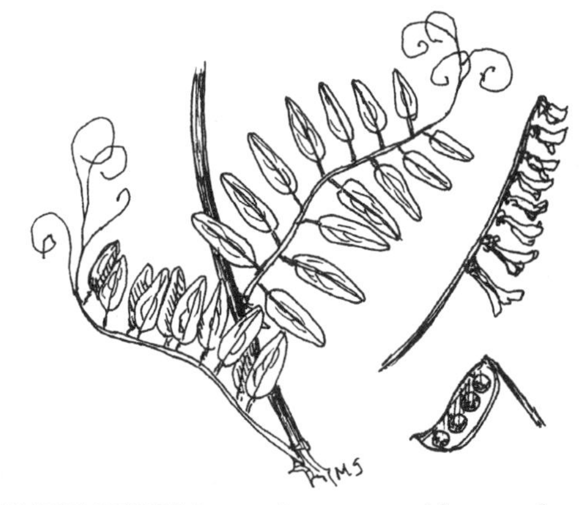

HAIRY VETCH *is an winter annual legume that grows slowly in fall, then fixes a lot of N in spring.*

Example 3. In a moderate rainfall region **after small grain harvest** in late summer, you want **a soil-protecting winter cover** that can **supply N** for no-till **corn** next spring. You want to kill the cover without herbicides. You need a legume that:

- can be drilled in late summer and put on a lot of fall growth
- will overwinter
- will fix a lot of N
- can be mow-killed shortly before (or after) corn planting
- could provide some weed-controlling, moisture-conserving residue

Hairy vetch works well in the Northeast, Midwest and parts of the mid-South. Mixing it with rye or another cereal improves its weed-management and moisture-conservation potential. Crimson clover may be an appropriate choice for the southeastern Piedmont. Austrian winter pea could be considered, alone or in a mix, in coastal plain environments, but will winterkill in Zone 7 and below. Where grain harvest occurs in late spring or early summer, LANA woollypod vetch might be a better choice.

Example 4. After a **spring broccoli** crop, you need a **weed-suppressing cover that adds N** and **organic matter**, and perhaps **mulch**, into which you will **no-till** seed fall lettuce or spinach. You want a cover that:

- is very versatile
- grows fast in hot weather
- can be overseeded into broccoli
- germinates on the soil surface under dry conditions
- fixes N
- persists until you're ready to kill it

Here, a quick-growing, warm-season legume such as cowpeas may work, especially if you can irrigate to hasten establishment during dry conditions.

5. Settle for the Best Available Cover. It's likely the "wonder crop" you want doesn't exist. One or more species could come close, as the above examples indicate. *Top Regional Cover Crop Species* (p. 66) can provide a starting point. Check

WINTER (cereal) RYE *is an annual grain that prevents soil and wind erosion. Its killed vegetation suppresses weeds for no-till planting.*

with regional experts. Keep in mind that you can mix two or more species, or try several options in small areas.

6. Or Build a Rotation Around Cover Crops. It's hard to decide in advance every field's crops, planting dates, fieldwork or management specifics. One alternative is to find out which cover crops provide the best results on your farm, then build a rotation around those covers, especially when trying to tackle some tough soil improvement or weed control issues. See *Full-Year Covers Tackle Tough Weeds* (p. 38).

With this "reverse" strategy, you plan covers according to their optimum field timing, and then determine the best windows for cash crops. A cover crop's strengths help you decide which cash crops would benefit the most.

For now, however, you probably want to fit one or more cover crops into your existing rotations. The charts and narratives in this book can help you select some of the most suitable species for your farming system and objectives. See *Crop Rotations with Cover Crops* (p. 34) to get you thinking more. When you've narrowed your choices, refer to Appendix A, *Testing Cover Crops on Your Farm* (p. 189) for some straightforward tips on what to do next.

Adapted from Northeast Cover Crop Handbook *by Marianne Sarrantonio, Rodale Institute, 1994.*

BUILDING SOIL FERTILITY AND TILTH WITH COVER CROPS

by Marianne Sarrantonio

Soil is an incredibly complex substance. It has physical and chemical properties that allow it to sustain living organisms—not just plant roots and earthworms, but hundreds of thousands of different insects, wormlike creatures and microorganisms. When these organisms are in balance, your soil cycles nutrients efficiently, stores water and drains the excess, and maintains an environment in which plants can thrive.

To recognize that a soil can be healthy, one has only to think of the soil as a living entity. It breathes, it transports and transforms nutrients, it interacts with its environment, and it can even purify itself and grow over time. If you view soil as a dynamic part of your farming system, unsustainable crop management practices amount to soil neglect. That neglect could worsen as the soil sickens and loses its life functions one by one.

Regardless of how healthy or alive your soil is right now, cover crops can play a vital role in ensuring that your soil provides a strong foundation for your farming system. While the most common reasons for including cover crops in a farming system may relate to the immediate short-term need, the continued practice of cover cropping becomes an investment in building healthy soil over the long term.

Cover crops improve soil in a number of ways. Protection against soil loss from erosion is perhaps the most obvious soil benefit of cover crops, but providing organic matter is a more long-term and equally important goal. Cover crops contribute indirectly to overall soil health by catching nutrients before they can leach out of the soil profile or, in the case of legumes, by adding nitrogen to the soil. Their roots can even help unlock some nutrients, converting them to more available forms. Cover crops provide habitat or a food source for some important soil organisms, break up compacted layers in the soil and help dry out wet soils.

EROSION PROTECTION

Erosion of topsoil occurs on many farms, depriving fields of the most fertile portion, that containing the highest percentage of organic matter and nutrients. Cover crops can play a major role in fighting soil erosion.

A raindrop falling at high speed can dislodge soil particles and cause them to move as far as 6 feet (42). Once a soil particle is loose, it is much more vulnerable to being carried away by running water. Any aboveground soil cover can take some of the punch out of a heavy rainfall simply by acting as a cushion for raindrops.

A cover crop also can:

- Slow the action of moving water, thus reducing its soil-carrying capacity, by creating an obstacle course of leaves, stems and roots through which the water must maneuver on its way downhill
- Increase the soil's ability to absorb and hold water, through improvement in pore structure, thereby preventing large quantities of water from moving across the soil surface
- Help stabilize soil particles in the cover crop root system

The reduction in soil erosion due to cover cropping will be roughly proportional to the amount of cover on the soil. The Revised Universal Soil Loss Equation developed by the Natural Resources Conservation Service predicts that a soil cover of just 40 percent when winter arrives can reduce erosion substantially until spring.

It's worthwhile to get covers established early, to ensure that maximum soil cover develops before winter rains. Consider overseeding covers at layby cultivation, aerial seeding or hand spreading before harvest, or planting as soon as possible after harvest. It's always a good idea to maintain year-round soil cover whenever possible.

ORGANIC MATTER ADDITIONS

The benefits of organic matter include improved **soil structure**, increased **infiltration and water-holding capacity**, increased **cation exchange capacity** (the ability of the soil to act as a short-term storage bank for positively charged plant nutrients) and more **efficient long-term storage** of nutrients. Without organic matter, you have no soil to speak of, only a dead mixture of ground-up and weathered rocks.

Organic matter includes thousands of different substances derived from decayed leaves, roots, microorganisms, manure and even groundhogs that died in their burrows. These substances function in different ways to build healthy soil. Different plants leave behind different kinds of organic matter as they decompose, so your choice of cover crop will largely determine which soil benefits you will receive.

Soil scientists may argue over how to classify the various soil organic components. Most will agree, however, that there is a portion that can be called the **"active" fraction**, and one that might be called the **"stable" fraction**, which is roughly equivalent to **humus**. There are many categories in between the active and stable fractions.

The active fraction represents the most easily decomposed parts of soil organic matter. It tends to be rich in simple sugars and proteins and consists largely of recently added fresh residues, microbial cells and the simpler waste products from microbial decay.

Because microorganisms, like human organisms, crave sweet stuff, compounds containing simple sugars disappear quickly. Proteins also are selected quickly from the menu of edible soil goodies. When these compounds are digested, many of the nutrients that they contain are released into the soil. Proteins are nitrogen-rich, so *the **active fraction** is responsible for the release of most N, as well as some K, P and other nutrients, from organic matter into the soil.* The easily decomposed proteins and sugars burn up almost completely as energy sources, and don't leave much behind to contribute to organic matter building.

After the microorganisms have devoured the portions of the active fraction that are easiest to digest, a more dedicated subset of these microorganisms will start munching on the more complex and tough material, such as celluloses and lignins, the structural materials of plants. Since cellulose is tougher than simple sugars, and lignin breaks down very slowly, they contribute more to the humus or stable fraction. Humus is responsible for giving the soil that rich, dark, spongy feeling and for properties such as water retention and cation exchange capacity.

Plant materials that are succulent and rich in proteins and sugars will release nutrients rapidly but leave behind little long-term organic matter. **Plant materials that are woodier or more fibrous** will release nutrients much more slowly, perhaps even tie up nutrients temporarily (see *Tillage, No Tillage and N Cycling*, p. 21), but will promote more stable organic matter, or humus, leading to better soil physical conditions, increased nutrient-holding capacity and higher cation exchange capacity.

In general, annual legumes are succulent. They release nitrogen and other nutrients quickly through the active fraction, but are not very effective at building up humus. Long-term use of annual legumes *can* increase soil humus, however, some research suggests (429).

Grains and other grasses and nonlegumes will contribute to humus production, but won't release nutrients very rapidly or in large quantities if incorporated as they approach maturity. Perennial legumes such as white and red clover may fall in both categories—their leaves will break down quickly, but their stems and root systems may become tough and fibrous and can contribute to humus accumulation.

Cover Crops Help "Glue" Soil

As soil microorganisms digest plant material, they produce some compounds in addition to the active and stable fractions of the organic matter. One group of these by-products is known as **polysaccharides**. These are complex sugars that act as glues in the soil to cement small soil particles into clusters or **aggregates**. Many farmers use the term "**crumb**" to describe soil clusters about the size of a grain of rice. A well-aggregated or "crumby" soil—not to be confused with crummy or depleted soil—has good aeration. It allows better infiltration and retention of water.

Cover crops can promote good aggregation in the soil through increased production of these and other microbial glues. See *Cover Crops Can Stabilize Your Soil* (p. 19). Well-aggregated soils also are less prone to compaction, which has been shown to reduce yields of vegetables such as snap beans, cabbage and cucumber by 50 percent or more (451).

As they decompose, leguminous cover crops seem to be better than grasses for production of polysaccharides (9). However, polysaccharides will decompose in a matter of months, so their aggregation effect is likely to last only the season after the use of the cover crop.

Grass species also promote good aggregation, but by a different mechanism. Grasses have a 'fibrous' root system—made of numerous fine roots spreading out from the base of the plant. These roots may release compounds that help aggregate the soil between roots.

Organic matter builds up very slowly in the soil. A soil with 3 percent organic matter might only increase to 4 percent after a decade or more of soil building. The benefits of increased organic matter, however, are likely to be apparent long before increased quantities are detectable. Some, such as enhanced aggregation, water infiltration rates and nutrient release, will be apparent the first season; others may take several years to become noticeable (429).

Your tillage method is an important consideration when using cover crops to build soil, because tillage will affect the rate of organic matter accumulation. *It is difficult to build up organic matter under conventional tillage regimes*. Tillage speeds up organic matter decomposition by exposing more surface area to oxygen, warming and drying the soil, and breaking residue into smaller pieces with more surfaces that can be attacked by decomposers. Like fanning a fire, tillage rapidly "burns up" or "oxidizes" the fuel, which in this case is organic matter. The resulting loss of organic matter causes the breakdown of soil aggregates and the poor soil structure often seen in overtilled soil.

When adding cover crops to a system, minimize tillage to maximize the long-term soil benefits. Many of the cover crops discussed in this book are ones you can seed into growing crops or no-till plant into crop residues. Otherwise, the gain in organic matter may be counteracted by higher decomposition rates.

TIGHTENING THE NUTRIENT LOOP

In addition to reducing topsoil erosion and improving soil structure, cover crops enhance nutrient cycling in your farming system by taking up nutrients that otherwise might leach out of the soil profile. These excess nutrients have the potential to pollute groundwater or local streams and ponds, not to mention impoverishing the soil they came from.

Of the common plant nutrients, nitrogen in the nitrate form is the most water-soluble and therefore the most vulnerable to leaching. Anytime soil is bare and appreciable rain falls, nitrates are on the move. Nitrate can be present in the soil at the end of a cropping season if the crop did not use all the N applied. Decomposing organic matter (including plant residues, compost and animal manures) also can supply nitrate-N, as long as the soil temperature is above freezing. Even in a field where the yearly application of N is well-suited to crop needs, nitrates can accumulate after crops are harvested and leach when it rains.

Cover crops reduce nitrate leaching in two ways. They soak up available nitrate for their own needs. They also use some soil moisture, reducing the amount of water available to leach nutrients.

The best cover crops to use for nitrate conservation are nonlegumes that form deep, extensive root systems quickly after cash crops are harvested. For much of the continental U.S., cereal rye is the best choice for catching nutrients after a summer crop. Its cold tolerance is a big advantage that allows rye to continue to grow in late fall and put down roots to a depth of three feet or more. Where winters are mild, rye can grow through the winter months.

Cover Crops Can Stabilize Your Soil

The more you use cover crops, the better your soil tilth, research continues to show. One reason is that cover crops, especially legumes, encourage populations of beneficial fungi and other microorganisms that help bind soil aggregates.

The fungi, called mycorrhizae, produce a water-insoluble protein known as **glomalin**, which catches and glues together particles of organic matter, plant cells, bacteria and other fungi (453). Glomalin may be one of the most important substances in promoting and stabilizing soil aggregates.

Most plant roots, not just those of cover crops, develop beneficial mycorrhizal relationships. The fungi send out rootlike extensions called **hyphae**, which take up water and soil nutrients to help feed plants. In low-phosphorus soils, for example, the hyphae can increase the amount of phosphorus that plants obtain. In return, the fungi receive energy in the form of sugars that plants produce in their leaves and send down to the roots.

Growing a cover crop increases the abundance of mycorrhizal spores. Legumes in particular can contribute to mycorrhizal diversity and abundance, because their roots tend to develop large populations of these beneficial fungi.

By having their own mycorrhizal fungi and by promoting mycorrhizal relationships in subsequent crops, cover crops therefore can play a key role in improving soil tilth. The overall increase in glomalin production also could help explain why cover crops can improve water infiltration into soil and enhance storage of water and soil nutrients, even when there has been no detectable increase in the amount of soil organic matter.

Research with soil high in residual N in the mid-Atlantic's coastal plain showed that cereal rye took up more than 70 lb. N/A in fall when planted by October 1. Other grasses, including wheat, oats, barley and ryegrass, were only able to take up about half that amount in fall. Legumes were practically useless for this purpose in the Chesapeake Bay study (46). Legumes tend to establish slowly in fall and are mediocre N scavengers, as they can fix much of their own N.

To maximize N uptake and prevent leaching, plant nonlegumes as early as possible. In the above study, rye took up only 15 lb. N/A when planting was delayed until November. It is important to give cover crops the same respect as any other crop in the rotation and plant them in a timely manner.

Not Just Nitrogen Cycling

Cover crops help bring other nutrients back into the upper soil profile from deep soil layers. Calcium and potassium are two macronutrients with a tendency to travel with water, though not generally on the express route with N. These nutrients can be brought up from deeper soil layers by any deep-rooted cover crop. The nutrients are then released back into the active organic matter when the cover crop dies and decomposes.

Although phosphorus (P) doesn't generally leach, as it is only slightly water-soluble, cover crops may play a role in increasing its availability in the soil. Some covers, such as buckwheat and lupins, are thought to secrete acids into the soil that put P into a more soluble, plant-usable form.

Some cover crops enhance P availability in another manner. The roots of many common cover crops, particularly legumes, house beneficial fungi known as **mycorrhizae**. The mycorrhizal fungi have evolved efficient means of absorbing P from the soil, which they pass on to their plant host. The filaments (hyphae) of these fungi effectively extend the root system and help the plants tap more soil P.

Keeping phosphorus in an **organic** form is the most efficient way to keep it cycling in the soil. So the return of any plant or animal residue to the soil helps maintain P availability. Cover crops also help retain P in your fields by reducing erosion.

Adding Nitrogen

One of nature's most gracious gifts to plants and soil is the way that legumes, with the help of rhizobial bacteria, can add N to enrich your soil. If you are not familiar with how this remarkable process works, see *Nodulation: Match Inoculant to Maximize N* (p. 122).

The nitrogen provided by N-fixation is used efficiently in natural ecosystems, thanks to the soil's complex web of interacting physical, chemical and biological processes. In an agricultural system, however, soil and crop management factors often interfere with nature's ultra-efficient use of organic or inorganic N. Learning a bit about the factors affecting N-use efficiency from legume plants will help build the most sustainable cropping system possible within your constraints.

How Much N is Fixed?

A number of factors determine how much of the N in your legume came from "free" N, fixed from N2 gas:

• Is the **symbiosis** (the interdependence of the rhizobia and the plant roots) effective? See *Nodulation: Match Inoculant to Maximize N* (p. 122). Use the correct rhizobial inoculant for the legume you're growing. Make sure it's fresh, was stored properly, and that you apply it with an effective sticking agent. Otherwise, there will be few nodules and N-fixation will be low.

• Is the soil fertile? N-fixation requires molybdenum, iron, potassium, sulfur and zinc to function properly. Soils depleted of these micronutrients will not support efficient fixation. Tissue testing your cash crops can help you decide if you need to adjust micronutrient levels.

• Is the soil getting enough air? N-fixation requires that N-rich air get to the legume roots. Waterlogging or compaction hampers the movement of air into the soil. Deep-rooted cover crops can help alleviate subsoil compaction (451).

• Is the pH adequate? Rhizobia generally will not live long in soils below pH 5.

• Does the legume/rhizobial pair have high fixation potential? Not all legumes were created equal—some are genetically challenged when it comes to fixation. Beans (*Phaseolus* spp.) are notoriously incapable of a good symbiotic relationship and are rarely able to fix much more than 40 lb. N/A in a whole season. Cowpeas (*Vigna unguiculata*) and vetches (*Vicia* spp.), on the other hand, are generally capable of high fixation rates. Check Chart 2 *Performances and Roles* (p. 67) and the sections on individual cover crops for information about their N-fixation potential.

Even under the best of conditions, legumes rarely fix more than 80 percent of the nitrogen they need to grow, and may only fix as much as 40 or 50 percent. The legume removes the rest of what it needs from the soil like any other plant. Legumes have to feed the bacteria to get them to work, so if there is ample nitrate already available in the soil, a legume will remove much of that first before expending the energy to get N-fixation going. In soils with high N fertility, legumes may fix little or no nitrogen. See *How Much N?* (p. 22).

While it is tempting to think of legume nodules as little fertilizer factories pumping N into the surrounding soil, that isn't what happens. The fixed N is almost immediately shunted up into the stems and leaves of the growing legume to form proteins, chlorophyll and other N-containing compounds. The fixed nitrogen will not become available to the next crop until the legume decomposes. Consequently, if the aboveground part of the legume is removed for hay, the majority of the fixed nitrogen also leaves the field.

What about the legume roots? Under conditions favoring optimal N fixation, a good rule of thumb is to think of the nitrogen left in the plant roots (15 to 30 percent of plant N) as being roughly equivalent to the amount the legume removed directly from the soil, and the amount in the stems and leaves as being equivalent to what was fixed.

Annual legumes that are allowed to flower and mature will transport a large portion of their biomass nitrogen into the seeds or beans. Also, once the legume has stopped actively growing, it will shut down the N-fixing symbiosis. In annual legumes this occurs at the time of flowering; no additional N gain will occur after that point. Unless you want a legume to reseed itself, it's generally a good idea to kill a legume cover crop in the early- to mid-blossom stage. You'll have obtained maximum legume N and need not delay planting of the following cash crop any further, aside from any period you may want for residue decomposition as part of your seedbed preparation.

How Nitrogen is Released

How much N will soil really acquire from a legume cover crop? Let's take it from the point of a freshly killed, annual legume, cut down in its prime at mid-bloom. The management and climatic events following the death of that legume will greatly affect the amount and timing of N release from the legume to the soil.

Most soil bacteria will feast on and rapidly decompose green manures such as annual legumes, which contain many simple sugars and proteins as energy sources. Soil bacteria love to party and when there is lots to eat, they do something that no party guest you've ever invited can do—they reproduce themselves, rapidly and repeatedly, doubling their population in as little as seven days under field conditions (306). Even a relatively inactive soil can come to life quickly with addition of a delectable green manure.

The result can be a very rapid and large release of nitrate into the soil within a week of the green manure's demise. This N release is more rapid when covers are plowed down than when left on the surface. As much as 140 lb. N/A has been measured 7 to 10 days after plowdown of hairy vetch (363). Green manures that are less protein-rich (N-rich) will take longer to release N. Those that are old and fibrous or woody are generally left for hard-working but somewhat sluggish fungi to convert slowly to humus over the years, gradually releasing small amounts of nutrients.

Other factors contribute significantly to how quickly a green manure releases its N. Weather has a huge influence. The soil organisms responsible for decomposition work best at warm temperatures and are less energetic during cool spring months.

Soil moisture also has a dramatic effect. Research shows that soil microbial activity peaks when 60 percent of the soil pores are filled with water, and declines significantly when moisture levels are higher or lower (244). This 60 percent water-filled pore space roughly corresponds to **field capacity**, or the amount of water left in the soil when it is allowed to drain for 24 hours after a good soaking rain.

Microbes are sensitive to soil chemistry as well. Most soil bacteria need a pH of between 6 and 8 to perform at peak; fungi (the slow decomposers) are still active at very low pH. Soil microorganisms also need most of the same nutrients that plants require, so low-fertility soils support smaller populations of primary decomposers, compared with high-fertility soils. Don't expect N-release rates or fertilizer replacement values for a given cover crop to be identical in fields of different fertility.

Many of these environmental factors are out of your direct control in the near term. Management factors such as fertilization, liming and tillage, however, also influence production and availability of legume N.

Tillage, No-Tillage and N-Cycling

Tillage affects decomposition of plant residues in a number of ways. First, any tillage increases soil contact with residues and increases the microbes' access to them. The plow layer is a hospitable environment for microbes, as they're sheltered from extremes of temperature and moisture. Second, tillage breaks the residue into smaller pieces, providing more edges for microbes to munch. Third, tillage will temporarily decrease the density of the soil, generally allowing it to drain and therefore warm up more quickly. All told, residues incorporated into the soil tend to decompose and release nutrients much faster than those left on the surface, as in a no-till system. That's not necessarily good news, however.

A real challenge of farming efficiently is to keep as much of the N as possible in a stable, storable form until it's needed by the crop. The best storage form of N is the organic form: the undecomposed residue, the humus or the microorganisms themselves.

Let's consider the N contained in the microbes. Nitrogen is a nutrient the microbes need for building proteins and other compounds. Carbon-containing compounds such as sugars are mainly

How Much N?

To find out if you might need more N than your green manure will supply, you need to estimate the amount of N in your cover crop. To do this, assess the total yield of the green manure and the percentage of N in the plants just before they die.

To estimate yield, take cuttings from several areas in the field, dry and weigh them. Use a yardstick or metal frame of known dimensions (1 ft. x 2 ft., which equals 2 ft^2 works well) and clip the plants at ground level within the known area. Dry them out in the sun for a few consecutive days, or use an oven at about 140° F for 24 to 48 hours until they are "crunchy dry." Use the following equation to determine per-acre yield of dry matter:

$$\text{Yield (lb.)/Acre} = \frac{\text{Total weight of dried samples (lb.)}}{\text{\# square feet you sampled}} \times \frac{\text{43,560 sq. ft.}}{\text{1 Acre}}$$

While actually sampling is more accurate, you can estimate your yield from the height of your green manure crop and its percent groundcover. Use these estimators:

At 100 percent groundcover and 6-inch height*, most nonwoody legumes will contain roughly 2,000 lb./A of dry matter. For each additional inch, add 150 lb. So, a legume that is 18 inches tall and 100 percent groundcover will weigh roughly:

Inches >6: 18 in.–6 in. = 12 in.

X 150 lb./in.: 12 in. X 150 lb./in. = 1,800 lb.

Add 2,000 lb.: 2,000 lb. + 1,800 lb. = 3,800 lb.

If the stand has less than 100 percent groundcover, multiply by (the percent ground cover / 100). In this example, for 60 percent groundcover, you would obtain:

3,800 x (60/100) = 2,280 lb.

Keep in mind that these are *rough estimates* to give you a quick guide for the productivity of your green manure. To know the exact percent N in your plant tissue, you would have to send it to a lab for analysis. Even with a delay for processing, the results could be helpful for the crop if you use split applications of N. Testing is always a good idea, as it can help you refine your N estimates for subsequent growing seasons.

The following rules of thumb may help here:

- Annual legumes typically have between 3.5 and 4 percent N in their aboveground parts prior to flowering (for young material, use the higher end of the range), and 3 to 3.5 percent at flowering. After flowering, N in the leaves decreases quickly as it accumulates in the growing seeds.

* For cereal rye, the height relationship is a bit different. Cereal rye weighs approximately 2,000 lb./A of dry matter at an 8-inch height and 100 percent groundcover. For each additional inch, add 150 lb., as before, and multiply by (percent groundcover/100). For most small grains and other annual grasses, start with 2,000 lb./A at 6 inches and 100 percent ground cover. Add 300 lb. for each additional inch and multiply by (percent groundcover/100).

energy sources, which the microorganisms use as fuel to live. The process of burning this fuel sends most of the carbon back into the atmosphere as carbon dioxide, or CO_2.

Suppose a lot of new food is suddenly put into the soil system, as when a green manure is plowed down. Bacteria will expand their populations quickly to tap the carbon-based energy that's available. All the new bacteria, though, will need some N, as well as other nutrients, for body building before they can even begin to eat. So any newly released or existing mineral N in soil gets scavenged by new bacteria.

Materials with a high carbon to nitrogen (C:N) ratio, such as mature grass cover crops, straw or any fibrous, woody residue, have a low N content. They can "tie up" soil N, keeping it **immobilized** (and unavailable) to crops until the carbon "fuel supply" starts depleting. Tie-up may last for several weeks in the early part of the growing season, and crop plants may show the yellowing characteristic of N deficiencies. That is why it often makes sense to wait one to three weeks after killing a low-N cover before planting the next crop, or to supplement with a more readily available N source when a delay is not practical.

• For perennial legumes that have a significant number of thick, fibrous or woody stems, reduce these estimates by 1 percent.

• Most cover crop grasses contain 2 to 3 percent N before flowering and 1.5 to 2.5 percent after flowering.

• Other covers, such as brassicas and buckwheat, will generally be similar to, or slightly below, grasses in their N content. To put it all together:

$$\text{Total N in green manure (lb./A)} = \text{yield (lb./A)} \times \frac{\% \text{ N}}{100}$$

To estimate what will be available to your crop this year, divide this quantity of N by:

- 2, if the green manure will be conventionally tilled;
- 4, if it will be left on the surface in a no-till system in Northern climates;
- 2, if it will be left on the surface in a no-till system in Southern climates.

Bear in mind that in cold climates, N will mineralize more slowly than in warm climates, as discussed above. So these are gross estimates and a bit on the *conservative* side.

Of course, cover crops will not be the only N sources for your crops. Your soil will release between 10 and 40 lb. N/A for each 1 percent organic matter. Cold, wet clays will be at the low end of the scale and warm, well-drained soils will be at the high end. You also may receive benefits from last year's manure, green manure or compost application.

Other tools could help you refine your nitrogen needs. On-farm test strips of cover crops receiving different N rates would be an example. Refer to Appendix A, *Testing Cover Crops on Your Farm* (p. 189) for some tips on designing an on-farm trial. In some regions, a pre-sidedress N test in spring could help you estimate if supplemental N will be cost-effective. Bear in mind that pre-sidedress testing does not work well when fresh plant residues have been turned in—too much microbial interference relating to N tie-up may give misleading results.

For more information on determining your N from green manures and other amendments, see the *Northeast Cover Crop Handbook* (361).

—Marianne Sarrantonio, Ph.D.

Annual legumes have low C:N ratios, such as 10:1 or 15:1. When pure stands of annual legumes are plowed down, the N tie-up may be so brief you will never know it occurred.

Mixed materials, such as legume-grass mixtures, may cause a short tie-up, depending on the C:N ratio of the mixture. Some N storage in the microbial population may be advantageous in keeping excess N tied up when no crop roots are there to absorb it.

Fall-planted mixtures are more effective in mopping up excess soil N than pure legumes and, as stated earlier, the N is mineralized more rapidly from mixtures than from pure grass. A fall-seeded mixture will adjust to residual soil N levels. When the N levels are high, the grass will dominate and when N levels are low, the legume will dominate the mixtures. This can be an effective management tool to reduce leaching while making the N more available to the next crop.

Potential Losses

A common misunderstanding about using green manure crops is that the N is used more efficiently because it's from a plant source. This is not necessarily true. Nitrogen can be lost from a green manure system almost as easily as from chemical fertilizers, and in comparable amounts. The reason is that the legume organic N may be converted to ammonium (NH_4), then to ammonia (NH_3) or nitrate (NO_3) before plants can take it up. Under no-till systems where killed cover crops remain on the surface, some ammonia (NH_3) gas can be lost right back into the atmosphere.

Nitrate is the form of N that most plants prefer. Unfortunately, it is also the most water-soluble form of N. Whenever there is more nitrate than plant roots can absorb, the excess may leach with heavy rain or irrigation water.

As noted earlier, nitrates in excess of 140 lb./A may be released into warm, moist soil within as little as seven to 10 days after plowing down a high-N legume, such as a hairy vetch stand. Since the following crop is unlikely to have much of a root system at that point, the N has a ticket for Leachville. Consider also that the green manure may have been plowed down to as deep as 12 inches—much deeper than anyone would consider applying chemical fertilizer. Moreover, green manures sometimes continue to decompose after the cash crop no longer needs N. This N also is prone to leaching.

To summarize, conventional plowing and aggressive disking can cause a rapid decomposition of green manures, which could provide too much N too soon in the cropping season. No-till systems will have a reduced and more gradual release of N, but some of that N may be vulnerable to gaseous loss, either by ammonia volatilization or by denitrification, which occurs when NO_3^- converts to gases under low O_2 (flooded) conditions. Thus, depending on management, soil and weather situations, N from legume cover crops may not be more efficiently used than N from fertilizer.

Some possible solutions to this cover crop nitrogen-cycling dilemma:

- A shallow incorporation of the green manure, as with a disk, may reduce the risk of gaseous loss.
- It may be feasible to no-till plant or transplant into the green manure, then mow or incorporate it between the rows 10-14 days later, when cash crop roots are more developed and able to take up N. This has some risk, especially when soil moisture is limiting, but can provide satisfactory results if seedling survival is assured.
- Residue from a grass/legume mix will have a higher C:N than the legume alone, slowing the release of N so it's not as vulnerable to loss.

Consider also that some portion of the N in the green manure will be conserved in the soil in an organic form for gradual release in a number of subsequent growing seasons.

OTHER SOIL-IMPROVING BENEFITS

Cover crops can be very useful as living plows to penetrate and break up compacted layers in the soil. Some of the covers discussed in this book, such as sweetclover and forage radish, have roots that reach as deep as three feet in the soil within one cropping season. The action of numerous pointy little taproots with the hydraulic force of a determined plant behind them can penetrate soil where plowshares fear to go. Grasses, with their tremendously extensive root systems, may relieve compacted surface soil layers. Sorghum-sudangrass can be managed to powerfully fracture subsoil. See *Summer Covers Relieve Compaction* (p. 110).

One of the less appreciated soil benefits of cover crops is an increase in the total numbers and diversity of soil organisms. As discussed earlier, diversity is the key to a healthy, well-functioning soil. Living covers help supply year-round food for organisms that feed off root by-products or that need the habitat provided on a residue-littered soil surface. Dead covers supply a more varied and increased soil diet for many organisms.

Of course, unwanted pests may be lured to the field. Effective crop rotations that include cover crops, however, tend to reduce rather than increase pest concerns. Pest-management considerations due to the presence of a cover crop are discussed in the next chapter, *Managing Pests with Cover Crops* (p. 25).

Finally, cover crops may have an added advantage of drying out and therefore warming soils during a cold, wet season. The flip side of this is that they may dry the soil out too much and rob the following crop of needed moisture.

There are no over-the-counter elixirs for renewing soil. A long-term farm plan that includes cover crops, however, can help ensure your soil's health and productivity for as long as you farm.

MANAGING PESTS WITH COVER CROPS

By Sharad C. Phatak and Juan Carlos Diaz-Perez

Cover crops are poised to play increasingly important roles on North American farms. In addition to slowing erosion, improving soil structure and providing fertility, we are learning how cover crops help farmers to manage pests (390). With limited tillage and careful attention to cultivar choice, placement and timing, cover crops can reduce infestations by insects, diseases, nematodes and weeds. Pest-fighting cover crop systems help minimize reliance on pesticides, and as a result cut costs, reduce your chemical exposure, protect the environment, and increase consumer confidence in the food you produce.

Farmers and researchers are using cover crops to design new strategies that preserve a farm's natural resources while remaining profitable. Key to this approach is to see a farm as an "agroecosystem"—a dynamic relationship of the mineral, biological, weather and human resources involved in producing crops or livestock. Our goal is to learn agricultural practices that are environmentally sound, economically feasible and socially acceptable.

Environmentally sustainable pest management starts with building healthy soils. Research in south Georgia (see *Georgia Cotton, Peanut Farmers*, p. 26) shows that crops grown on biologically active soils resist pest pressures better than those grown on soils of low fertility, extreme pH, low biological activity and poor soil structure.

There are many ways to increase biological activity in soil. Adding more organic material by growing cover crops or by applying manure or compost helps. Reducing or eliminating pesticides favors diverse, healthy populations of beneficial soil flora and fauna. So does reducing or eliminating tillage that causes losses of soil structure, biological life or organic matter. These losses make crops more vulnerable to pest damage.

Farming on newly cleared land shows the process well. Land that has been in a "cover crop" of trees or pastures for at least 10 years remains productive for row crops and vegetables for the first two to three years. High yields of agronomic and horticultural crops are profitable, with comparatively few pesticide and fertilizer inputs. After that period—under conventional systems with customary clean tillage—annual crops require higher inputs. The first several years of excessive tillage destroys the food sources and micro-niches on which the soil organisms that help suppress pests depend. When protective natural biological systems are disrupted, pests have new openings and crops are much more at risk.

Cover crop farming is different from clean-field monocropping, where perfection is rows of corn or cotton with no thought given to encouraging biological diversity. Cover crops bring more forms of life into the picture and into your management plan. By working with a more diverse range of crops, some growing at the same time in the same field, you've got a lot more options. Here's a quick overview of how these systems work.

Insect Management

In balanced ecosystems, insect pests are kept in check by their natural enemies (409). These natural pest control organisms —called beneficials in agricultural systems—include predator and parasitoid insects and diseases. Predators kill and eat other insects; parasitoids spend their larval stage inside another insect, which then dies as the invader's larval stage ends. However, in conventional agricultural systems, synthetic chemical treatments that kill insect pests also typically kill the natural enemies of the insects. Conserving and encouraging beneficial organisms is a key to achieving sustainable pest management.

You should aim to combine strategies that make each farm field more hospitable to beneficials. Reduce pesticide use, and, when use is essential, select materials that are least harmful to beneficials. Avoid or minimize cultural practices

Georgia Cotton, Peanut Farmers Use Cover Crops to Control Pests

TIFTON, Ga.—Here in southwestern Georgia, I'm working with farmers who have had dramatic success creating biologically active soil in fields that have been conventionally tilled for generations. We still grow the traditional cash crops of cotton and peanuts, but with a difference. We've added cover crops, virtually eliminated tillage, and added new cash crops that substitute for cotton and peanuts some years to break disease cycles and allow for more biodiversity.

Our strategies include no-till planting (using modified conventional planters), permanent planting beds, controlled implement traffic, crop rotation and annual high-residue winter cover crops. We incorporate fertilizer and lime prior to the first planting of rye in the conversion year. This is usually the last tillage we plan to do on these fields for many years. Together, these practices give us significant pest management benefits within three years.

Growers are experimenting with a basic winter cover crop>summer cash crop rotation. Our cover crops are ones we know grow well here. Rye provides control of disease, weed and nematode threats. Legume crops are crimson clover, subterranean clover or cahaba vetch. They are planted with the rye or along field borders, around ponds, near irrigation lines and in other non-cropped areas as close as possible to fields to provide the food needed to support beneficials at higher populations.

When I work with area cotton and peanut growers who want to diversify their farms, we set up a program that looks like this:

- **Year 1.** *Fall*—Adjust fertility and pH according to soil test. Deep till if necessary to relieve subsurface soil compaction. Plant a cover crop of rye, crimson clover, cahaba vetch or subterranean clover.

  Spring—Strip-till rows 18 to 24 inches wide, leaving the cover crop growing between the strips. Three weeks later, plant cotton.
- **Year 2.** *Fall*—Replant cereal rye or cahaba vetch, allow crimson or subclover hard seed to germinate.

  Spring—Strip-till cotton.
- **Year 3.** *Fall*—Plant rye.

  Spring—Desiccate rye with herbicides. No-till plant peanuts.
- **Year 4.** The cycle starts again at year 1.

Vegetable farmers frequently use fall-planted cereal rye plowed down before vegetables, or crimson clover strip-tilled before planting vegetables. The crimson clover matures, drops hard seed, then dies. Most of the seed germinates in the fall. Cereal/legume mixes have not been more successful than single-crop cover crop plantings in our area.

Some vegetable farmers strip-till rows into rye in April. The strips are planted in early May to Southern peas, lima beans or snap beans. Rye in row middles will be dead or nearly dead. Rye or crimson clover can continue the rotation.

Vegetable farmers also broadcast crimson clover in early March. They desiccate the cover, strip-till rows, then plant squash in April. The clover in the row middles will set seed then die back through summer. The crimson strips will begin to regrow in the fall from the dropped seed, and fall vegetables may be planted in the tilled areas after the July squash harvest.

Insecticide and herbicide reduction begins the first year, with no applications needed by the third or fourth year in many cases.

The farmers get weed control by flail mowing herbicide-killed, fall-planted rye, leaving about 6 inches of stubble. Alternatively, they

such as tilling and burning that kill beneficials and destroy their habitat. Build up the sustenance and habitat that beneficials need. Properly managed cover crops supply moisture, physical niches and food in the form of insects, pollen, honeydew and nectar.

By including cover crops in your rotations and not spraying insecticides, beneficials often are already in place when you plant spring or summer crops. However, if you fully incorporate cover crops into the soil, you destroy or disperse most could use a roller to kill the rye crop. One or two post-emerge herbicide applications should suffice in the first few years. I don't recommend cultivation for weed control because it increases risks of soil erosion and damages the protective outer leaf surface layer (cuticle) that helps prevent plant diseases.

We see changes on farms where the rotations stay in place for three or more years:

- **Insects. Insecticide costs under conservation tillage are $50 to $100/A less** than conventional crop management in the area for all kinds of crops. The farmers using the alternative system often substitute with insect control materials such as *Bacillus thuringiensis* (Bt), pyrethroids, and insect growth regulators that have less severe environmental impact than chemical pesticides. These products are less persistent in the field environment, more targeted to specific pests and do less harm to beneficials. By planting cover crops on field edges and in other non-crop areas, these farmers are increasing the numbers of beneficials in the field environments.

Pests that are *no longer a problem* on the cover-cropped farms include thrips, bollworm, budworm, aphids, fall armyworm, beet armyworm and white flies. On my no-till research plots with cover crops and long rotations, I've not used insecticides for six years on peanuts, for eight years on cotton and for 12 years on vegetables. I'm working with growers who use cover crops and crop rotations to economically produce cucumbers, squash, peppers, eggplant, cabbage, peanuts, soybeans and cotton with only one or two applications of insecticide—sometimes with none.

- **Weeds.** Strip-tilling into over-wintered cover crops provides acceptable weed control for relay-cropped cucumbers (325). Conventional management of rye in our area is usually to disk or kill it with broad-spectrum herbicides such as paraquat or glyphosate. Rye can also be killed with a roller, providing an acceptable level of weed control for the subsequent cash crop.
- **Diseases.** I've been strip-tilling crimson clover since 1985 to raise tomatoes, peppers, eggplant, cucumbers, cantaloupes, lima beans, snap beans, Southern peas and cabbages. I'm using no fungicides. Our research staff has raised peanuts no-tilled into cereal rye for the past six years, also without fungicides.
- **Nematodes.** If we start on land where pest **nematodes** are not a major problem, this system keeps them from becoming a limiting factor.

Even though the conventional wisdom says you can't build organic matter in our climate and soils, we have top-inch readings of 4 percent organic matter in a field that tested 0.5 percent four years ago.

We are still learning, but know that we can rotate crops, use cover crops and cut tillage to greatly improve our sustainability. In our experience, we've reduced total costs by as much as $200 per acre for purchased inputs and tillage. Parts of our system will work in many places. Experiment on a small scale to look more closely at what's really going in your soil and on your crops. As you compare insights and share information with other growers and researchers in your area, you'll find cover crops that help you control pests, too.

—Sharad C. Phatak

of the beneficials that were present. Conservation tillage is a better option because it leaves more of the cover crop residue on the surface. No-till planting only disturbs an area 2 to 4 inches wide, while strip-tilling disturbs an area up to about 24 inches wide between undisturbed row middles.

Cover crops left on the surface may be living, temporarily suppressed, dying or dead. In any event, their presence protects beneficials and their habitat. The farmer-helpful organisms are hungry, ready to eat the pests of cash crops that are planted into the cover-crop residue. The ultimate goal is to provide year-round food and habitat for beneficials to ensure their presence within or near primary crops.

We're just beginning to understand the effects of cropping sequences and cover crops on beneficial and insect pest populations. Researchers have found that **generalist predators**, which feed on many species, may be an important biological control. During periods when pests are scarce or absent, several important generalist predators can subsist on nectar, pollen and alternative prey afforded by cover crops. This suggests you can enhance the biological control of pests by using cover crops as habitat or food for the beneficials in your area.

This strategy is important for farmers in the South, where pest pressure can be especially heavy. In south Georgia, research showed that populations of beneficial insects such as insidious flower bugs (*Orius insidiosus*), bigeyed bugs (*Geocoris* spp.) and various lady beetles (*Coleoptera coccinellidae*) can attain high densities in various vetches, clovers and certain cruciferous crops. These predators subsisted and reproduced on nectar, pollen, thrips and aphids, and were established before key pests arrived. Research throughout Georgia, Alabama and Mississippi showed that when summer vegetables were planted amid "dying mulches" of cool-season cover crops, some beneficial insects moved in to attack crop pests.

When crops are attacked by pests, they send chemical signals that attract beneficial insects. The beneficials move in to find their prey (420).

Maximizing natural predator-pest interaction is the primary goal of biologically based Integrated Pest Management (IPM), and cover crops can play a leading role. For example:

- Colorado potato beetles were observed at 9 a.m. attacking eggplant that had been strip-till planted into crimson clover. By noon, assassin bugs had clustered around the feeding beetles. The beneficial bugs destroyed all the beetles by evening.
- Cucumber beetles seen attacking cucumber plants were similarly destroyed by beneficials within a day.
- Lady beetles in cover crop systems help to control aphids attacking many crops.

Properly selected and managed, cover crops can enhance the soil and field environment to favor beneficials. Success depends on properly managing the cover crop species matched with the cash crops and anticipated pest threats. While we don't yet have prescription plantings guaranteed to bring in all the needed beneficials—and only beneficials—for long lists of cash crops, we know some associations:

- We identified 13 known beneficial insects associated with cover crops during one growing season in south Georgia vegetable plantings (53, 55, 57).
- In cotton fields in south Georgia where residues are left on the surface and insecticides are not applied, more than 120 species of beneficial arthropods, spiders and ants have been observed.
- Fall-sown and spring-sown insectory mixes with 10 to 20 different cover crops work well under orchard systems. These covers provide habitat and alternative food sources for beneficial insects. This approach has been used successfully by California almond and walnut growers participating in the Biologically Intensive Orchard Systems (BIOS) project of the University of California (184).

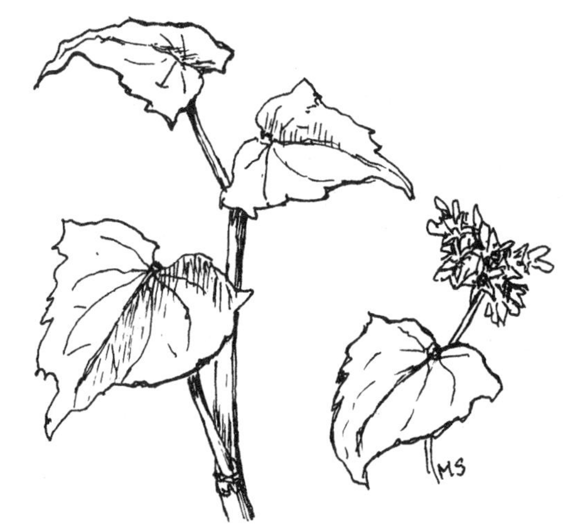

BUCKWHEAT *grows quickly in cool, moist weather.*

The level of ecological sustainability depends on the grower's interests, management skills and situation. Some use no insecticides while others have substantially reduced insecticide applications on peanut, cotton and vegetable crops.

• In Georgia, Mississippi and South Carolina, minimally tilled crimson clover or cahaba vetch before cotton planting have been successful in reducing fertilizer N up to 50 percent and insecticide inputs by 30 to 100 percent.

• Many farmers are adopting a system of transplanting tomatoes, peppers and eggplant into a killed hairy vetch or vetch/rye cover crop. Benefits include weed, insect and disease suppression, improved fruit quality and overall lower production cost.

• Leaving "remnant strips" of a cover when most of the crop is mowed or incorporated provides a continuing refuge and food source for beneficials, which might otherwise leave the area or die. This method is used in orchards when continued growth of cover crops would cause moisture competition with trees.

• Insect movement is orchestrated in a system developed by Oklahoma State University for pecan growers. As legume mixtures senesce, beneficials migrate into trees to help suppress harmful insects. Not mowing the covers from August 1 until shuck split of the developing pecans lessens the unwanted movement of stink bugs, a pest which can damage green pecans (261). In California, lygus bugs on berseem clover or alfalfa are pests of cash crops. Be careful that cover crop maturity or killing a cover doesn't force pests into a neighboring cash crop.

Disease Management

Growers traditionally have been advised to turn under plant debris by moldboard plowing to minimize crop losses due to diseases (321, 322, 403, 405, 406). Now we realize that burying cover crop residues and disrupting the entire soil profile eliminates beneficial insect habitats and the benefits of weed control by crop residues. The increased use of conservation tillage increases the need to manage crop disease without burying cover crops.

In the field, although plants are exposed to a wide diversity of microorganisms, plant infection by microorganisms is rare (314). A pathogen has to cross many plant barriers before it can cause disease to roots, stem or leaves. You can use cover crops to reinforce two of these barriers.

Cover crops can enhance the soil and field environment to favor beneficial insects.

Plant cuticle layer. This often waxy surface layer is the first physical barrier to plant penetration. Many pathogens and all bacteria enter the plant through breaks, such as wounds, or natural openings, such as stomata, in this cuticle layer. This protective layer can be physically damaged by cultivation, manipulation, spraying and sand-blasting from wind erosion, as well as by the impact and soil splashing from raindrops and overhead irrigation. Spray adjuvants may also damage the waxes of the cuticle resulting in more disease, as with *Botritys cinerea* rot in grapes (356, 367). In well-developed minimum-till or no-till crop systems with cover crops, you may not need cultivation for weed control (see below) and you can minimize spraying. Organic mulches form living, dying or killed covers that hold soil, stop soil splashing and protect crops from injury to the cuticle.

Plant surface microflora. Many benign organisms are present on the leaf and stem surface. They compete with pathogens for a limited supply of nutrients. Some of these organisms produce natural antibiotics. Epiphytic bacteria adhere to plant surfaces forming multicellular structures known as biofilms (339). These biofilms play an important role in plant disease. Pesticides, soaps, surfactants, spreaders and sticking agents can kill or disrupt these beneficial microorganisms, weakening the plant's defenses against pathogens (356, 367). Cover crops can help this natural protection process work by reducing the need for application of synthetic crop protection materials. Further, cover crop plant surfaces can support healthy populations of beneficial microorgan-

Select Covers that Balance Pests, Problems of Farm

Many crops can be managed as cover crops, but only a few have been studied specifically for their pest-related benefits on cash crops and field environments.

Learn all you can about the impacts of a cover crop species to help you manage it in your situation. Here are several widely used cover crops described by their effects **under conservation tillage** in relation to insects, diseases, nematodes and weeds.

• **Cereal Rye** (*Secale cereale*)—This winter annual grain is perhaps the most versatile cover crop used in the continental United States. Properly managed under conservation tillage, rye has the ability to reduce soil-borne diseases, nematodes and weeds. Rye is a non-host plant for root-knot nematodes and soil-borne diseases. It produces significant biomass that smothers weeds when it is left on the surface and also controls weeds allelopathically through natural weed-suppressing compounds.

As it grows, rye provides habitat, but not food, for beneficial insects. Thus, only a small number of beneficial insects are found on rye.

Fall-planted rye works well in reducing soil-borne diseases, root-knot nematodes and broadleaf weeds in all cash crops that follow, including cotton, soybean and most vegetables. Rye will not control weedy grasses. Because it can increase numbers of cut worms and wire worms in no-till planting conditions, rye is not the most suitable cover where those worms are a problem ahead of grass crops like corn, sweet corn, sorghum or pearl millet.

• **Wheat** (*Triticum aestivum*)—A winter annual grain, wheat is widely adapted and works much like rye in controlling diseases, nematodes and broadleaf weeds. Wheat is not as effective as rye in controlling weeds because it produces less biomass and has less allelopathic effect.

• **Crimson Clover** (*Trifolium incarnatum*)—Used as a self-reseeding winter annual legume throughout the Southeast, fall-planted crimson clover supports and increases soil-borne diseases, such as the pythium-rhizoctonia complex, and root-knot nematodes. It suppresses weeds effectively by forming a thick mulch. Crimson clover supports high densities of beneficial insects by providing food and habitat. Because some cultivars produce "hard seed" that resists immediate germination, crimson clover can be managed in late spring so that it reseeds in late summer and fall.

isms, including types of yeasts that can migrate onto a cash crop after planting or transplanting.

Soilborne pathogenic **fungi** limit production of vegetables and cotton in the southern U.S. (404, 405, 406, 407). *Rhizoctonia solani, Pythium myriotylum, Pythium phanidermatum* and *Pythium irregulare* are the most virulent pathogenic fungi that cause damping-off on cucumbers, snap beans, and other vegetables. *Sclerotium rolfsii* causes rot in all vegetables and in peanuts and cotton. Infected plants that do not die may be stunted because of lesions caused by fungi on primary or secondary roots, hypocotyls and stems, and may have reduced yields of low quality. But after two or three years in cover cropped, no-till systems, damping-off is not a serious disease, as experience on south Georgia farms and research plots shows. Increased soil organic matter levels may help in reducing plant disease incidence and severity by enhancing natural disease suppression (252, 424).

In soils with high levels of disease inoculum, however, it takes time to reduce population levels of soil pathogens using only cover crops. After tests in Maine with oats, broccoli, white lupine (*Lupinus albus*) and field peas (*Pisum sativum*) researchers cautioned it may take three to five years to effectively reduce stem lesion losses on potatoes caused by *R. solani* (240). Yet there are single-season improvements, too. For example, in an Idaho study, *Verticillium* wilt of potato was reduced by 24 to 29 percent following sudangrass

• **Subterranean Clover** (*Trifolium subterraneum*)—A self-reseeding annual legume, fall-planted subterranean clover carries the same risks as crimson clover with soil-borne diseases and nematodes. It suppresses weeds more effectively in the deep South, however, because of its thick and low growth habit. Subclover supports a high level of beneficial insects.

• **Cahaba White Vetch** (*Vica sativa X V. cordata*)—This cool-season annual legume is a hybrid vetch that increases soilborne diseases yet suppresses root-knot nematodes. It supports beneficial insects, yet attracts very high numbers of the tarnished plant bug, a serious pest.

• **Buckwheat** (*Fagopyrum esculentum*)— A summer annual non-legume, buckwheat is very effective in suppressing weeds when planted thickly. It also supports high densities of beneficial insects. It is suitable for sequential planting around non-crop areas to provide food and habitat for beneficial insects. It is very attractive to honeybees.

A well-planned crop rotation maximizes benefits and compensates for the risks of cover crops and cash crops. Planting rye in a no-till system substantially reduces root-knot nematodes, soil-borne diseases and broadleaf weeds. By using clovers and vetches in your fields and adding beneficial habitat in non-cultivated areas, you can increase populations of beneficial insects that help to keep insects pests under control. Mixed plantings of small grains and legumes combine benefits of both while reducing their shortcomings.

As pesticides of all types (fungicides, herbicides, nematicides and insecticides) are reduced, the field environment becomes increasingly resilient in keeping pest outbreaks in check. Plantings to further increase beneficial habitat in non-cultivated areas can help maintain pollinating insects and pest predators, but should be monitored to avoid build-ups of potential pests. Researchers are only beginning to understand how to manage these "insectary plantings."

Editor's Note: Each cover crop listed here, except for cahaba vetch, is included in the charts (p. 62 and following) and is fully described in its respective section. Check the *Table of Contents* (p. 4) for location.

—Sharad C. Phatak

green manure. Yield of U.S. No.1 potatoes increased by 24 to 38 percent compared with potatoes following barley or fallow (394).

Nematode Management

Nematodes are minute roundworms that interact directly and indirectly with plants. Some species feed on roots and weaker plants, and also introduce disease through feeding wounds. Most nematodes are not plant parasites, but feed on and interact with many soil-borne microorganisms, including fungi, bacteria and protozoa. Damage to the crop from plant-parasitic nematodes results in a breakdown of plant tissue, such as lesions or yellow foliage; retarded growth of cells, seen as stunted growth or shoots; or excessive growth such as root galls, swollen root tips or unnatural root branching.

If the community of nematodes contains diverse species, no single species will dominate.

This coexistence would be the case in the undisturbed field or woodland described above.

In conventional crop systems, pest nematodes have abundant food and the soil environment is conducive to their growth. This can lead to rapid expansion of plant parasitic species, plant disease and yield loss. Cropping systems that increase biological diversity over time usually prevent the onset of nematode problems. Reasons may include a dynamic soil ecological balance and improved, healthier soil structure with higher organic matter (5, 245, 424). In Michigan, to limit nematodes between potato crops, some potato growers report that two years of radish improves potato production and lowers pest control costs (270, 271).

Once a nematode species is established in a field, it is usually impossible to eliminate it. Some covers can enhance a resident parasitic nematode population if they are grown before or after another crop that hosts a plant-damaging nematode species.

If a nematode pest species is absent from the soil, planting a susceptible cover crop will not give rise to a problem, assuming the species is not introduced on seed, transplants or machinery (357). Iowa farmer Dick Thompson reports that researchers analyzing his fields found no evidence that hairy vetch, a host for soybean cyst nematode, caused any problem with the pest in his soybeans. This may be due to his use of compost in strip-cropped fields with an oats/hairy vetch>corn>soybean rotation.

You can gradually reduce a field's nematode pest population or limit nematode impact on crops by using specific cover crops. Nematode control tactics involving covers include:

- Manipulating soil structure or soil humus
- Rotating with non-host crops
- Using crops with nematicidal effects, such as brassicas

Cover crops may also improve overall plant vitality to lessen the nematode impact on yield. But if you suspect nematode trouble, send a soil sample for laboratory analysis to positively identify the nematode species. Then be sure any cover crops you try aren't alternate hosts for that pest species. Area IPM specialists can help you.

Using brassicas and many grasses as cover crops can help you manage nematodes. Cover crops with documented nematicidal properties against at least one nematode species include sorghum-sudangrass hybrids (*Sorghum bicolor X S. bicolor var. sudanese*), marigold (*Tagetes patula*), hairy indigo (*Indigofera hirsuta*), showy crotalaria (*Crotolaria spectabilis*), sunn hemp (*Crotalaria juncea*), velvetbean (*Mucuna deeringiana*), rapeseed (*Brassica rapa*), mustards and radish (*Raphanus satiuus*).

You must match specific cover crop species with the particular nematode pest species, then manage it correctly. For example, cereal rye residue left on the surface or incorporated to a depth of several inches suppressed Columbia lance nematodes in North Carolina cotton fields better than if the cover was buried more deeply by moldboard plowing. Associated greenhouse tests in the study showed that incorporated rye was effective against root-knot, reniform and stubby root nematodes, as well (20).

Malt barley, corn, radishes and mustard sometimes worked as well as the standard nematicide to control sugar beet nematode in Wyoming sugar beets, a 1994 study showed. Increased production more than offset the cover crop cost, and lamb grazing of the brassicas increased profit without diminishing nematode suppression. The success is conditional upon a limited nematode density. The cover crop treatment was effective only if there were fewer than 10 eggs or juveniles per cubic centimeter of soil. A moderate sugar beet nematode level was reduced 54 to 75 percent in about 11 weeks, increasing yield by nearly 4 tons per acre (231).

Weed Management

Cover crops are widely used as smother crops to shade and out-compete weeds (412). Cereal grains establish quickly as they use up the moisture, fertility and light that weeds need to survive. Sorghum-sudangrass hybrids and buckwheat are warm-season crops that suppress weeds through these physical means and by plant-produced natural herbicides (allelopathy).

Cereal rye is an overwintering crop that suppresses weeds both physically and chemically. If rye residue is left on the soil surface, it releases allelochemicals that inhibit seedling growth of many annual small-seeded broadleaf weeds, such as pigweed and lambsquarters. The response of grassy weeds to rye is more variable. Rye is a major component in the killed organic mulches used in no-till vegetable transplanting systems.

Killed cover crop mulches last longer if the stalks are left intact, providing weed control well into the season for summer vegetables. Two implements have been modified specifically to enhance weed suppression by cover crops. The undercutter uses a wide blade to slice just under the surface of raised beds, severing cover crop plants from their root mass. An attached rolling harrow increases effectiveness (95, 96, 97). A

Buffalo rolling stalk chopper does no direct tillage, but aggressively bends and cuts crops at the surface (303). Both tools work well on most legumes when they are in mid-bloom stage or beyond.

Killed mulch of a cover crop mix of rye, hairy vetch, crimson clover and barley kept processing tomatoes nearly weed-free for six weeks in an Ohio test. This length of time is significant, because other research has shown that tomato fields kept weed-free for 36 days yield as much as fields kept weed-free all season (97, 150). The roller is another method used to terminate the cover crop (13). The roller flattens and crimps the cover crop, forming a flat mat of cover crop residue that effectively control weeds.

Cover crops can also serve as a "living mulch" to manage weeds in vegetable production. Cover crops are left to grow between rows of the cash crop to suppress weeds by blocking light and outcompeting weeds for nutrients and water. They may also provide organic matter, nitrogen (if legumes) and other nutrients mined from underneath the soil surface, beneficial insect habitat, erosion prevention, wind protection and a tough sod to support field traffic.

To avoid competition with the cash crop, living mulches can be chemically or mechanically suppressed. In the Southeast, some cool-season cover crops such as crimson clover die out naturally during summer crop growth and do not compete for water or nutrients. However, cover crops that regrow during spring and summer—such as subterranean clover, white clover and red clover—can compete strongly for water with spring-planted crops unless the covers are adequately suppressed.

In New York, growing cover crops overseeded within three weeks of potato planting provided good weed suppression, using 70 percent less herbicide. Yield was the same as, or moderately reduced from, the standard herbicide control plots in the two-year study. Hairy vetch, woollypod vetch, oats, barley, red clover and an oats/hairy vetch mix were suppressed as needed with fluazifop and metribuzin (341).

Cover crops often suppress weeds early, then prevent erosion or supply fertility later in the season. For example, shade-tolerant legumes such as red clover or sweetclover that are planted with spring grains grow rapidly after grain harvest to prevent weeds from dominating fields in late summer. Overseeding annual ryegrass or oats at soybean leaf yellowing provides a weed-suppressing cover crop before frost and a light mulch to suppress winter annuals, as well.

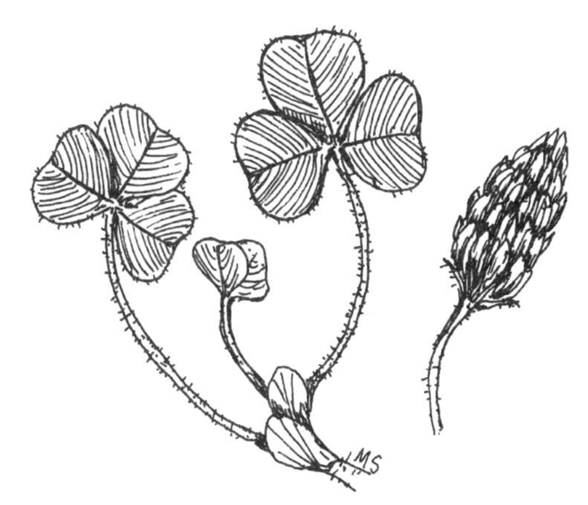

CRIMSON CLOVER, *a winter annual legume, grows rapidly in spring to fix high levels of nitrogen.*

Healthy soils grow healthy weeds as well as healthy crops, making it difficult to manage weeds in conservation tillage without herbicides. Long term strategies for weed management should include:

- Reducing the weed seed bank
- Preventing weeds from going to seed
- Cleaning equipment before moving to different fields and farms
- Planting cover crops to help manage weeds in conservation tillage

Cover crops can play a pest-suppressing role on virtually any farm. As we find out more about the pest management benefits of cover crop systems, they will become even more attractive from both an economic and an environmental perspective. Traditional research will identify some new pieces of these biologically based systems. However, growers who understand how all the elements of their farm fit together will be the people who will really bring cover crops into the prominence they deserve in sustainable farming.

CROP ROTATION WITH COVER CROPS

Readers' note: > indicates progression to another crop; / indicates a mixture of crops growing at the same time.

One of the biggest challenges of cover cropping is to fit cover crops into your current rotations, or to develop new rotations that take full advantage of their benefits. This section will explore some of the systems used successfully by farmers in different regions of the U.S. One might be easily adapted to fit your existing crops, equipment and management. Other examples may point out ways that you can modify your rotation to make the addition of cover crops more profitable and practical.

Whether you add cover crops to your existing rotations or totally revamp your farming system, you should devote as much planning and attention to your cover crops as you do to your cash crops: Failure to do so can lead to failure of the cover crop and cause problems in other parts of your system. Also remember that there is likely no single cover crop that is right for your farm.

Before you start:

- Review *Benefits of Cover Crops* (p. 7) and *Selecting the Best Cover Crops for Your Farm* (p. 9).
- Decide which benefits are most important to you.
- Read the examples below, then consider how these cover crop rotations might be **adapted to your particular conditions**.
- Talk to your neighbors and the other "experts" in your area, including the contact people listed in *Regional Experts* (p. 202).
- Start small on an easily accessible plot that you will see often.
- Be an opportunist—and an optimist. If your cropping plans for a field are disrupted by weather or other conditions outside of your control, this may be the ideal window for establishing a cover crop.
- Consider using an early-maturing cash crop to allow for timely planting of the cover crop.
- Cover crops can be used for feed. Consider harvesting or grazing for forage or alternative livestock such as sheep and goats.

The ideas in this book will help you see cover crop opportunities, no matter what your system. For more in-depth scientific analysis of cover crops in diverse cropping systems, see several comprehensive reviews listed in Appendix F (77, 106, 362, 390).

COVER CROPS FOR CORN BELT GRAIN AND OILSEED PRODUCTION

In addition to providing winter cover and building soil structure, nitrogen (N) management will probably be a major factor in your cover crop decisions for the corn>soybean rotation. A fall-planted grass or small grain will scavenge leftover N from the previous corn or soybean crop. Legumes are much less efficient at scavenging N, but will add N to the system for the following crop. Legume/grass mixtures are quite good at both.

Corn>Soybean Systems

Keep in mind that corn is a heavy N feeder, soybeans benefit little, if at all, from cover crop N and that you have a shorter time for spring cover crop growth before corn than before soybeans.

▼ **Precaution.** If you use herbicides, be sure to check labels for plant-back or rotation intervals to ensure that your cover crop isn't adversely affected.

Cover crop features: rye provides winter cover, scavenges N after corn, becomes a long-lasting (6 to 12 week) residue to hold moisture and suppress weeds for your soybeans; **hairy vetch** provides spring ground cover, abundant N and a moderate-term (4 to 8 week) mulch for a no-till corn crop; **field peas** are similar to vetch, but residue breaks down faster; **red clover** is also similar, but produces slightly less N and has less vigorous spring regrowth; **berseem clover** grows quickly to provide several cuttings for high-N green manure.

Here are some options to consider adapting to your system:

Corn>Rye>Soybeans>Hairy Vetch.
In Zone 7 and warmer, you can grow a cover crop every year between your corn and full-season beans. Also, you can use wheat or another small grain to replace the cover crop before beans, in a three-crop, two-year rotation (corn>wheat>doublecrop beans). In all cases, another legume or a grass/legume mixture can be used instead of a single species cover crop. Where it is adapted, you can use crimson clover or a crimson/grass mixture instead of vetch.

In cooler areas, plant rye as soon as possible after corn harvest. If you need more time in the fall, try overseeding in rowed beans at drydown "yellow leaf" stage in early fall, or in early summer at the last cultivation of corn. Seeding options include aerial application where the service is economical, using a specialty high-body tractor with narrow tires, or attaching a broadcast seeder, air seeder or seed boxes to a cultivator.

▼ **Precaution.** Broadcast seeding of cover crops into standing crops is less dependable than other seeding methods. Success will depend on many factors, including adequate rainfall amount and distribution after seeding.

Kill the rye once it is about knee-high, or let it go a bit longer, killing it a couple of weeks before planting beans. Killing the rye with herbicides and no-tilling beans in narrow rows allows more time for cover crop growth, since you don't have to work the ground. If soil moisture is low, consider killing the rye earlier. Follow the beans with hairy vetch or a vetch/small grain mixture.

Legumes must be seeded at least 6 weeks before hard frost to ensure winter survival. Seed by drilling after soybean harvest, or by overseeding before leaf drop. Allow the vetch (or mixture) to grow as long as possible in spring for maximum N fixation.

Harvesting sweet corn, seed corn or silage corn opens a window for timely cover crop planting. Harvest or graze the small grain or legume / small grain mixture in spring if needed for feed. See individual cover crop chapters for management details.

In Pennsylvania, Ed Quigley seeds rye or spring oats after corn silage harvest. The oats can be cut for silage in fall if planted by early September. Rye can be made into rylage or sprayed before no-till corn the following spring.

Worried about planting your corn a bit late because you're waiting for your cover crop to mature? Research in Maryland, Illinois and elsewhere suggests that no-tilling corn towards the end of the usual window when using a legume cover crop has its rewards. The delay can result in greater yields than earlier planting, due to greater moisture conservation and more N produced by the cover crop, or due to the timing of summer drought (82, 84, 300). In Pennsylvania, however, delayed planting sometimes reduced corn yields following rye (118).

Growers are looking to add a small grain to their corn>soybean rotation.

Check your state variety trial data for a shorter season corn hybrid that yields nearly as well as slightly longer season corn. The cover crop benefit should overcome many yield differences.

Worried about soil moisture? There's no question that growing cover crops may consume soil moisture needed by the next crop. In humid regions, this is a problem only in an unusually dry spring. Time permitting, allow 2 to 3 weeks after killing the cover crop to alleviate this problem. While spring rainfall may compensate for the moisture demand of most cover crops by normal planting dates, cover crops can quickly dry out a field. Later in the season, killed cover crop residues in conservation tillage systems can conserve moisture and increase yields.

In dryland areas of the Southern Great Plains, lack of water limits cover crop use. (See *Dryland Cereal Cropping Systems*, p. 42).

In any system where you are using accumulated soil moisture to grow your cash crop, you need to be extra careful. However, as noted in this section and elsewhere in the book, farmers and researchers are finding that water-thrifty cover crops may be able to replace even a fallow year without adversely affecting the cash crop.

Corn>Rye>Soybeans>Small Grain>Hairy Vetch. This rotation is similar to the corn>rye>soybeans rotation described above, except you add a year of small grains following the beans. In crop rotation research from different areas, many benefits accrue as the rotation becomes longer. This is because weed, disease and insect pest problems generally decrease with an increase in years between repeat plantings of the same crop.

Residue from small grains provides good organic matter for soil building, and in the case of winter grains, the plants help to prevent erosion over winter after soybeans loosen up the soil.

The length of the growing season will determine how you fit in cover crops after full-season soybeans in the rotation. Consider using a short-season bean if needed in order to achieve timely planting after soybean harvest. Calculate whether cover crop benefits will compensate for a possible yield loss on the shorter season beans. If there is not enough time to seed a legume after harvest, use a small grain rather than no cover crop at all.

The small grain scavenges leftover N following beans. Legume cover crops reduce fertilizer N needed by corn, a heavy N feeder. If you cannot seed the legume at least six weeks before a hard frost, consider overseeding before leaf drop or at last cultivation.

▼ **Precaution.** Because hairy vetch is hard seeded, it will volunteer in subsequent small grain crops.

An alternate rotation for the lower mid-South is corn>crimson clover (allowed to go to seed) > soybeans > crimson clover (reseeded) > corn. Allow the crimson clover to go to seed before planting beans. The clover germinates in late summer under the beans. Kill the cover crop before corn the next spring. If possible, choose a different cover crop following the corn this time to avoid potential pest and disease problems with the crimson clover.

▼ **Precaution.** In selecting a cover crop to interseed, do not jeopardize your cash crop if soil moisture is usually limiting during the rest of the corn season! Banding cover crop seed in row middles by using insecticide boxes or other devices can reduce cover crop competition with the cash crop.

3 Year: Corn>Soybean>Wheat/Red Clover. This well-tested Wisconsin sequence provides N for corn as well as weed suppression and natural control of disease and insect pests. It was more profitable in recent years as the cost of synthetic N increased. Corn benefits from legume-fixed N, and from the improved cation exchange capacity in the soil that comes with increasing organic matter levels.

Growers in the upper Midwest can add a small grain to their corn>bean rotation. The small grain, seeded after soybeans, can be used as a cover crop, or it can be grown to maturity for grain. When growing wheat or oats for grain, frost-seed red clover or sweet-clover in March, harvest the grain, then let the clover grow until it goes dormant in late fall. Follow with corn the next spring. Some secondary tillage can be done in the fall, if conditions allow. One option is to attach sweeps to your chisel plow and run them about 2 inches deep, cutting the clover crowns.

Alternatively, grow the small grain to maturity, harvest, then immediately plant a legume cover crop such as hairy vetch or berseem clover in July or August. Soil moisture is critical for quick germination and good growth before frost. For much of the northern U.S., there is not time to plant a legume after soybean harvest, unless it can be seeded aerially or at the last cultivation. If growing spring grains, seed red clover or sweet-clover directly with the small grain.

An Iowa study compared no-till and conventional tillage corn>soybean>wheat/clover rotation with annual applications of composted swine manure. Berseem clover or red clover was frostseeded into wheat in March. Corn and soybean yields were lower in no-till plots the first year, while wheat yield was not affected by tillage. With yearly application of composted swine manure, however, yield of both corn and soybean were the same for both systems beginning in year two of the 4-year study (385).

Adding a small grain to the corn>soybean rotation helps control white mold on soybeans, since two years out of beans are needed to reduce pathogen populations. Using a grain/legume mix will scavenge available N from the bean crop, hold soil over winter and begin fixing N for the corn. Clovers or vetch can be harvested for seed, and red or yellow clover can be left for the second year as a green manure crop.

Using a spring seeding of oats and berseem clover has proved effective on Iowa farms that also have livestock. The mix tends to favor oat grain production in dry years and berseem production in wetter years. Either way the mixture provides biomass to increase organic matter and build soil. You can clip the berseem several times before flowering for green manure.

▼ **Precaution.** Planting hairy vetch with small grains may make it difficult to harvest a clean grain crop. Instead, seed vetch after small grain harvest. Be sure to watch for volunteer vetch in subsequent small grain crops. It is easily controlled with herbicides but will result in significant penalties if found in wheat grain at the elevator.

COVER CROPS FOR VEGETABLE PRODUCTION

Vegetable systems have many windows for cover crops. Periods of one to two months between harvest of early planted spring crops and planting of fall crops can be filled using fast-growing warm-season cover crops such as buckwheat, cowpeas, sorghum-sudangrass hybrid, or another crop adapted to your conditions. As with other cropping systems, plant a winter annual cover crop on fields that otherwise would lie fallow.

Where moisture is sufficient, many vegetable crops can be overseeded with a cover crop, which will then be established and growing after vegetable harvest. Select cover crops that tolerate shade and harvest traffic, especially where there will be multiple pickings.

Cover crop features: Oats add lots of biomass, are a good nurse crop for spring-seeded legumes, and winterkill, doing away with the need for spring killing and tilling. **Sorghum-sudangrass hybrid** produces deep roots and tall, leafy stalks that die with the first frost. **Yellow sweetclover** is a deep rooting legume that provides cuttings of green manure in its second year. **White clover** is a persistent perennial and good N source. **Brassicas and mustards** can play a role in pest suppression and nutrient management. Mixtures of **hairy vetch and cereal rye** are increasing used in vegetable systems to scavenge nutrients and add N to the system.

Residue from small grains provides organic matter for soil building. . . to prevent erosion over winter.

In Zone 5 and cooler, plant rye, oats or a summer annual (in August) after snap bean or sweet corn harvest for organic matter production and erosion control, especially on sandy soils. Spray or incorporate the following spring, or leave unkilled strips for continued control of wind erosion.

If you have the option of a full year of cover crops in the East or Midwest, plant hairy vetch in the spring, allow to grow all year, and it will die back in the fall. Come back with no-till sweet or field corn or another N-demanding crop the following spring. Or, hairy vetch planted after about August 1 will overwinter in most zones with adequate snow coverage. Allow it to grow until early flower the following spring to achieve full N value. Kill for use as an organic mulch for no-till transplants or incorporate and plant a summer crop.

Full-Year Covers Tackle Tough Weeds

TROUT RUN, Pa.—Growing cover crops for a full year between cash crops, combined with intensive tillage, helps Eric and Anne Nordell control virtually every type of weed nature throws at their vegetable farm—even quackgrass. They are also manipulating the system to address insects.

The couple experimented with many different cover crops on their north-central Pennsylvania farm while adapting a system to battle quackgrass. Originally modeled on practices developed on a commercial herb farm in the Pacific Northwest, the Nordells continue to make modifications to fit their ever-changing conditions.

In the fallow year between cash crops, the Nordells grow winter cover crops to smother weeds and improve soil. Combined with summer tillage, the cover crops keep annual weeds from setting seed. Cognizant of the benefits of reduced tillage, they continue to modify their tillage practices—reducing tillage intensity whenever possible.

Regular use of cover crops in the year before vegetables also improves soil quality and moisture retention while reducing erosion. "Vegetable crops return very little to the soil as far as a root system," says Eric, a frequent speaker on the conference circuit. "You cut a head of lettuce and have nothing left behind. Growing vegetables, we're always trying to rebuild the soil."

Continual modification to their system is the name of the game. When they set up their original 4-year rotation in the early 80's, tarnished plant bugs were not an issue on their farm, but in the 90's they became a major problem in lettuce. The problem—and the solution!—was in their management of yellow sweetclover.

In their original rotation, sweetclover was overseeded into early cash crops such as lettuce. After overwintering, the sweetclover was mowed several times the following year before plowing it under and planting late vegetable crops. When the tarnished plant bugs began moving in—possibly attracted by the flowering sweetclover—the Nordells realized that mowing the sweetclover caused the plant bugs to move to the adjacent lettuce fields. It was time to change their system.

Fully committed to the use of cover crops, they first tried to delay mowing of the sweetclover until after lettuce harvest. Eventually, they decided to revamp their clover management completely. They now plant sweetclover in June of the second or fallow year of the rotation. This still gives the sweetclover plenty of time to produce a soil-building root system before late vegetables. It flowers later, so they are no longer mowing it and forcing tarnished plant bugs into lettuce fields.

Yellow blossom sweetclover—one of the best cover crop choices for warm-season nitrogen production—puts down a deep taproot before winter if seeded in June or July, observes Eric. "That root system loosens the soil, fixes nitrogen, and may even bring up minerals from the subsoil with its long tap root."

You can sow annual ryegrass right after harvesting an early-spring vegetable crop, allow it to grow for a month or two, then kill, incorporate and plant a fall vegetable.

Some farmers maximize the complementary weed-suppressing effects of various cover crop species by orchestrating peak growth periods, rooting depth and shape, topgrowth differences and species mixes. See *Full-Year Covers Tackle Tough Weeds* (above).

3 Year: Winter Wheat/Legume Interseed> Legume>Potatoes. This eastern Idaho rotation conditions soil, helps fight soil disease and provides N. Sufficient N for standard potatoes depends on rainfall being average or lower to prevent leaching that would put the soil N below the shallow-rooted cash crop.

2 Year Options: For vegetable systems in the Pacific Northwest and elsewhere, plant a winter wheat cover crop followed by sweet corn or

Originally part of their weed management program, Eric points out that the clover alone would not suppress weeds. It works on their farm because of their successful management efforts over a decade to suppress overall weed pressure using intensive tillage, crop rotation and varied cover crops. The same concept applies to the tarnished plant bug. Never satisfied with a single strategy rather than a whole-system approach, the Nordells also began interseeding a single row of buckwheat into successive planting of short term cash crops like lettuce, spinach and peas.

The idea was to create a full-season insectiary in the market garden, moderating the boom and bust cycle of good and bad insects. They also hoped that the buckwheat would provide an alternate host for the plant bugs. The strategy seems to be working. Data collected as part of a research project with the Northeast Organic Network (NEON) found very few tarnished plant bugs in their lettuce but lots in the buckwheat insectiary.

The two pronged cover crop approach using buckwheat and a different management regime for sweetclover seems to be doing the trick. The next step, currently being evaluated, is to mix Italian ryegrass with the sweetclover to increase root mass and sod development between June planting and frost.

Rye and vetch are a popular combination to manage nitrogen. The rye takes up excess N from the soil, preventing leaching. The vetch fixes additional nitrogen which it releases after it's killed the following spring. With the August seeding, the Nordells' rye/vetch mixture produces "a tremendous root system" and much of its biomass in fall.

The Nordells plow the rye/vetch mix after it greens up in late March to early April, working shallowly so as not to turn up as many weed seeds. They understand that such early kill sacrifices some biomass and N for earlier planting of their cash crop—tomatoes, peppers, summer broccoli or leeks—around the end of May.

Thanks to their weed-suppressing cover crops, the Nordells typically spend less than 10 hours a season hand-weeding their three acres of cash crops, and never need to hire outside weeding help. "Don't overlook the cover crops' role in improving soil tilth and making cultivation easier," adds Eric. Before cover cropping, he noticed that their silty soils deteriorated whenever they grew two cash crops in a row. "When the soil structure declines, it doesn't hold moisture and we get a buildup of annual weeds," he notes.

The Nordells can afford to keep half their land in cover crops because their tax bills and land value are not as high as market gardeners in a more urban setting. "We take some land out of production, but in our situation, we have the land," Eric says. "If we had to hire people for weed control, it would be more costly."

To order a video describing this system ($10 postpaid) or a booklet of articles from the Small Farmers' Journal ($12 postpaid), write to Eric and Anne Nordell, Beech Grove Farm, 3410 Route 184, Trout Run, PA 17771.

Updated in 2007 by Andy Clark

onions. Another 2-yr. option is green peas > summer sorghum-sudangrass cover crop > potatoes (in year 2). Or, seed mustard green manure after winter or spring wheat. Come back with potatoes the following year. For maximum biofumigation effect, incorporate the mustard in the fall (see *Brassicas and Mustards*, pp. 81).

1 Year: Lettuce>Buckwheat>Buckwheat>Broccoli>White Clover/Annual Ryegrass. The Northeast's early spring vegetable crops often leave little residue after their early summer harvest. Sequential buckwheat plantings suppress weeds, loosen topsoil and attract beneficial insects. Buckwheat is easy to kill by mowing in preparation for fall transplants. With light tillage to incorporate the relatively small amount of fast-degrading buckwheat residue, you can then sow a winter grass/legume cover mix to hold soil throughout the fall and over winter. Planted at least 40 days before frost, the white clover should overwinter and provide green manure or a living mulch the next year.

California Vegetable Crop Systems Innovative work in California includes rotating cover crops as well as cash crops, adding diversity to the system. This was done in response to an increase in *Alternaria* blight in LANA vetch if planted year after year.

4 Year: LANA Vetch>Corn>Oats/Vetch> Dry Beans>Common Vetch>Tomatoes>S-S Hybrid/ Cowpea>Safflower. The N needs of the cash crops of sweet corn, dry beans, safflower and canning tomatoes determine, in part, which covers to grow. Corn, with the highest N demand, is preceded by LANA vetch, which produces more N than other covers. Before tomatoes, common vetch works best. A mixture of purple vetch and oats is grown before dry beans, and a mix of sorghum-sudangrass and cowpeas precedes safflower.

In order to get maximum biomass and N production by April 1, LANA vetch is best planted early enough (6 to 8 weeks before frost) to have good growth before "winter." Disked in early April, LANA provides all but about 40 lb. N/A to the sweet corn crop. Common vetch, seeded after the corn, can fix most of the N required by the subsequent tomato crop, with about 30 to 40 lb.N/A added as starter.

A mixture of sorghum-sudangrass and cowpeas is planted following tomato harvest. The mixture responds to residual N levels with N-scavenging by the grass component to prevent winter leaching. The cowpeas fix enough N for early growth of the subsequent safflower cash crop, which has relatively low initial N demands. The cover crop breaks down fast enough to supply safflower's later-season N demand.

▼ **Precaution.** If you are not using any herbicides, vetch could become a problem in the California system. Earlier kill sacrifices N, but does not allow for the production of hard seed that stays viable for several seasons.

ANNUAL *and* PERENNIAL MEDIC *cultivars can fix N on low moisture and can reduce erosion in dryland areas compared with bare fallow between crop seasons.*

COVER CROPS FOR COTTON PRODUCTION

In what would otherwise be continuous cotton production, any winter annual cover crop added to the system can add rotation benefits, help maintain soil productivity, and provide the many other benefits of cover crops highlighted throughout this book.

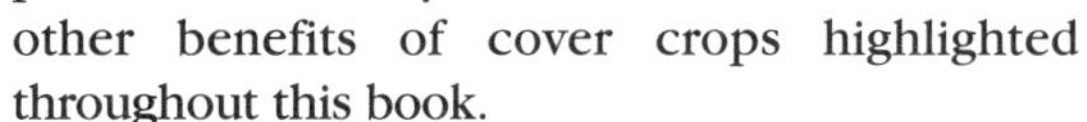

Hairy vetch, crimson clover, or mixtures with rye or another small grain can reduce erosion, add N and organic matter to the system. Drill after shredding stalks in the fall and kill by spraying or mowing prior to no-till seeding of cotton in May. Or, aerially seed just before application of defoliant. The dropping leaves mulch the cover crop seed, aiding germination. Rye works better than wheat. Yields are usually equal to, or greater than yields in conventional tillage systems with winter fallow.

Balansa clover, a promising cover crop for the South, reseeds well in no-till cotton systems (see *Up-and-Coming Cover Crops*, p. 191).

1 Year: Rye/Legume>Cotton. Plant the rye/legume mix in early October, or early enough to allow the legume to establish well before cooler winter temperatures. Kill by late April, and if soil moisture permits, no-till plant cotton within three to five days using tined-wheel row cleaner attachments to clear residue. Band-spray normal preemergent herbicides over the cleaned and planted row area. Cotton will need additional

Start Where You Are

In many instances, you can begin using cover crops without substantially altering your cash crop mix or planting times or buying new machinery. Later, you might want to change your rotation or other practices to take better advantage of cover crop benefits.

We'll use a basic Corn Belt situation as a model. From a corn>soybean rotation, you can expand to:

Corn>Cover>Soybean>Cover. Most popular choices are rye or rye/vetch mixture following corn; vetch or rye/vetch mixture following beans. Broadcast or drill covers immediately after harvest. Hairy vetch needs at least 15 days before frost in 60° F soil. Rye will germinate as long as soil is just above freezing. Drill for quicker germination. Consider overseeding at leaf-yellowing if your post-harvest planting window is too short.

If you want to make certain the legume is well established for maximum spring N and biomass production, consider adding a small grain to your rotation.

Corn>Soybean>Small Grain/Cover. Small grains could be oats, wheat or barley. Cover could be vetch, field peas or red clover. If you want the legume to winterkill to eliminate spring cover crop killing, try a non-hardy cultivar of berseem clover or annual alfalfa.

If you have livestock, a forage/hay market option or want more soil benefits, choose a longer-lived legume cover.

Corn>Soybean>Small Grain/Legume> Legume Hay, Pasture or Green Manure. Yellow sweetclover or red clover are popular forage choices. An oats/berseem interseeding provides a forage option the first year. Harvesting the cover crop or terminating it early in its second season opens up new options for cash crops or a second cover crop.

Late-season tomatoes, peppers, vine crops or sweet corn all thrive in the warm, enriched soil following a green manure. Two heat-loving covers that could be planted after killing a cool-season legume green manure are buckwheat (used to smother weeds, attract beneficial insects or for grain harvest) and sorghum-sudangrass hybrid (for quick plow-down biomass or to fracture compacted subsoil).

These crops would work most places in the Corn Belt. To get started in your area, check *Top Regional Cover Crop Species* (p. 66) to fill various roles, or *Cultural Traits and Planting* (p. 69) to find which cover crops fit best in your system.

weed control toward layby using flaming, cultivation or directed herbicides. Crimson clover, hairy vetch, Cahaba vetch and Austrian winter peas are effective legumes in this system.

Multiyear: Reseeding Legume>No-Till Cotton> Legume>No-Till Cotton. Subterranean clover, Southern spotted burclover, balansa clover and some crimson clover cultivars set seed quickly enough in some areas to become perpetually reseeding when cotton planting dates are late enough in spring. Germination of hard seed in late summer provides soil erosion protection over winter, N for the following crop and an organic mulch at planting.

Strip planting into reseeding legumes works for many crops in the South, including cotton, corn, sweet potatoes, peanuts, peppers, cucumbers, cabbage and snap beans. Tillage or herbicides are used to create strips 12 to 30 inches wide. Wider killed strips reduce moisture competition by the cover crop before it dies back naturally, but also reduce the amount of seed set, biomass and N produced. Wider strips also decrease the mulching effect from the cover crop residue. The remaining strips of living cover crop act as in-field insectary areas to increase overall insect populations, resulting in more beneficial insects to control pest insects.

▼ Precautions

- Watch for moisture depletion if spring is unusually dry.
- Be sure to plant cotton by soil temperature (65° F is required), because cover crops may keep soil cool in the spring. Don't plant too early!
- A delay of two to three weeks between cover crop kill and cotton planting reduces these problems, and reduces the chance of stand losses due to insects (cutworm), diseases or allelopathic chemicals.
- Additional mid-summer weed protection is needed during the hot-season "down time" for the reseeding legumes.
- Reseeding depends on adequate hard seed production by the clovers. Dry summer weather favors hard seed production while wet summers reduce the percentage of hard seed.

DRYLAND CEREAL-LEGUME CROPPING SYSTEMS

Soil moisture availability and use by cover crops are the dominant concerns in dryland production systems. Yet more and more innovators are finding that carefully managed and selected cover crops in their rotations result in increased soil moisture availability to their cash crops. They are finding ways to incorporate cover crops into flexible rotations that can be modified to capitalize on soil moisture when available while preventing adverse effects on cash crops. This delicate balance between water use by the cover crop and water conservation—particularly in conservation tillage systems—will dictate, in part, how cover crops work in your rotation. See also *Managing Cover Crops in Conservation Tillage Systems* (p. 44).

Perennial legumes provide numerous benefits to grain cropping systems in the Northern Plains, including increased grain yield, nutrient scavenging, carbon sequestration, breaking weed and insects cycles and for use as feed (129).

Cover crop features: **perennial medics** persist due to hard seed (of concern in some systems), providing green manure and erosion control; **field peas** and **lentils** (grain legumes) are shallow-rooted yet produce crops and additional N in years of good rainfall.

An excellent resource describing these rotations in detail is *Cereal-Legume Cropping Systems: Nine Farm Case Studies in the Dryland Northern Plains, Canadian Prairies and Intermountain Northwest* (258).

7 to 13 Years: Flax>Winter Wheat>Spring Barley>Buckwheat>Spring Wheat>Winter Wheat>Alfalfa (up to 6 years) >Fallow

System sequences are:

- Flax or other spring crops (buckwheat, wheat, barley) are followed by fall-seeded wheat (sometimes rye), harvested in July, leaving stubble over the winter;
- Spring-seeded barley or oats, harvested in August, leaving stubble over the winter;
- Buckwheat, seeded in June and harvested in October, helps to control weeds;
- A spring small grain, which outcompetes any volunteer buckwheat (alternately, fall-seeded wheat, or fall-seeded sweetclover for seed or hay).

The rotation closes with up to 6 years of alfalfa, plowdown of sweetclover seeded with the previous year's wheat or an annual legume green manure such as Austrian winter peas or berseem clover.

There are many points during this rotation where a different cash crop or cover crop can be substituted, particularly in response to market conditions. Furthermore, with cattle on the ranch, many of the crops can be grazed or cut for hay.

Moving into areas with more than 12 inches of rain a year opens additional windows for incorporating cover crops into dryland systems.

9 Year: Winter Wheat>Spring Wheat> Spring Grain/Legume Interseed>Legume Green Manure/Fallow>Winter Wheat> Spring Wheat>Grain/Legume Interseed> Legume> Legume. In this rotation, one year of winter wheat and two years of spring-seeded crops follow a two or three-year legume break. Each legume sequence ends with an early summer incorporation of the legume to save moisture followed by minimal surface tillage to control weeds. Deep-rooted winter wheat follows sweetclover, which can leave surface soil layers fairly dry. Spring-seeded grains prevent weeds that show up with successive winter grain cycles and have shallower roots that allow soil moisture to build up deeper in the profile.

In the second spring-grain year, using a low-N demanding crop such as kamut wheat reduces the risk of N-deficiency. Sweetclover seeded with the kamut provides regrowth the next spring that helps to take up enough soil water to prevent saline seep. Black medic, INDIANHEAD lentils and field peas are water-efficient substitutes for the deep-rooted—and water hungry—alfalfa and sweetclover. These peas and lentils are spring-sown, providing back-up N production if the forage legumes fail to establish.

While moisture levels fluctuate critically from year to year in dryland systems, N levels tend to be more stable than in the hot, humid South, and adding crop residue builds up soil organic matter more easily. Careful management of low-water use cover crops can minimize soil water loss while adding organic matter and N. Consequently, dryland rotations can have a significant impact on soils and the field environment when used over a number of years.

These improved soils have higher organic matter, a crumbly structure, and good water retention and infiltration. They also resist compaction and effectively cycle nutrients from residue to subsequent crops.

Remember, the benefits of cover crops accrue over several years. You will see improvements in crop yield, pest management and soil tilth if you commit to cover crop use whenever and wherever possible in your rotations.

MANAGING COVER CROPS IN CONSERVATION TILLAGE SYSTEMS

by Kipling Balkcom, Harry Schomberg, Wayne Reeves and Andy Clark
with Louis Baumhardt, Hal Collins, Jorge Delgado, Sjoerd Duiker, Tom Kaspar and Jeffrey Mitchell

INTRODUCTION

Conservation tillage is defined as a system that leaves enough crop residue on the soil surface after planting to provide 30% soil cover, the amount needed to reduce erosion below tolerance levels (SSSA). Today, however, most conservation tillage practitioners aim for greater soil cover because of additional benefits of crop residue. Cover crops are critical to producing this residue and have the potential to maximize conservation tillage benefits.

Benefits of *conservation tillage* systems include:

- reduced soil erosion
- decreased labor and energy inputs
- increased availability of water for crop production
- improved soil quality

Cover crops further benefit conservation tillage systems by:

- producing crop residues that increase soil organic matter and help control weeds
- improving soil structure and increasing infiltration
- protecting the soil surface and dissipating raindrop energy
- reducing the velocity of water moving over the soil surface
- anchoring soil and adding carbon deep in the soil profile (via roots)

Conservation tillage has been adopted on more and more acres since the 1970's thanks to improvements in equipment, herbicides and other technologies. Several long-term, incremental benefits of conservation tillage have emerged. The most important benefits have been attributed to the accumulation of organic matter at the soil surface.

This accumulation of surface organic matter results in:

- increased aggregate stability, which helps to increase soil water infiltration and resist erosion
- improved nutrient cycling and water quality, due to keeping nutrients in the field
- increased biological activity, which improves nutrient cycling and can influence diseases and pests

Additional benefits from conservation tillage systems compared to intensive or conventional tillage systems (89) include:

- Reduced labor and time—one or two trips to prepare land and plant compared to three or more reduces labor and fuel costs by 50% or more.
- Reduced machinery wear—fewer trips mean fewer repairs.
- Increased wildlife—crop residues provide shelter and food for wildlife, such as game birds and small animals, which can result in additional farm revenue.
- Improved air quality—by reducing wind erosion (amount of dust in the air), fossil fuel emissions from tractors (fewer trips) and release of carbon dioxide into the atmosphere (tillage stimulates the release of carbon from organic matter).

In an Iowa study comparing no-till and conventional tillage in a corn>soybean>wheat/clover rotation, corn and soybean yields were lower in no-till plots the first year. With yearly application of composted swine manure, however, yield of both corn and soybean were the same for both systems beginning in year two of the study. Wheat yields were not affected by tillage, but increased with compost application (385).

Cover crop contributions to conservation tillage systems

Biomass. Conservation tillage systems depend on having crop residues on the soil surface for most of the year. Cover crops help provide the additional biomass needed to meet this year-round requirement. A typical high residue cover crop should contain at least 4000 lb. biomass/A.

In low-fertility soils, you can increase biomass production of grass cover crops with the addition of a small amount of N fertilizer. Cover crops grown in soils with higher levels of organic matter, or following a legume summer crop like soybean, may not need additional N fertilizer. Remember, minimal cover crop residue or biomass translates into minimal benefits.

Soil improvement. Cover crop biomass is a source of organic matter that stimulates soil biological activity. Soil organic matter and cover crop residues improve soil physical properties, resulting in:

- greater water infiltration, due to direct effects of the residue coverage or to changes in soil structure
- greater soil aggregation or tilth, resulting in better nutrient and moisture management
- less surface sealing, because residue intercepts rain drops, reducing the dispersal of clay particles during a rainfall or irrigation event
- greater soil porosity, due to the macropores that are formed as roots die and decompose

Improvements in soil physical properties depend on soil type, crops grown and residue management system, as well as temperature and rainfall. Regardless of soil type, however, tillage will very quickly negate cover crop benefits associated with increased soil organic matter. Simply put, tillage breaks down organic matter much faster than no-till.

Improvements in soil physical properties due to cover crops have been documented widely in conservation tillage systems (25, 52, 106, 114, 115, 119, 238, 318, 419).

Erosion control. Cover crops and conservation tillage combine to reduce soil erosion and wind erosion (26, 115, 119, 223, 267).

In Kentucky, on a Maury silt loam soil with a 5% slope, soil loss was 8 tons/A for conventionally tilled corn with the corn residue and cover crop turned under in the spring. In contrast, for no-tillage corn with 3 tons/A of corn residue remaining on the soil surface, soil loss was 1 ton/A without a cover crop and 0.9 tons/A with a winter cover crop (91, 151).

Tillage breaks down organic matter much faster than no-till.

In Missouri, on a Mexico series silt loam soil with a claypan, inclusion of a rye or wheat cover crop reduced soil loss in no-tillage silage corn from 9.8 to 0.4 tons/A/year (437).

Rotation effects. Crop rotation provides numerous benefits to any cropping system. It is critical to reducing the incidence of diseases and pests, and is also credited with improving nutrient use and reducing weeds. Cover crops increase the complexity and intensity of crop rotations, effectively increasing crop rotation benefits. Note, however, as addressed throughout this book, that cover crops can adversely affect other crops in the rotation.

Cover crop management in conservation tillage systems

Nutrient management. Nitrogen and phosphorus are the two macronutrients most likely to be lost from cropping systems. Cover crops help reduce losses of these nutrients by:

- increasing infiltration—thus reducing surface runoff and erosion
- taking up nutrients—or acting as a 'catch crop'
- using water for cover crop growth during peak leaching season (late fall through early spring)—reducing the amount of water available to leach nutrients

Grasses and brassicas are better than legumes at reducing N leaching (106, 234, 265). Cereal rye is very effective at reducing N leaching because it is cold tolerant, has rapid growth, and produces a large quantity of biomass (111). Winter annual weeds do not effectively reduce N losses.

Cover crops may reduce the efficiency of N fertilizer in no-till systems, depending on the method of application. Surface applications of urea-containing fertilizers to soils with large amounts of cover crop residues can result in large losses of ammonia N. When applied to the soil surface, urea and urea-ammonium nitrate (UAN) solutions volatilize more than ammonium nitrate and subsequently lose more N to the atmosphere.

Injecting urea-containing fertilizers into the soil eliminates volatilization losses. Banding urea-containing fertilizers also reduces volatilization losses because banding minimizes fertilizer and residue contact, while increasing fertilizer and soil contact.

Nitrogen dynamics with nonlegume cover crops. Differences between nonlegumes and legume cover crops are mostly related to nitrogen management. Legumes fix N while nonlegumes can only use N already in the soil. Legume residues usually contain more total N that is more readily available to subsequent crops.

The addition of fresh crop residues stimulates growth of soil microbes and increases microbial demand for nutrients, particularly N. Microorganisms use C, N and other nutrients as a food source in order to break down the residues. If the amount of N in the residues is too low, the microorganisms use soil N instead, reducing N availability to the cash crop. This is called *N immobilization*. If the amount of N in the residues is greater than microbial demand, N is released and N availability for plant growth is increased, a process called *N mineralization*.

Small grain and other grass cover crops usually result in an initial, if not persistent, immobilization of N during the cash crop season. The N *content* of small grain cover crop residues varies greatly, but generally ranges from 20 to 50 lb./A for the aboveground biomass and 8 to 20 lb./A in the roots. The N *contribution* from small grain cover crops depends on N availability during the cover crop growing period, the total amount of biomass produced and the growth stage when the cover is terminated.

The **carbon to nitrogen ratio** (C:N ratio) of cover crop residue is a good indicator of whether immobilization or mineralization will occur. Values exceeding 30 parts carbon to one part nitrogen (C:N ratio of 30:1) are generally expected to immobilize N during the early stages of the decomposition process. For more information about C:N ratios and cover crop nutrient dynamics, see *Building Soil Fertility and Tilth with Cover Crops* (pp. 16).

The C:N ratio of small grain residues is mostly dependent on time of termination. Early termination of grass cover crops results in a narrower C:N ratio, typical of young plant tissue. If killed too early, this narrower C:N ratio results in rapid decomposition of a smaller amount of residue, reducing ground coverage.

Because of the need for residue in conservation tillage systems, small grain cover crops are often allowed to grow as long as possible. Termination date depends on crop rotation and climate. When small grain cover crops are killed at flowering, the C:N ratio is usually greater than 30:1.

In Pennsylvania, delaying rye termination date from early to late boot stage increased average above-ground dry matter accumulation from 1200 to 3700 lb./A with no negative effect on corn yield (118).

In Alabama, rye, black oat and wheat cover crops were terminated at different growth stages with a roller or roller-herbicide combinations. Biomass production was about 2.0 - 2.6 tons/A at the flag leaf stage and corresponding C:N ratio averaged 25:1, regardless of cover crop species.

At flowering, biomass averaged 4.2, 3.8, and 3.3 tons/A for black oat, rye and wheat, respectively, and the C:N ratio for all covers was 36:1. Killing at soft dough stage did not increase biomass production for any of the covers, but did increase the C:N ratios, which would increase N immobilization (13).

The wide C:N ratio of small grain residues must be taken into account for best N management. Nitrogen fertilizer rates for cash crops may need to be increased 25 to 30 lb./A following a high residue cereal cover crop.

In N-limited soils, early-season growth of the cash crop is usually enhanced if this N is applied as starter fertilizer. Although yield increases from starter N applications are dependent on rainfall and crop, they occur frequently enough to justify the practice. Starter fertilizer promotes more rapid canopy development, which reduces weed competition and can offset the negative effects of cool, wet soils often experienced with conservation tillage systems. Ideally, starter fertilizers should be placed near the seeding row in a 2 X 2 band, i.e. 2 inches to the side and 2 inches below the seed.

Legumes add N. Legume cover crops obtain nitrogen from the atmosphere through a symbiotic relationship with nitrogen fixing bacteria. The N content of legume cover crops and the amount of N available to subsequent crops is affected by:

- Legume species and adaptation to soil and climatic conditions
- residual soil N
- planting date
- termination date

Cover crop management affects the N content of legume cover crops and the contribution of N to the following cash crop. Early establishment of legume cover crops results in greater biomass production and N production. The nitrogen content of legume cover crops is optimized at the flowering stage. Legumes can contribute from 15 to 200 lb. N/A to the subsequent crop, with typical values of 50 to 100 lb./A.

In North Carolina, delaying the kill date of crimson clover 2 weeks beyond 50% bloom, and hairy vetch 2 weeks beyond 25% bloom increased the biomass of clover by 41% and vetch by 61%. Corresponding increases in N content were 23% for clover and 41% for vetch (427). In Maryland, hairy vetch fixed about 2 lb. N/acre/day from April 10 to May 5, resulting in an additional 60 lb. N/A in aboveground biomass (82, 83, 86).

The C:N ratio of mature legume residues varies from 25:l to 9:1 and is typically well below 20:1, the guideline threshold where rapid mineralization of the N in the residue occurs. Residues on the soil surface decompose more slowly than those incorporated in conventional tillage systems. Consequently, in conservation tillage systems, legume-residue N may not be readily available during the early part of the growing season.

Due to the initial lag in availability of N from legume cover crop residues, any additional fertilizer N should be applied to cash crops at planting in conservation-tillage winter annual legume systems. Splitting N applications to corn grown in these systems, as is generally recommended for conventional-tilled corn grown without legume cover crops, is not necessary (347).

Grass-legume mixtures. Mixtures of grass and legume cover crops provide the same benefits to conservation tillage but often mitigate the nitrogen immobilization of pure grass cover crops. The grass component scavenges residual nitrogen effectively, while the legume adds fixed nitrogen that is more readily available to the cash crop (86, 343, 344, 345).

The C:N ratio of grass-legume mixtures is usually intermediate to that of pure stands. In several studies in Maryland, the C:N ratio of mixtures of hairy vetch and rye never exceeded 25:1; the C:N ratio of pure rye ranged from 30:1 to 66:1 across several spring kill dates (81, 83, 84, 85, 86).

Water availability. Cover crops use soil water while they are growing. This can negatively affect cash crop yields. Once killed, however, cover crop residues may increase water availability by increasing infiltration and reducing evaporation losses.

Short-term soil water depletion at planting may or may not be offset by soil water conservation later in the growing season. This is dependent on rainfall distribution in relation to crop development. A rainfall event following cover crop termination enables soil surface water recharge, which usually provides adequate soil moisture in humid regions to facilitate cash crop planting. Time of termination becomes more critical as the probability of precipitation decreases (423).

Cover crops increase water availability by:

- decreasing evaporation due to a mulching effect
- increasing infiltration of rainfall by decreasing runoff velocity
- increasing organic matter, which increases water-holding capacity
- improving soil structure and consequently increasing root interception of soil water

- protecting the soil surface from raindrop impact, thus reducing development of a surface seal or crust, which can greatly reduce infiltration

In Alabama, cover crop residue left on the soil surface reduced runoff and increased infiltration by 50 to 800% compared to removing or incorporating the residues (418, 419). In Georgia, infiltration rates were 100% greater *even after removal of crop residues* for a Cecil sandy loam when grain sorghum was no-till planted into crimson clover compared to a tilled seedbed without a cover crop (52).

In Maryland, pure and mixed stands of hairy vetch and rye did not deplete soil water or adversely affect corn yield. Rather, the additional residue from cover crops killed in early May conserved soil moisture and contributed to greater corn yield (84, 85).

In Kentucky, surface evaporation from May to September was five times less under no-till (which leaves a surface mulch) than with conventional tillage. Because less water was lost to evaporation, more water was available for the crop (91).

Cover crop use in *dryland systems* is often limited by moisture availability. A literature review of dryland cover crop studies on the Great Plains concluded that use of cover crops on dryland cropping systems of the Great Plains reduced yields of subsequent crops. However, in semi-arid Texas, 5 tons/A of wheat straw increased available soil water by 73% and more than doubled grain sorghum yields from 26 to 59 bushels/A (423).

The risk for early-season soil water depletion by cover crops is the same regardless of the tillage system. However, the full potential of cover crops to increase infiltration and conserve soil water can only be achieved in a conservation system where cover crop residues are left on the surface. Conservation tillage increased water use efficiency compared to a traditional wheat>fallow system with tillage (319, 135).

One way to reduce the risk of early-season soil water depletion by cover crops is to desiccate the cover some time prior to planting the cash crop. For example, yield reductions due to early-season depletion of soil water can be reduced by killing the cover crop 2 to 3 weeks before planting the cash crop (428, 290, 348). Depending on your situation, you could extend this window to terminate cover crops in conservation systems from 4 to 6 weeks prior to planting the cash crop.

Cover crops can sometimes be used to deplete soil water on poorly drained soils, allowing an earlier planting date for the cash crop, but the practical advantage of this practice is not certain.

Soil temperature. Cover crop residues keep the soil cooler, reduce daily fluctuations of soil temperature, and reduce soil temperature maximums and minimums. The cooler soil temperatures, which benefit the cash crop throughout the summer, can delay spring planting compared to a system without a cover crop.

Spring soil temperature is particularly important in cover crop/conservation tillage systems. Where possible, plant your cash crop according to soil temperature rather than the calendar. Follow local recommendations about the appropriate soil temperature for your cash crop. As noted below, the use of row cleaners will allow faster soil warmup.

The harmful effects of planting when the soil temperature is too low were demonstrated in Colorado for conservation tillage with continuous corn (but not cover crops). Low soil temperatures contributed to reduced corn yields over 5 years (171, 172).

Insects and diseases. Conservation tillage systems alter pest dynamics, due in large part to residues left on the soil surface. Conservation tillage systems with surface residues create a more diverse plant/soil ecosystem than conventional tillage systems (137, 185, 416).

Cover crops may harbor insects, diseases, and nematodes that could be harmful to the cash crop. Before planting a cover crop, be sure to investigate specific pest/crop interactions that may become a problem (100). Understanding these interactions and the conditions that favor them helps you make good management decisions. For example:

- Cereal rye, orchardgrass and crimson clover may attract armyworms.

- Clover root curculio, a common pest of red clover, can attack alfalfa.
- Chickweed can attract black cutworm or slugs.
- Johnsongrass is a host to maize dwarf mosaic virus (MDMV), which also infects corn.

Conversely, cover crops can be used in conservation tillage systems to attract beneficial insects. One approach is to allow a live strip of cover crops to remain between crop rows to serve as habitat and a food source until the main crop is established. This approach resulted in one less insecticide application in conservation-tilled cotton compared to conventional cotton in South Georgia (368, 416).

For more information about cover crops and beneficial insects, see *Manage Insects on Your Farm: A Guide to Ecological Strategies* (409, http://www.sare.org/publications/insect.htm).

Cover crop residues have been shown to reduce the incidence of several diseases in many different cash crop systems by reducing splash dispersal of pathogens. Small grain cover crops in conservation tillage have also been shown to reduce peanut yield losses from Tomato Spotted Wilt Virus (TSWV), with greater residue amounts resulting in lower incidence of TSWV. This benefit was directly related to less incidence of damage from thrips, the vector of TSWV (49).

Some cover crops can serve as an overwintering host for nematodes and may thus increase the severity of nematode damage. This may be a greater concern where crops are not rotated, like continuous cotton in some areas of the South. On the other hand, cover crops such as brassicas can reduce nematode populations (48, 231, 283, 284, 285, 353, 430).

On a Maryland sandy soil, winterkilled forage radish increased bacteria-eating nematodes, rye and rapeseed increased the proportion of fungal feeding nematodes, while nematode communities without cover crops were intermediate. The Enrichment Index, which indicates a greater abundance of opportunistic bacteria-eating nematodes, was 23% higher in soils that had brassica cover crops than the unweeded control plots. These samples, taken in November, June (a month after spring cover crop kill), and August (under corn), suggest that the cover crops, living or dead, increased bacterial activity and may have enhanced nitrogen cycling through the food web (432).

Consult local experts to minimize the potential for pest problems

The need for sound crop rotation is greater in conservation systems than in conventional systems. Cover crops should be a key component of any conservation rotation system. With the vast number of potential combinations of crops, cover crops, and diseases, consult local experts to ensure that you manage cover crops in conservation tillage systems to minimize the potential for pest problems.

Weed management. Cover crops affect weeds and weed management in conservation tillage systems in several ways:

- Cover crops compete with weeds for light, water and nutrients.
- Cover crop residue can suppress weed seed germination; the more residue the better.
- Grass cover crops (high C:N ratio) usually provide longer-lasting residue than legumes.
- Some cover crops release weed-suppressing allelopathic compounds.
- Conservation tillage does not continually turn up new weed seeds for germination.
- Cover crops can become weeds.

Some legume, cereal and brassica cover crops release allelopathic compounds that can reduce weed populations and/or suppress weed growth (39, 45, 176, 177, 178, 336, 359, 410, 422). Unfortunately, these same allelopathic compounds can also stunt and/or kill cash crop seedlings, particularly cotton (24) and some small seeded vegetable crops. Allowing time between the termination date and the cash crop planting date reduces the risk to cash crops because these chemicals leach out of the cover crop residue and are decomposed by soil microorganisms.

Cereal rye is known to release phenolic and benzoic acids that can inhibit weed seed germination and development. In Arkansas, the concentration of these allelopathic chemicals varied 100-fold among ten varieties of rye in the boot stage, with the cultivar BONEL having the greatest concentration and PASTAR the least. Factoring in the yield of each cultivar with the concentration and activity of the inhibitors, BONEL, MATON and ELBON were considered the best rye cultivars for allelopathic use (66).

Conservation tillage and the allelopathic effects of cover crop residue can both contribute to the suppression of weeds in these systems (452). In Alabama, a conservation tillage system using rye or black oat cover crops eliminated the need for post-emergence herbicides in soybean and cotton (335, 349). Including rye or black oat increased yields of non-transgenic cotton in 2 of 3 years, compared to conservation tillage without a cover crop.

Economics of cover crop establishment and use

Using cover crops in any tillage system usually costs more time and money than not using cover crops. Depending on your particular system, you may or may not be repaid for your investment over the short term. If you are already using cover crops but are considering switching to conservation tillage, the economics are similar to using cover crops in conventional tillage systems, but the benefits may be expressed more in the conservation system (51).

Factors affecting the economics of cover crop use include:

- the cash crop grown
- the cover crop selected
- time and method of establishment
- method of termination
- the cash value applied to the environment, soil productivity and soil protection benefits derived from the cover crop.
- the cost of nitrogen fertilizer and the fertilizer value of the cover crop
- the cost of fuel

The economic picture is most affected by seed costs, energy costs and nitrogen fertility dynamics in cover crop systems. Cover crop seed costs vary considerably from year to year and from region to region, but historically, legume cover crops cost about twice as much to establish as small grain covers. The increased cost of the legume cover crop seed can be offset by the value of N that legumes can replace.

Depending on your system, legume cover crops can replace 45 to 100 lb. N/A. On the other hand, a rye cover crop terminated at a late stage of growth might require 20-30 lb. more N/A due to N immobilization by the wide C:N ratio rye residue. Thus, the difference in cost between a rye cover crop and a legume cover crop would be offset by the value of 65 to 125 lb. N/A. At a price of $0.45/lb. N, this would be worth $29 to $56/A.

Cover crop establishment in conservation tillage systems

The major challenges to cover crop adoption in both tilled and conservation tillage systems include seeding time and method, killing time and method, and cover crop residue management to ensure good stands of the cash crop. Success with cover crops requires adequate attention to each.

Plant cover crops on time. In order to maximize benefits—or to work at all—cover crops need to be planted early, sometimes before the summer crop is harvested. Timely planting results in:

- good root establishment and topgrowth before the crops go dormant
- reduced chance of winter kill
- more biomass production compared to later planting dates
- greater uptake of residual soil N

Timely fall planting is particularly important before early vegetables or corn. Corn is typically planted early in the spring, which forces an early cover crop termination date. A late planted cover crop that must be terminated early will not produce sufficient biomass to provide adequate soil protection and enhance soil quality.

Planting methods. Cover crops in conservation tillage systems are usually planted with a drill or broadcast on the soil surface, but several alternate methods can be used. Good soil-seed contact is required for germination and emergence. Most small seeded legumes require shallow seed placement (1/4 inch), while larger seeded legumes and small grains are generally planted up to 1.5 inches deep (see CHARTS, p. 62).

Conservation tillage drills can handle residue and provide uniform seeding depth and adequate seed-soil contact, even with small seeded cover crops. In some situations, preplant tillage can be used to control weeds and disrupt insect and disease life cycles.

Broadcast seeding requires an increase in the seeding rate compared to other methods (see CHARTS, p. 62). Broadcasting is often the least successful seeding method. Small-seeded species such as clovers tend to establish better by broadcasting than larger seed species. A drop-type or cyclone-type seeder can be used on small acreage and provides a uniform distribution of seed. Conventional drills work adequately in some conservation tillage systems—depending on the amount of residue—and may be more successful than broadcast seeding.

On larger areas, *aerial seeding* by fixed-wing aircraft or helicopter in late summer during crop die-down can be effective. As the leaves of the summer crop drop off, they aid germination by covering the seed, retaining moisture and protecting the soil.

In colder climates, *frost-seeding* can be used for some cover crop species (see *individual cover crop chapters in this book*). Seed is broadcast during late fall or early spring when the ground has been "honeycombed" by freezing and thawing. The seed falls into the soil cracks and germinates when the temperature rises in the spring.

Some legumes can be managed to reseed the following year. This reduces economic risks and seeding costs. Reseeding systems generally depend on well-planned rotations such as that reported in Alabama (311), where crimson clover was followed with strip-tilled soybean planted late enough to let the clover reseed. Corn was grown the next year in the reseeded clover. In this system, the cover crop is planted every other year rather than annually. Grain sorghum can be planted late enough in the South to allow crimson clover to reseed in a conservation-tillage system.

The introduction of legume cover crops that bloom and set seed earlier also improves their utility for reseeding in conservation-tillage systems. Auburn University in cooperation with USDA-NRCS has released several legume cover crops that flower early, including AU ROBIN and AU SUNRISE crimson clover, and AU EARLY COVER hairy vetch (288). Leaving 25 to 50% of the row area alive when desiccating the cover crop allows reseeding without reducing corn grain yields. However, the strips of live cover crop may compete with the cash crop for water, a potential problem during a spring drought.

Decisions about when to kill the cover crop must be site- and situation-specific.

Spring management of cover crops in conservation tillage systems

Kill date. Timing of cover crop termination affects soil temperature, soil moisture, nutrient cycling, tillage and planting operations, and the effects of allelopathic compounds on the subsequent cash crop. Because of the many factors involved, decisions about when to kill the cover crop must be site- and situation-specific.

There are a number of pros and cons of killing a cover crop early vs. late. Early killing:

- allows time to replenish soil water
- increases the rate of soil warming
- reduces phytotoxic effects of residues on cash crops
- reduces survival of disease inoculum
- speeds decomposition of residues, decreasing potential interference with planter operation
- increases N mineralization from lower C:N ratio cover crops

Advantages for later kill include:

- more residue available for soil and water conservation
- better weed control from allelopathic compounds and mulch affect
- greater N contribution from legumes
- better potential for reseeding of the cover crop

After 25 Years, Improvements Keep Coming

By Pat Sheridan, Sheridan Farms, Fairgrove, Mich as interviewed by Ron Ross for the No-Till Farmer

Talk to 10 no-tillers and you'll probably hear 10 different viewpoints on why it pays to quit disturbing and start building the soil. At Sheridan Farms, we've got our list, too. We are able to better time planting, weed control and other production chores. And we've got the potential for sediment and nutrient runoff into Saginaw Bay on Lake Huron under control.

Like a lot of no-tillers would testify, however, these changes didn't come quickly, nor without some reluctance and skepticism along the way. In our first years of no-tilling, starting in 1982, we did just about everything wrong and had an absolute train wreck. We overcame a few hurdles early on, started adding more no-till acres and were 100% continuous no-till by 1990.

Cover Crop Success

We started working with cover crops about 20 years ago. We deal with about a dozen different soil types, 80 percent of which are clay loam. And much of our land is poorly drained, low-organic-matter lake bed soils.

Cereal rye has been a good cover crop year in and year out for this mixture of soils. We like the AROOSTOCK variety from Maine because it provides fast fall and spring growth and its smaller seed size makes it more economical to plant.

In late August, we fly rye into standing corn and also into soybeans if we're coming back with soybeans the following year. We learned that rye is easier to burn down when it's more than 2 feet tall than when it has grown only a foot or less.

The rye crop also helps us effectively manage soil moisture. If it looks like we're going to get a dry spring, we burn down rye with Roundup as soon as we can; but if it's wet, we let the rye grow to suck up excess moisture. We can be very wet in the spring, but Michigan also receives less rain during the growing season than any other Great Lakes state, on average. Moisture management is critical to us.

We've seen less white mold in no-tilled soybeans wherever we have heavy residue. We've had years with zero white mold when our conventional till neighbors faced a costly problem. It's become a simple equation: the heavier the residue mat, the less white mold.

An open mind welcomes a lot of ideas that, with a little tweaking, can deliver even more success to your fields.

Deep-Rooted Crops

We're looking for a cover crop that will help establish a more diverse rotation, so we can always follow a broadleaf crop behind a grass crop and vice-versa. Oilseed radish is beginning to show real promise. It has about the same tremendous appetite for nitrogen as wheat, and it develops a very deep root mass. It's an excellent nutrient scavenger.

This combination enables the cover crop to capture maximum nitrogen from deep in the soil profile to feed the following corn crop. No one has ever proven to me that we need nutrients down deep. It sounds good to have a plant food layer at 16 to 18 inches, but I much prefer the nitrogen and other nutrients near the surface where the crop can use them.

Deep-rooted cover crops like oilseed radish can help reverse the traditional theory of nitrogen stratification. Nitrogen allowed to concentrate deep in the soil scares us because it is more likely to leach into the tile lines and reach Lake Huron.

We've also tried wheat, hairy vetch, crimson clover and a dry bean and soybean mix for cover crops, and we'll keep experimenting. Recently, I traded oilseed radish seed to Kansas no-tillers Red and David Sutherland for Austrian winter pea seed. The Sutherlands have reported good moisture retention and nitrogen fixation with the peas. We like what we've seen with the peas, as well.

Less Nitrogen, More Corn

We partly credit the cover crop program with sharply reducing our fertilizer bills. In fact, the first time I hit a 200-bushel corn yield, I did it with only 140 to 150 pounds of nitrogen per acre, or about 0.7 to 0.8 pounds of nitrogen per bushel. As anyone who has been growing corn knows, the typical nitrogen recommendation has been about a pound-plus per bushel. Oversupplying nitrogen has absolutely no value. I think the whole nutrient cycle concept is intriguing; no-till in conjunction with cover crops really makes it work.

Organic Matter Boost

When we started no-tilling, we had heard stories from farmers and others that we could expect to see increased organic matter content in our soils after a few years. But some soil experts cautioned that this likely wouldn't happen. Fortunately, we've triggered significant humus development during the past 20 years, with organic matter increasing from about 0.5 to as much as 2.5 percent. This is a real bonus in addition to all the other benefits from no-tilling, and we expect to see even more improvement as we include more cover crops in our rotation.

What Works At Home?

Our county is part of the Saginaw Bay watershed, the largest in the state with more than 8,700 square miles. Everything we do as farmers can affect the water quality of the bay, and we're very conscious of that.

A group of about 150 farmers from three counties formed the Innovative Farmers of Michigan in 1994. Our objectives have been to reduce the amount of sediment entering the bay and change our farming practices to reduce nutrient and pesticide runoff. We don't want our soils in the bay. After a 3-year study, financed with an EPA 319 grant in 1996, we came up with some pretty dramatic results. We found that conservation tillage does not reduce yields; in fact we saw significant yield increases in corn.

Also, reduced tillage increases the soil's capacity to supply nitrogen and phosphorus to a growing crop. Water-holding capacity and water infiltration rates were higher on no-till fields. We reduced the potential for soil erosion from water by up to 70 percent and from wind by up to 60 percent, compared to conventional tillage. At the end of the project, we were getting a lot better handle on what no-till systems work best in our three-county area.

At Sheridan Farms, we'll keep looking for more diversity and hope to get back to a four- or five-crop continuous no-till system. The most valuable lesson we learned is there is no universal truth or no-till game plan that will apply for everyone. Eventually, we adapted a no-till system that fits our particular soil types, crops, climate, long-range goals and farming style.

—Adapted with permission from "The No-Till Farmer," May 2006. www.no-tillfarmer.com

As a general rule, cover crops, particularly cereals, need to be terminated 2-3 weeks ahead of planting to allow plant material to dry out and become brittle. Dry brittle cover crop residue allows tillage and planting equipment to cut through the residue more easily, as opposed to semi-dry cover crop residue. Semi-dry residue is tough and hard to cut, which can result in considerable dragging of the residue as implements traverse the field.

Allelopathic compounds can be a greater problem with crop establishment when fresh residues become trapped in the seed furrow, a condition known as "hairpinning." Hairpinning can be a problem even for residues that have been on the surface for a number of weeks if planting in the morning when residues are still moist from precipitation or dew. Hairpinning can reduce seed to soil contact and cash crop stands.

You can sometimes plant the cash crop directly into standing (live) cover crop, then kill the cover crop. This allows more time for cover crop growth and biomass production, and usually side-steps the problem of planting into tough cover crop residue. However, planting into standing green residue can increase the risk of allelopathic chemicals affecting sensitive cash crop seedlings, and can make it difficult to align rows when planting.

Killing methods

Many kill methods have been developed and tested. Some are described below. Be sure to check with Extension or other farmers for recommended methods for your area and crop system.

Killing with an herbicide. Killing cover crops with a non-selective herbicide is the standard method used by conservation tillage growers. They favor this option because they can cover many acres quickly and herbicides are relatively cheap. Herbicides can be applied at any time or growth stage to terminate the cover crop.

Killing with a roller-crimper. Cover crops can be killed using a mechanical roller (often called a roller-crimper). The roller kills the cover crop by breaking (crimping) the stems. The crimping action aids in cover crop desiccation.

The cover crop is rolled down parallel to the direction of planting to form a dense mat on the soil surface, facilitating planter operation and aiding in early season weed control. When using a roller alone for cover crop termination, best results are obtained when rolling is delayed until flowering stage or later.

Roller-crimpers work best with tall-growing cover crops. Small weeds are not killed by rolling. Weed suppression by the mat of rolled cover crop residue depends on cover crop, weed species and height, and the density (thickness) of the cover crop mat.

Rollers can be front- or rear-mounted. They usually consist of a round drum with equally spaced blunt blades around the drum. Blunt blades are used to crimp the cover crop. This is preferable to sharp blades that would cut the cover crop and dislodge residue that might interfere with seed soil contact at planting.

The roller-crimper is a viable way to kill cover crops without using herbicides. It also helps prevent planter problems that can occur when tall-growing cover crops lodge in many different directions after chemical termination

In Alabama, a mechanical roller was used to kill black oat, rye and wheat cover crops. The roller combined with glyphosate at one-half the recommended rate was as effective as using glyphosate at the full recommended rate to kill all cover crops. The key was to use the roller at flowering. Herbicides can be eliminated if the roller operation occurs at the soft dough stage or later, a good option for organic growers (13).

▼ **Precaution:** Applying non-selective herbicides at reduced rates could lead to weed resistance. The half rate of herbicide may not completely eradicate the weed, increasing the chance that the weed will produce seed. Under these circumstances, such seeds are more likely to be resistant to the herbicide. Therefore, it is safer to completely eliminate the use of the non-selective herbicide with a roller or use the non-selective herbicide at the labeled rate, with or without the roller.

Growers and researchers are addressing several barriers to the use of rollers:

- Operation speed was hampered by vibration. Using curved blades on the roller drum alleviates this problem.

- Most rollers are 8 rows or smaller, but growers have built wider rollers that can be folded for transportation.
- Rolling and planting can be done in one operation by using a front-mounted roller and rear-mounted drill, saving time and energy.

For more information about cover crop rollers, see ATTRA (11) and *Cover Crop Roller Design Holds Promise for No-Tillers*, p. 146.

Mowing/chopping. Mowing and chopping are quick methods to manage large amounts of cover crop residue by cutting it into smaller pieces. An alternative to the use of herbicides, it is more energy intensive.

In humid climates, mowed residues break down faster, negating some of the residue benefits of conservation tillage. In drier climates, cover crop residues do not decompose as fast, but wind and water may cause residue to accumulate in low areas or remove it from the field altogether.

Cutting residue into smaller pieces may adversely affect the performance of tillage and planting equipment because coulters designed to cut through residue may instead push small residue pieces into the soil. Use "row cleaners" or "trash whippers" to prevent this problem.

Living mulch. Living mulches are cover crops that co-exist with the cash crop during the growing season and continue to grow after the crop is harvested. Living mulches do not need to be reseeded each year (182). They can be chosen and managed to minimize competition with the main cash crop yet maximize competition with weeds. The living mulch can be an annual or perennial plant established each year, or it can be an existing perennial grass or legume stand into which a crop is planted.

Living mulch systems are dependent on adequate moisture for the cash crop. They can be viable for vineyards, orchards, agronomic crops, such as corn, soybean, and small grains, and many vegetables. Legumes are often used because they fix nitrogen, a portion of which will be available for the companion crop. If excess nitrogen is a problem, living mulches (especially grasses) can serve as a sink to tie up some of this excess nitrogen and hold it until the next growing season.

In conservation tillage systems, living mulches can improve nitrogen budgets, provide weed and erosion control, and may contribute to pest management and help mitigate environmental problems.

Living mulch systems are feasible in Midwest alfalfa-corn rotations (386). Use in corn-soybean rotations was also feasible but more challenging because soybean is more susceptible to competition from the living mulch. With adequate suppression, living mulches can be managed to minimize competition with corn with little or no reduction in yield. The system requires close monitoring and careful control of competition between the living mulch and grain crop to maintain crop yields.

Cash crop establishment. Cash crop establishment can be complicated by the use of cover crops in conservation-tillage systems. Cover crops can reduce cotton, corn and soybean stands if not managed well. Possible causes of stand reductions include:

- poor seed-soil contact due to residue interference with planter operations
- soil water depletion
- wet soils due to residue cover
- cold soils due to residue cover
- allelopathic effects of cover crop residues
- increased levels of soilborne pathogens
- increased predation by insects and other pests
- free ammonia (in the case of legume covers)

To prevent stand problems following cover crops:

- Check for good seed-soil contact and seed placement, particularly seeding depth.
- Be sure that coulters are cutting through cover crop residue rather than pushing it into the soil along with the seed.
- Desiccate the cover crop at least 2 to 3 weeks before planting the cash crop.
- Monitor the emerging crop for early season insect problems such as cutworms.

Small seeded crops like vegetables and cotton are especially susceptible to stand reductions following cover crops. Winter annual legumes may cause more problems due to allelopathic effects and/or increased populations of plant pathogens.

Residue management systems that leave cover crop residue on the surface can reduce the risk of stand problems provided the residue does not interfere with planter operation.

Good seed placement is more challenging where residues remain on the soil surface. However, improvements in no-till planter design have helped. Equipment that removes crop residue from the immediate seeding area can help to reduce stand losses (see equipment discussion, below).

Surface residues reduce soil temperature. The relative influence of this temperature reduction on crop growth is greater in northern areas of a crop's adapted zone. Removal of residue from the zone of seed placement will increase soil temperature in the seed zone and also decrease the amount of residue that comes in contact with the seed. This will result in better seed-soil contact and less allelopathic effects from residue to the developing seedling.

No-till planters. The key to successful no-till cash crop establishment in cover crop residues is adequate seed to soil contact at a desired seeding depth. No-till planters are heavier than conventional planters. The additional weight allows the planter to maintain desired seeding depth in rough soil conditions and prevents the planter from floating across the soil surface and creating uneven seed placement. Individual planter row units are typically equipped with heavy-duty down-pressure springs that allow the operator to apply down pressure in uneven soil conditions to maintain depth control.

Row cleaners are designed to operate in heavy cover crop residue. Manufacturers have developed different types of row cleaners that can be matched to various planters. All row cleaners are designed to sweep residue away from the opening disks of the planter units. Removing this residue reduces the chance of pushing residue into the seed furrow (hairpinning).

All row cleaners can be adjusted to match specific field conditions. Row cleaners should be adjusted to move residue but not soil. If too much soil is disturbed in the row, the soil will dry out and can crust over, which will hinder emergence. In addition, disturbed soil can promote weed emergence in the row creating unnecessary competition between weeds and the cash crop.

Spoked closing wheels improve establishment in poorly drained or fine-textured soils. On these soils, traditional cast-iron or smooth rubber closing wheels can result in soil crusting. Spoked closing wheels crumble the seed trench closed for adequate seed to soil contact, but leave the soil loose and friable for plant emergence.

Additional planter attachments to ensure adequate seed to soil contact in rough soil conditions include V-slice inserts and seed firmers. V-slice inserts clean the seed trench created by the opening disks. Seed firmers press the seed into the soil at the bottom of the seed trench.

Strip-tillage equipment. Strip-tillage equipment is designed to manage residue and perform some non-inversion tillage in the row. In the South, strip-tillage refers to in-row subsoiling (14-16 inches deep) to reduce compaction, with minimal disturbance of residue on the soil surface. In the Midwest, zone-tillage typically refers to shallow tillage within the row designed to remove residue and enhance soil warming in the seed zone.

Regardless of manufacturer, strip tillage implements typically consist of a coulter that runs ahead of a shank, followed by such attachments as additional coulters, rolling baskets, drag chains, or press wheels. Depending on conditions, these attachments are used alone or in various combinations to achieve different degrees of tillage.

When strip-tilling in cover crop residue, the coulter should be positioned as far forward of the shank as possible and centered on the shank. This allows the coulter to operate in firm soil enabling it to cut residue ahead of the shank. By cutting the residue ahead of the shank, the residue can flow through the shanks more easily and not wrap up or drag behind the implement.

Fine-textured soils sometimes stick to the shank and may accumulate there, disturbing too much soil and making the slit too wide. This can impede planter operations and is referred to as "blowout." Plastic shields that fit over the shank help prevent blowout. Another way to reduce blowout is to install splitter points on the subsoil shanks. The splitter points look like shark fins that attach vertically upright to the tips of the shank points. They fracture the soil at the bottom of the trench, preventing soil upheaval to the soil surface. The soil fracture created is analogous to stress cracks in concrete.

Row cleaners can be used on cool, poorly drained soils to enable faster soil-warming in spring. This may allow earlier planting and helps ensure optimal plant emergence conditions. Available for most strip-tillage implements, row cleaners function much like row cleaners for planters, sweeping cover crop residue away from the row. Adjustments for strip tillage row cleaners are not as flexible as those on planters.

Vegetable establishment. Adoption of no-tillage systems for transplanted vegetable crops was limited by equipment and stand establishment problems. This problem was overcome in the 1990's with the development of the Subsurface Tiller-Transplanter (287). The SST-T is a "hybrid," combining subsurface soil loosening to alleviate soil compaction and effective setting of transplants, in one operation with minimum disturbance of surface residues or surface soil.

The spring-loaded soil-loosening component of the subsurface tiller tills a narrow strip of soil ahead of the double disk shoe of the transplanter. The double-disk shoe moves through the residue and the tilled strip with relatively little resistance. In addition, the planter can be equipped with fertilizer and pesticide applicators to reduce the number of trips required for a planting operation.

Regional Roundup: Cover Crop Use in Conservation Tillage Systems

Midwest—*Tom Kaspar*

Soils. Soils of the Midwest contain high levels of organic matter compared to other regions. Research has yet to confirm if cover crops can increase soil organic matter contents beyond current levels. The possibility of using corn stover as a bioenergy source would leave the soil unprotected and much more vulnerable to degradation, but cover crops could offset any detrimental effects associated with corn stover removal. The degree to which cover crops could protect the soil following corn stover removal has not been investigated.

Farm systems. Midwest farms are large, averaging 350 acres. Cover crops and conservation tillage are most common in corn and soybean systems, with or without livestock. Cover crops are also commonly used in vegetable systems.

Cover crop species. Rye and other small grains are the primary cover crops used in the Midwest. Legume cover crop include red clover, hairy vetch and sweetclovers.

Cover crop benefits. Advantages of cover crops in the Midwest include reducing erosion, anchoring residues in no-till systems, suppressing winter annual weeds and nutrient management. The ability of cover crops to scavenge nitrates is particularly beneficial in the Midwest, where the majority of United States corn is produced, because the high N requirement of corn increases the potential for nitrate loss.

Drawbacks. Cover crops have reduced corn (but not soybean) yields when they are terminated at planting. Earlier termination helps reduce this problem, but residue benefits are reduced. The potential biomass production is complicated by the already short, cold cover crop growing season between harvest and planting of corn and soybean crops. Cash crop planting and harvest coincide with cover crop kill and planting dates.

Management. Cover crops need to be planted at the same time farmers are harvesting corn and soybean to ensure adequate biomass production. Producers would benefit from alternative cover crop establishment methods, such as overseeding before harvest, seeding at weed cultivation with delayed emergence, or frost seeding after harvest. Environmental payments or incentives may entice growers to try alternative practices.

Northeast—*Sjoerd Duiker*

Cover crops are becoming an integral part of crop production in the Northeast. This is due in large part to the increasing adoption of no-tillage systems, because cover crops can be managed more easily than with tillage, while cover crop residues in no-till systems lead to multiple benefits.

Soils. There are many soil types in the Northeast, including soils developed in *glacial* deposits or from melt water lakes; sedimentary soils formed from the *sedimentary* rocks sandstone, shale and limestone; *Piedmont* soils, remnants of a coastal mountain range with complex geology, characterized by a gently to strongly undulating landscape; and *coastal plain* soils, developed in unconsolidated material deposited by rivers and the ocean, often sandy with a shallow water table.

Soil and nutrient management in the region aim to address soil erosion, clay subsoils, fragipans, shallow water tables and the nutrient enrichment caused by the high density of animal production.

Farm systems. Farms in the Northeast are diverse, tend to be small, and include cash grain, perennial forage, dairy, hog, poultry, fruit and vegetable operations. Nutrient management regulations in some states encourage the use of cover crops and conservation tillage practices, particularly for the application of manure. Farmer organizations such as the Pennsylvania No-Till Alliance actively promote cover crops for their soil-improving benefits, while government programs such as the 2006 Maryland cover crop subsidy of $30-$50 per acre led to a dramatic increase in cover crop acreage.

Cover crop species. Cover crop options and niches are as diverse as the farming systems in the region. Rye, wheat, oats and ryegrass are the most common grass cover crops; hairy vetch, crimson clover and Austrian winter pea are important legumes; buckwheat finds a place in many vegetable systems and brassica crops such as forage radish are increasingly being tested and used in the region.

Cover crop benefits. Cover crops are planted for erosion control, soil improvement, moisture conservation, forage and nutrient management, particularly the nitrogen and phosphorus from intensive animal agriculture. Cover crops can fit into many different niches in the region, particularly fruit and vegetable systems (1, 2, 3, 4). Recent work with forage radish (*Raphanus sativus* L.) suggests that its large taproot can penetrate deep soil layers and alleviate compaction (446).

Drawbacks. Barriers to the adoption of cover crops include the time and cost of establishment and management, water use, and, for some systems, the length of the growing season.

Management. Farmers and other researchers fit cover crops into many different niches using:

- timely seeding, overseeding into standing crops, or interseeding, including some use of living mulches
- various termination methods, including mowing or rolling standing cover crops
- manipulation of cover crop kill and cash crop planting dates to maximize cover crop benefits

Southeast—*Kipling Balkcom*
High-residue cover crops are essential to the success of conservation systems in the Southeast.

Soils. Soils in the Southeast are highly weathered, acidic, and often susceptible to erosion due to their low organic matter content. Decades of conventional tillage practices have exacerbated their poor physical and chemical condition.

Farm systems. Southeast farms raise various combinations of cotton, soybeans, corn, peanuts and small grains. Some include livestock, have access to irrigation or raise fruit and vegetables.

Cover crop species. Rye, wheat, oats, hairy vetch and crimson clover are the cover crop mainstays for grain and oil crop systems.

Cover crop benefits. Cover crop biomass is needed on the weathered soils of the Southeast to add organic matter and improve soil physical, chemical, and biological properties. Cover crop residues reduce soil erosion and runoff, increase infiltration and conserve soil moisture, particularly beneficial in dry years or on drought-prone soils.

Drawbacks. Major concerns are:

- water management
- integration of different cover crop species into southeastern crop production
- reduced effectiveness of pre-emergence herbicides in high-residue systems

Producers are also concerned about residue interference with efficient equipment operation, adequate soil moisture at planting, and stand establishment problems. In addition, some are not willing to commit to the additional management level or perceived costs.

Management. Producers like the idea of reducing trips across the field, which reduces fuel and labor costs and saves time. Significant increases in the use of cover crops and conservation tillage systems in the Southeast have paralleled the adoption of new genetic varieties of corn, soybean and cotton that are herbicide resistant or have incorporated genes for improved insect resistance. These genetic changes reduced some of the challenges associated with weed and insect management, making the conservation tillage systems easier to manage. The relatively longer growing season usually allows ample time to plant cover crops after cash crops.

Northern Plains—*Jorge Delgado*

Rainfall and moisture availability are the major factors affecting the use of cover crops in conservation tillage systems.

Soils. Soils of the Northern Great Plains are exposed to high wind conditions with enough force to move soil particles off site in minimum tillage conditions where soil cover is low. Crop systems do not, in general, leave substantial residue on the soil surface, due in part to low annual rainfall in non-irrigated systems.

Farm systems. Farms in the Northern Plains tend to be large and can be divided into irrigated and non-irrigated systems. Crops rotations include potato, safflower, dry bean, sunflower, canola, crambe, flax, soybean, dry pea, wheat and barley.

Cover crop species. Rye, field pea (Austrian winter pea, trapper spring pea), sweetclover and sorghum-sudangrass are commonly grown.

Cover crop benefits. Cover crop residues improve water retention, helping to increase soil water content and yields. Cover crops reduce wind erosion and nutrient loss, and increase soil carbon. High crop residue and winter cover crops also sequester carbon and nitrogen and increase the availability of other macro- and micronutrients (7, 113).

Drawbacks. Rainfall amount, the availability of irrigation and water use by cover crops are critical considerations for the region. Cool, wet spring weather is exacerbated by cover crop residues that delay soil-warming. Cover crops and conservation tillage often reduce cash crop yields, even in irrigated systems (171, 172).

Management. Management is key to increasing nutrient use efficiencies and reduce nutrient losses to the environment (112, 113, 114, 371). Management is also the key to increasing water use efficiency.

Southern Plains —*Louis Baumhardt*

Conservation tillage was first introduced for soil erosion control on the Great Plains. It followed inversion tillage that incorporated crop residue and degraded the soil's natural cohesiveness and aggregation. Combined with the 1930's dry and windy conditions, this intensive tillage produced catastrophic wind erosion known historically as the "Dust Bowl" (26). Use of conservation tillage practices for much of the Southern Great Plains seems to lag behind other regions, but may be underestimated, in part, because insufficient residue is produced in dryland areas to qualify as conservation tillage acres.

Soils. Soils of the Southern Great Plains were formed from a range of materials including, for example, an almost flat aeolian mantle in the north (Texas High Plains and western Kansas) and reworked Permian sediments of the Texas Rolling Plains extending towards western Oklahoma. These soils have varied mineralogy, are frequently calcareous, and generally have poor structure and low organic matter content (37). All Southern Great Plains soils are managed for wind erosion control and water conservation.

Farm systems. Farm systems on the Southern Great Plains vary with irrigation. They are larger and more diverse as irrigation declines to distribute risk and meet production requirements. Principal crops include cotton, corn, peanut, grain sorghum, soybean, and sunflower. Grain and forage crops support the regional cattle industry.

Wheat-sorghum-fallow is a common rotation and permits cattle grazing on wheat forage and sorghum stubble (27). This and similar rotations may include additional years of sorghum or a wheat green fallow before cotton.

Cover crop species. Water governs cover crop species selection, but wheat, rye, and oats are most common. Wheat is commonly grown for grain or forage and as a green fallow crop between annual cotton crops (29).

Cover crop benefits. Cover crop residue helps meet the 30% cover requirement for conservation tillage, helps control wind erosion in low residue crops, and provides other water infiltration and storage benefits.

Drawbacks. Cover crop use in the region depends on precipitation or the availability and economy of irrigation to produce residue. Some crops such as cotton produce insufficient residue for soil cover, but establishing cover crops competes for water needed by the subsequent cotton crop (28). Grazing crop residues and cover crops also limits the amount of crop residue left on the soil surface and must be balanced against the value of the forage.

Management. Southern Great Plains producers often use cover crops to control wind erosion in annual crops like cotton that produce insufficient cover to protect the soil. During years with limited precipitation, cover crops compete for water resources needed to establish primary cash crops (28). Nevertheless, producers wishing to grow cotton on soils subject to wind erosion have successfully introduced residue producing winter cereal crops with minimum irrigation input.

Conservation tillage increases storage of precipitation in the soil through increased infiltration and reduced evaporation. This additional water supplements growing season precipitation and irrigation to meet crop water needs on the semi-arid Southern Great Plains.

Pacific Northwest —*Hal Collins*

Under dryland conservation tillage systems in the Pacific Northwest (PNW), winter precipitation and limited water availability are major factors affecting the use of cover crops. With irrigation, heavy crop residues from previous grain crops can negatively impact cover crop stand establishment. Annual precipitation in agricultural regions of the PNW ranges from 15 to 76 cm, due to orographic effects of the Cascade and Blue Mountain Ranges that strongly influence total precipitation and distribution patterns in Washington, Oregon and Idaho.

Soils. Soils of the PNW have developed from aeolian and flood deposits originating from volcanic activity and the last continental glaciations (~12,000 years BP) under shrub-steppe vegetation. Soils that developed on wind blown loessal deposits are typically silt loams with moderate to strong structure and soil organic C contents ranging from 1-2%. Soils developing on the flood deposits of Glacial Lakes' Missoula and Bonneville in the Columbia Basin of Washington, Oregon and Idaho are predominately sands to silt loams with weak soil structure and low soil organic C (<1%). Cultivated soils of the region are exposed to severe soil erosion from water and snow melt in the higher precipitation zones and due to high wind conditions in low rainfall areas (Columbia Basin).

Farm systems. Farms in the dryland and irrigated regions of the PNW tend to be large (2,000+ acres on average). Dryland regions are commonly cropped to wheat, barely, canola, oats, grass seed and dry peas. Crop rotations under irrigation are diverse, vegetable based rotations that include potato, onion, carrots, field corn, sweet corn, fresh beans and peas, sugar beets, mint, canola, mustards, safflower, dry pea, grass seed, alfalfa, wheat and barley.

Cover crop species. Field pea (Austrian winter pea), sweetclover, hairy vetch, sudangrass, small grains (wheat, triticale) and a variety of brassica species are used in the region.

Cover crop benefits. Cover crop residues improve water retention, infiltration and storage, soil structure, soil carbon reserves, microbial activity and crop yields. Cover crop residues have been shown to reduce water and wind erosion and nutrient loss from leaching and overland flow of sediments. High crop residues and the use of winter cover crops under irrigation sequester carbon and nitrogen and increase the availability of other macro and micronutrients. Cover crop residues can meet or exceed the 30% cover requirement for conservation tillage in low rainfed areas.

Drawbacks. Rainfall amount and distribution, the availability of irrigation and water use by cover crops are critical considerations for the dryland regions. Heavy residues under irrigation inhibit stand establishment. Cool, wet spring weather exacerbated by cover crop residues delay soil warming and seedling emergence of cash crops. Absentee land owners combined with the diversity of cropping under irrigation of high value vegetable cropping has limited adoption of conservation tillage and cover crop use. Cover crops and conservation tillage can reduce economic benefits and crop yields under some situations.

Management. Management of cover crops is complex and differs in dryland and irrigated systems. Cover crops are managed to reduce nutrient losses, increase nutrient use efficiencies and reduce severity of soil pathogens (88, 111, 115, 430). Management is also key to increasing water use efficiency and can affect protein content of small grains.

California —*Jeffrey Mitchell*

Despite the many benefits of cover cropping and conservation tillage, adoption by row crop producers in California has been limited. Cover crops are used on less than 5% of California's annual crop acreage and conservation tillage practices are used on less than 2% of annual cropland.

Soils. A wide range of soil types are used for agricultural production in California. Cover crops and conservation tillage are used predominantly on finer-textured clay loams or loam soils. More recently, conservation tillage is increasingly used in dairy forage production systems on coarser soil types.

Farm systems. Most cover crop use in conservation tillage systems in California has been for processing and fresh market commercial tomato production systems (187). Research is underway evaluating cover crops in CT corn and cotton systems (281).

Cover crop species. In tomato systems, the most successful and manageable cover crops are mixtures of triticale, rye and pea. Vetches are used for field corn.

Cover crop benefits. California farmers use cover crops to reduce intercrop tillage, suppress winter weeds, reduce pathogen buildup and manage nutrients.

Drawbacks. Producers are most concerned about cooler temperatures above and below mulch, slower maturing crops, cover crop regrowth and specialized management required. In-season weed management options may be limited in conservation tillage systems.

Management. Cover crops are normally grown from mid-October to mid-March in California's Central Valley. Aboveground biomass production can reach 11,000 lb. of aboveground dry matter/A without irrigation (279, 280). The cover crops are mowed or chopped in March using ground-driven stalk choppers, or merely allowed to collapse following herbicide application.

Tomatoes can be no-till transplanted directly into the mulch or transplanted following a strip-till pass using either narrow PTO-driven rotary mulchers or ground-driven strip-till implements modified for tomato beds (250). Because of inadequate weed control by the cover crop mulch itself, high residue cultivators that effectively slice through residues while cultivating weeds are necessary for in-season weed control.

Field corn has also been successfully direct seeded into flail mowed vetch cover crops in the Sacramento Valley. Corn yield is similar to "green manure" systems in which winter cover crops are incorporated.

SUMMARY AND RECOMMENDATIONS

Cover crops benefit conservation tillage systems by:

- decreasing soil erosion
- providing crop residues to increase soil organic matter
- improving soil structure and increasing infiltration
- increasing availability of water for crop production
- improving soil quality
- aiding in early season weed control
- breaking disease cycles

To enhance the beneficial effects of cover crops:

- Plant in a timely fashion.
- Consider additional N fertilizer for small grain covers only if residual N is low.
- Terminate covers 2-3 weeks ahead of anticipated planting date to allow soil moisture recharge and reduce problems associated with allelopathy, pests, and planter operation.
- Take advantage of equipment modifications designed for tillage and/or planter operations in heavy residue.

INTRODUCTION TO CHARTS

The four comprehensive charts that follow can help orient you to the major cover crops most appropriate to your needs and region. Bear in mind that choice of cultivar, weather extremes and other factors may affect a cover crop's performance in a given year.

CHART 1: TOP REGIONAL COVER CROP SPECIES

This chart lists up to five cover crop recommendations per broad bioregion for six different major purposes: N Source, Soil Builder, Erosion Fighter, Subsoil Loosener, Weed Fighter and Pest Fighter. If you know your main goal for a cover crop, Chart 1 can suggest which cover crop entries to examine in the charts that follow and help you determine which major cover narrative(s) to read first.

Disclaimer. The crops recommended here will not be the most successful in all cases within a bioregion, and others may work better in some locations and in some years. The listed cover crops are, however, thought by reviewers to have the best chance of success in most years under current management regimes.

CHART 2: PERFORMANCE AND ROLES

This chart provides relative ratings (with the exception of two columns having quantitative ranges) of what the top covers do best, such as supply or scavenge nitrogen, build soil or fight erosion.

Seasonality has a bearing on some of these ratings. A cover that grows best in spring could suppress weeds better than in fall. Unless otherwise footnoted, however, the chart would rate a cover's performance (relative to the other covers) for the entire time period it is likely to be in the field. Ratings are general for the species, based on measured results and observations over a range of conditions. The individual narratives provide more seasonal details. The added effect of a nurse crop is included in the "Weed Fighter" ratings for legumes usually planted with a grain or grass nurse crop.

Column headings

Legume N Source. Rates legume cover crops for their *relative* ability to supply fixed N. (*Nonlegumes have not been rated* for their biomass nitrogen content, so this column is left blank for nonlegumes.)

Total N. A *quantitative* estimate of the reasonably expected range of total N provided by a legume stand (from all biomass, above- and below ground) in lb. N/A, based mostly on published research. This is total N, not the fertilizer replacement value. *Grasses have not been rated* for their biomass nitrogen content because mature grass residues tend to immobilize N. *Brassicas* are less likely to immobilize N than grasses.

Dry Matter. A *quantitative* estimate of the range of dry matter in lb./A/yr., based largely on published research. As some of this data is based on research plots, irrigated systems or multicut systems, your on-farm result probably would be in the low to midpoint of the dry matter range cited. This estimate is based on fully dry material. "Dry" alfalfa hay is often about 20 percent moisture, so a ton of hay would only be 1,600 lb. of "dry matter."

N Scavenger. Rates a cover crop's ability to take up and store excess nitrogen. Bear in mind that the sooner you plant a cover after main crop harvest—or overseed a cover into the standing crop—the more N it will be able to absorb.

Soil Builder. Rates a cover crop's ability to produce organic matter and improve soil structure. The ratings assume that you plan to use cover crops regularly in your cropping system to provide ongoing additions to soil organic matter.

Erosion Fighter. Rates how extensive and how quickly a root system develops, how well it holds soil against sheet and wind erosion and the influence the growth habit may have on fighting wind erosion.

Weed Fighter. Rates how well the cover crop outcompetes weeds by any means through its life cycle, including killed residue. Note that ratings for the legumes assume they are established with a small-grain nurse crop.

Good Grazing. Rates relative production, nutritional quality and palatability of the cover as a forage.

Quick Growth. Rates the speed of establishment and growth.

Lasting Residue. Rates the effectiveness of the cover crop in providing a long-lasting mulch.

Duration. Rates how well the stand can provide long-season growth.

Harvest Value. Rates the cover crop's economic value as a forage (F) or as a seed or grain crop (S), bearing in mind the relative market value and probable yields.

Cash Crop Interseed. Rates whether the cover crop would hinder or help while serving as a companion crop.

CHART 3A: CULTURAL TRAITS

This chart shows a cover crop's characteristics such as life cycle, drought tolerance, preferred soils and growth habits.The ratings are general for the species, based on measured results and observations over a range of conditions. Choice of cultivar, weather extremes and other factors may affect a cover crop's performance in a given year.

Column headings

Aliases. Provides a few common names for the cover crop.

Type. Describes the general life cycle of the crop.

B = Biennial. Grows vegetatively during its first year and, if it successfully overwinters, sets seed during its second year.

CSA = Cool-Season Annual. Prefers cool temperatures and depending on which Hardiness Zone it is grown in, could serve as a fall, winter or spring cover crop.

SA = Summer Annual. Germinates and matures without a cold snap and usually tolerates warm temperatures.

WA = Winter Annual. Cold-tolerant, usually planted in fall and often requires freezing temperatures or a cold period to set seed.

LP = Long-lived Perennial. Can endure for many growing seasons.

SP = Short-lived Perennial. Usually does not persist more than a few years, if that long.

Hardy Through Zone. Refers to the standard USDA Hardiness Zones. See map on inside front cover. Bear in mind that regional microclimate, weather variations, and other near-term management factors such as planting date and companion species can influence plant performance expectations.

Tolerances. How well a crop is likely to endure despite stress from heat, drought, shade, flooding or low fertility. The best rating would mean that the crop is expected to be fully tolerant.

Habit. How plants develop.
C = Climbing
U = Upright
P = Prostrate
SP = Semi-Prostrate
SU = Semi-Upright

pH Preferred. The pH range in which a species can be expected to perform reasonably well.

Best Established. The season in which a cover crop is best suited for planting and early growth. Note that this can vary by region and that it's important to ascertain local planting date recommendations for specific cover crops.

Season: F = Fall ; Sp = Spring; Su = Summer; W = Winter

Time: E = Early; L = Late; M = Mid

Minimum Germination Temperature. The minimum soil temperature (F) generally required for successful germination and establishment.

CHART 3B: PLANTING

Depth. The recommended range of seeding depth (in inches), to avoid either overexposure or burying too deeply.

Rate. Recommended seeding rate for drilling and broadcasting a pure stand in lb./A, bu/A. and oz./100 sq. ft., assuming legal standards for germination percentage. Seeding rate will depend on the cover crop's primary purpose and other factors. See the narratives for more detail about establishing a given cover crop. Pre-inoculated ("rhizo-coated") legume seed weighs about one-third more than raw seed. Increase seeding rate by one-third to plant the same amount of seed per area.

Cost. Material costs (seed cost only) in dollars per pound, based usually on a 50-lb. bag as of fall 2006. Individual species vary markedly with supply and demand. Always confirm seed price and availability before ordering, and before planning to use less common seed types.

Cost/A. Seed cost per acre based on the midpoint between the high and low of reported seed prices as of fall 1997 and the midpoint recommended seeding rate for drilling and broadcasting. Your cost will depend on actual seed cost and seeding rate. Estimate excludes associated costs such as labor, fuel and equipment.

Inoculant Type. The recommended inoculant for each legume. Your seed supplier may only carry one or two common inoculants. You may need to order inoculant in advance. See *Seed Suppliers*, p. 195.

Reseeds. Rates the likelihood of a cover crop re-establishing through self-reseeding if it's allowed to mature and set seed. Aggressive tillage will bury seed and reduce germination. Ratings assume the tillage system has minimal effect on reseeding. Dependable reseeding ability is valued in some orchard, dryland grain and cotton systems, but can cause weed problems in other systems. See the narratives for more detail.

CHARTS 4A AND 4B

These charts provide relative ratings of other management considerations—benefits and possible drawbacks—that could affect your selection of cover crop species.

The till-kill rating assumes tillage at an appropriate stage. The mow-kill ratings assume mowing at flowering, but before seedheads start maturing. See sectional narratives for details.

Ratings are based largely on a combination of published research and observations of farmers who have grown specific covers. Your experience with a given cover could be influenced by site-specific factors, such as your soil condition, crop rotation, proximity to other farms, weather extremes, etc.

CHART 4A: POTENTIAL ADVANTAGES

Soil Impact. Assesses a cover's relative ability to loosen subsoil, make soil P and K more readily available to crops, or improve topsoil.

Soil Ecology. Rates a cover's ability to fight pests by suppressing or limiting damage from nematodes, soil disease from fungal or bacterial infection, or weeds by natural herbicidal (allelopathic) or competition/smothering action. Researchers report difficulty in conclusively documenting allelopathic activity distinct from other cover crop effects, and nematicidal impacts are variable, studies show. These are general, tentative ratings in these emerging aspects of cover crop influence.

Other. Indicates likelihood of attracting beneficial insects, of accommodating field traffic (foot or vehicle) and of fitting growing windows or short duration.

CHART 4B: POTENTIAL DISADVANTAGES

Increase Pest Risks. Relative likelihood of a cover crop becoming a weed, or contributing to a likely pest risk. Overall, growing a cover crop rarely causes pest problems, but certain cover crops may contribute to particular pest, disease or nematode problems in localized areas, for example by serving as an alternate host to the pest. See the narratives for more detail.

▼ Readers note the shift in meaning for symbols on this chart only.

Management Challenges. Relative ease or difficulty of establishing, killing or incorporating a stand. "Till-kill" refers to killing by plowing, disking or other tillage. "Mature incorporation" rates the difficulty of incorporating a relatively mature stand. Incorporation will be easier when a stand is killed before maturity or after some time elapses between killing and incorporating.

Chart 1 TOP REGIONAL COVER CROP SPECIES[1]

Bioregion	N Source	Soil Builder	Erosion Fighter	Subsoil Loosener	Weed Fighter	Pest Fighter
Northeast	red cl, hairy v, berseem, swt cl	ryegrs, swt cl, sorghyb, rye	rye, ryegrs, sub cl, oats	sorghyb, swt cl, forad	sorghyb, ryegrs, rye, buckwheat	rye, sorghyb, rape
Mid-Atlantic	hairy v, red cl, berseem, crim cl	ryegrs, rye, swt cl, sorghyb	sub cl, cowpeas, rye, ryegrs	sorghyb, swt cl, forad	rye, ryegrs, oats, buckwheat	rye, sorghyb, rape
Mid-South	hairy v, sub cl, berseem, crim cl	ryegrs, rye, sub cl, sorghyb	sub cl, cowpeas, rye, ryegrs	sorghyb, swt cl	buckwheat, ryegrs, sub cl, rye	rye, sorghyb
Southeast Uplands	hairy v, red cl, berseem, crim cl	ryegrs, rye, sorghyb, swt cl	sub cl, cowpeas, rye, ryegrs	sorghyb, swt cl	buckwheat, ryegrs, sub cl, rye	rye, sorghyb
Southeast Lowlands	winter peas, sub cl, hairy v, berseem, crim cl	ryegrs, rye, sorghyb, sub cl	sub cl, cowpeas, rye, ryegrs, sorghyb	sorghyb	berseem, rye, wheat, cowpeas, oats, ryegrs	rye, sorghyb
Great Lakes	hairy v, red cl, berseem, crim cl	ryegrs, rye, sorghyb, ryegrs, swt cl	oats, rye, ryegrs	sorghyb, swt cl, forad	berseem, ryegrs, rye, buckwht, oats	rye, sorghyb, rape
Midwest Corn Belt	hairy v, red cl, berseem, crim cl	rye, barley, sorghyb, swt cl	wht cl, rye, ryegrs, barley	sorghyb, swt cl, forad	rye, ryegrs, wheat, buckwht, oats	rye, sorghyb
Northern Plains	hairy v, swt cl, medics	rye, barley, medic, swt cl	rye, barley	sorghyb, swt cl	medic, rye, barley	rye, sorghyb
Southern Plains	winter peas, medic, hairy v	rye, barley, medic	rye, barley	sorghyb, swt cl	rye, barley	rye, sorghyb
Inland Northwest	winter peas, hairy v	medic, swt cl, rye, barley	rye, barley	sorghyb, swt cl	rye, wheat, barley	rye, mustards, sorghyb
Northwest Maritime	berseem, sub cl, lana v, crim cl	ryegrs, rye, sorghyb, lana v	wht cl, rye, ryegrs, barley	sorghyb, swt cl	ryegrs, lana v, oats, wht cl	rye, mustards
Coastal California	berseem, sub cl, lana v, medic	ryegrs, rye, sorghyb, lana v	wht cl, cowpeas, rye, ryegrs	sorghyb, swt cl	rye, ryegrs, berseem, wht cl	sorghyb, crim cl, rye
Calif. Central Valley	winter peas, lana v, sub cl, medic	medic, sub cl	wht cl, barley, rye, ryegrs	sorghyb, swt cl	ryegrs, wht cl, rye, lana v	sorghyb, crim cl, rye
Southwest	medic, sub cl	sub cl, medic, barley	barley, sorghyb		medic, barley	

[1]ryegrs=annual ryegrass. buckwht=buckwheat. forad=forage radish. rape=rapeseed. sorghyb=sorghum-sudangrass hybrid. berseem=berseem clover. winter peas=Austrian winter pea. crim cl=crimson clover. hairy v=hairy vetch. red cl=red clover. sub cl=subterranean clover. swt cl=sweetclover. wht cl=white clover. lana v=LANA woollypod vetch.

Chart 2 **PERFORMANCE AND ROLES**

Species	Legume N Source	Total N (lb./A)[1]	Dry Matter (lb./A/yr.)	N Scavenger[2]	Soil Builder[3]	Erosion Fighter[4]	Weed Fighter	Good Grazing[5]	Quick Growth
Annual ryegrass *p. 74*			2,000–9,000	◕	◕	◕	◕	◕	◕
Barley *p. 77*			2,000–10,000	◕	◕	●	◕	◕	◕
Oats *p. 93*			2,000–10,000	◕	◑	◕	●	◑	●
Rye *p. 98*			3,000–10,000	●	●	●	●	◑	●
Wheat *p. 111*			3,000–8,000	◕	◕	◕	◕	◕	◕
Buckwheat *p. 90*			2,000–4,000	○	◑	◔	●	○	●
Sorghum-sudan. *p. 106*			8,000–10,000	●	●	●	◕	◕	●
Mustards *p.81*		30–120	3,000–9,000	◑	◕	◕	◕	◑	◕
Radish *p. 81*		50–200	4,000–7,000	●	◕	◕	●	◑	◕
Rapeseed *p. 81*		40–160	2,000–5,000	◕	◑	◕	◕	◑	◕
Berseem clover *p. 118*	●	75–220	6,000–10,000	◕	◕	◕	●	●	●
Cowpeas *p. 125*	●	100–150	2,500–4,500	◔	◑	●	●	◑	◕
Crimson clover *p. 130*	◕	70–130	3,500–5,500	◑	◕	◕	◕	●	◑
Field peas *p. 135*	●	90–150	4,000–5,000	◔	◑	◕	◑	◕	◕
Hairy vetch *p. 142*	●	90–200	2,300–5,000	◔	◕	◑	◑	◑	◔
Medics *p. 152*	◑	50–120	1,500–4,000	◔	◑	◑	◕	◕	●
Red clover *p. 159*	◕	70–150	2,000–5,000	◑	◕	◑	◕	●	◔
Subterranean clovers *p.164*	●	75–200	3,000–8,500	◔	◕	◕	●	◕	◑
Sweetclovers *p. 171*	●	90–170	3,000–5,000	◔	●	◕	◕	◕	◑
White clover *p. 179*	●	80–200	2,000–6,000	◔	◑	◕	◕	●	◔
Woollypod vetch *p. 185*	●	100–250	4,000–8,000	◑	●	◑	●	◑	◕

[1]**Total N**—Total N from all plant. Grasses not considered N source. [2]**N Scavenger**—Ability to take up/store excess nitrogen. [3]**Soil Builder**—Organic matter yield and soil structure improvement. [4]**Erosion Fighter**—Soil-holding ability of roots and total plant. [5]**Good Grazing**—Production, nutritional quality and palatability. Feeding pure legumes can cause bloat.

○=Poor; ◔=Fair; ◑=Good; ◕=Very Good; ●=Excellent

Chart 2 **PERFORMANCE AND ROLES** continued

	Species	Lasting Residue[1]	Duration[2]	Harvest Value[3] F*	Harvest Value[3] S*	Cash Crop Interseed[4]	Comments
NONLEGUMES	Annual ryegrass	◕	◕	◑	◔	●	Heavy N and H_2O user; cutting boosts dry matter significantly.
	Barley	●	◑	◕	◑	◕	Tolerates moderately alkaline conditions but does poorly in acid soil < pH 6.0.
	Oats	◑	◔	◑	◑	●	Prone to lodging in N-rich soil.
	Rye	●	◕	◔	◔	◕	Tolerates triazine herbicides.
	Wheat	◕	◕	◑	◕	◔	Heavy N and H_2O user in spring.
	Buckwheat	○	◔	○	◔	◕	Summer smother crop; breaks down quickly.
	Sorghum-sudangrass	◕	●	●	○	○	Mid-season cutting increases yield & root penetration
BRASSICAS	Mustards	◔	◑	○	◔	○	Suppresses nematodes and weeds.
	Radish	◔	◑	◕	◔	◔	Good N scavenging and weed control; N released rapidly.
	Rapeseed	◑	◕	◔	●	○	Suppresses *Rhizoctonia.*
LEGUMES	Berseem clover	◑	●	●	◕	◑	Very flexible cover crop, green manure, forage.
	Cowpeas	◔	●	◑	◑	◑	Season length, habit vary by cultivar.
	Crimson clover	◑	◔	●	◕	●	Established easily, grows quickly if planted early in fall; matures early in spring.
	Field peas	◔	◑	●	◕	●	Biomass breaks down quickly.
	Hairy vetch	◔	◕	◔	◕	◑	Bi-culture with small grain expands seasonal adaptability.
	Medics	◑	◕	◑	◑	◑	Use annual medics for interseeding.
	Red clover	◔	◑	●	◕	●	Excellent forage, easily established; widely adapted.
	Subterranean clover	◕	◕	◕	○	●	Strong seedlings, quick to nodulate.
	Sweetclovers	◕	◕	◕	◑	◑	Tall stalks, deep roots in second year.
	White clover	◔	●	◕	◑	◕	Persistent after first year.
	Woollypod vetch	◔	◕	◔	◕	◑	Reseeds poorly if mowed within 2 months of seeddrop; overgrazing can be toxic.

[1]**Lasting Residue**—Rates how long the killed residue remains on the surface. [2]**Duration**—Length of vegetative stage. [3]**Harvest Value**—Economic value as a forage (F) or as seed (S) or grain. [4]**Cash Crop Interseed**—Rates how well the cover crop will perform with an appropriate companion crop.

○ =Poor; ◔ =Fair; ◑ =Good; ◕ =Very Good; ● =Excellent

Chart 3A **CULTURAL TRAITS**

Group	Species	Aliases	Type[1]	Hardy through Zone[2]	Tolerances: heat	drought	shade	flood	low fert	Habit[3]	pH (Pref.)	Best Established[4]	Min. Germin. Temp.
NONLEGUMES	Annual ryegrass *p. 74*	Italian ryegrass	WA	6	Fair	Fair	Very Good	Very Good	Fair	U	6.0–7.0	ESp, LSu, EF, F	40F
NONLEGUMES	Barley *p. 77*		WA	7	Very Good	Very Good	Good	Fair	Very Good	U	6.0–8.5	F, W, Sp	38F
NONLEGUMES	Oats *p. 93*	spring oats	CSA	8	Fair	Fair	Fair	Good	Good	U	4.5–7.5	LSu, ESp W in 8+	38F
NONLEGUMES	Rye *p. 98*	winter, cereal, or grain rye	CSA	3	Good	Very Good	Very Good	Good	Excellent	U	5.0–7.0	LSu, F	34F
NONLEGUMES	Wheat *p. 111*		WA	4	Good	Good	Good	Poor	Good	U	6.0–7.5	LSu, F	38F
NONLEGUMES	Buckwheat *p. 90*		SA	NFT	Good	Poor	Fair	Fair	Fair	U/SU SU	5.0–7.0	Sp to LSu	50F
NONLEGUMES	Sorghum-sudan. *p. 106*	Sudax	SA	NFT	Excellent	Excellent	Good	Good	Good	U	6.0–7.0	LSp, ES	65F
BRASSICAS	Mustards *p.81*	brown, oriental white, yellow	WA, CSA	7	Good	Very Good	Good	Fair	Fair	U	5.5–7.5	Sp, LSu	40F
BRASSICAS	Radish *p. 81*	oilseed, Daikon, forage radish	CSA	6	Good	Fair	Good	Fair	Fair	U	6.0–7.5	Sp, LSu, EF	45F
BRASSICAS	Rapeseed *p. 81*	rape, canola	WA	7	Fair	Good	Good	Fair	Fair	U	5.5–8	F, Sp	41F
LEGUMES	Berseem clover *p. 118*	BIGBEE, multicut	SA, WA	7	Very Good	Good	Very Good	Good	Good	U/SU SU	6.2–7.0	ESp, EF	42F
LEGUMES	Cowpeas *p. 125*	crowder peas, southern peas	SA	NFT	Excellent	Very Good	Good	Fair	Excellent	SU/C	5.5–6.5	ESu	58F
LEGUMES	Crimson clover *p. 130*		WA, SA	7	Good	Fair	Very Good	Fair	Good	U/SU	5.5–7.0	LSu/ESu	
LEGUMES	Field peas *p. 135*	winter peas, black peas	WA	7	Fair	Good	Fair	Fair	Fair	C	6.0–7.0	F, ESp	41F
LEGUMES	Hairy vetch *p. 142*	winter vetch	WA, CSA	4	Fair	Good	Good	Fair	Fair	C	5.5–7.5	EF, ESp	60F
LEGUMES	Medics *p. 152*		SP, SA	4/7	Excellent	Very Good	Very Good	Fair	Good	P/Su	6.0–7.0	EF, ESp, ES	45F
LEGUMES	Red clover *p. 159*		SP, B	4	Fair	Fair	Very Good	Good	Fair	U	6.2–7.0	LSu; ESp	41F
LEGUMES	Subterranean cl. *p. 164*	subclover	CSA	7	Good	Very Good	Very Good	Good	Excellent	P/SP	5.5–7.0	LSu, EF	38F
LEGUMES	Sweetclovers *p. 171*		B, SA	4	Very Good	Excellent	Fair	Fair	Excellent	U	6.5–7.5	Sp/S	42F
LEGUMES	White clover *p. 179*	white dutch ladino	LP, WA	4	Good	Good	Very Good	Very Good	Good	P/SU	6.0–7.0	LW, E to LSp, EF	40F
LEGUMES	Woollypod vetch *p. 185*	Lana	CSA	7	Very Good	Very Good	Good	Good	Very Good	SP/C	6.0–8.0	F	

[1]B=Biennial; CSA=Cool season annual; LP=Long-lived perennial; SA=Summer annual; SP=Short-lived perennial; WA=Winter annual
[2]See USDA Hardiness Zone Map, inside front cover. NFT=Not frost tolerant. [3]C=Climbing; U=Upright; P=Prostrate; SP=Semi-prostrate; SU=Semi-upright. [4]E=Early; M=Mid; L=Late; F=Fall; Sp=Spring; Su=Summer; W=Winter

○=Poor; ◔=Fair; ◑=Good; ◕=Very Good; ●=Excellent

Chart 3B **PLANTING**

Group	Species	Depth	Seeding Rate: Drilled lb./A	Seeding Rate: Drilled bu/A	Seeding Rate: Broadcast lb./A	Seeding Rate: Broadcast bu/A	Seeding Rate: Broadcast oz./100 ft^2	Cost ($/lb.)[1]	Cost/A (median)[2] drilled	Cost/A (median)[2] broadcast	Inoc. Type	Reseeds[3]
NONLEGUMES	Annual ryegrass	0-½	10-20	.4-.8	20-30	.8-1.25	1	.70-1.30	12	24		U
NONLEGUMES	Barley	¾-2	50-100	1-2	80-125	1.6-2.5	3-5	.17-.37	20	27		S
NONLEGUMES	Oats	½-1½	80-110	2.5-3.5	110-140	3.5-4.5	4-6	.13-.37	25	33		S
NONLEGUMES	Rye	¾-2	60-120	1-2	90-160	1.5-3.0	4-6	.18-.50	25	35		S
NONLEGUMES	Wheat	½-1½	60-120	1-2	60-150	1-2.5	3-6	.10-.30	18	22		S
NONLEGUMES	Buckwheat	½-1½	48-70	1-1.4	50-90	1.2-1.5	3-4	.30-.75	32	38		R
NONLEGUMES	Sorghum-sudangrass	½-1½	35	1	40-50	1-1.25	2	.40-1.00	26	34		S
BRASSICAS	Mustards	¼-¾	5-12		10-15		1	1.50-3.00	16	24		U
BRASSICAS	Radish	¼-½	8-13		10-20		1	1.50-2.50	22	32		S
BRASSICAS	Rapeseed	¼-¾	5-10		8-14		1	1.00-2.00	11	16		S
LEGUMES	Berseem clover	¼-½	8-12		15-20		2	1.70-2.50	22	39	crimson, berseem	N
LEGUMES	Cowpeas	1-1½	30-90		70-120		5	.85-1.50	71	113	cowpeas, lespedeza	S
LEGUMES	Crimson clover	¼-½	15-20		22-30		2-3	1.25-2.00	27	40	crimson, berseem	U
LEGUMES	Field peas	1½-3	50-80		90-100		4	.61-1.20	50	75	pea, vetch	S
LEGUMES	Hairy vetch	½-1½	15-20		25-40		2	1.70-2.50	35	65	pea, vetch	S
LEGUMES	Medics	¼-½	8-22		12-26		2/3	2.50-4.00	58	75	annual medics	R
LEGUMES	Red clover	¼-½	8-10		10-12		3	1.40-3.30	23	28	red cl, wht cl	S
LEGUMES	Subterranean clover	¼-½	10-20		20-30		3	2.50-3.50	45	75	clovers, sub, rose	U
LEGUMES	Sweetclovers	¼-1.0	6-10		10-20		1.5	1.00-3.00	16	32	alfalfa, swt cl	U
LEGUMES	White clover	¼-½	3-9		5-14		1.5	1.10-4.00	19	30	red cl, wht cl	R
LEGUMES	Woollypod vetch	½-1	10-30		30-60		2-3	1.25-1.60	30	65	pea, vetch	S

[1]Per pound in 50-lb. bags as of summer/fall 2006; To locate places to buy seed, see *Seed Suppliers* (p. 166).
[2]Mid-point price at mid-point rate, seed cost only. [3]**R**=Reliably; **U**=Usually; **S**=Sometimes; **N**=Never (reseeds).

Chart 4A **POTENTIAL ADVANTAGES**

	Species	Soil Impact			Soil Ecology				Other		
		subsoiler	free P&K	loosen topsoil	nematodes	disease	allelopathic	choke weeds	attract beneficials	bears traffic	short windows
NONLEGUMES	Annual ryegrass *p. 74*	Good	Good	Excellent	Good	Good	Good	Excellent	Fair	Excellent	Excellent
	Barley *p. 77*	Good	Good	Very Good	Fair	Good	Very Good	Very Good	Good	Good	Excellent
	Oats *p. 93*	Poor	Fair	Very Good	Poor	Good	Very Good	Excellent	Poor	Good	Excellent
	Rye *p. 98*	Fair	Very Good	Excellent	Good	Very Good	Excellent	Excellent	Fair	Very Good	Excellent
	Wheat *p. 111*	Good	Very Good	Very Good	Fair	Fair	Fair	Very Good	Fair	Good	Good
	Buckwheat *p. 90*	Poor	Excellent	Very Good	Fair	Poor	Very Good	Excellent	Excellent	Poor	Excellent
	Sorghum-sudangrass *p. 106*	Excellent	Good	Good	Very Good	Very Good	Excellent	Excellent	Good	Good	Excellent
BRASSICAS	Mustards *p.81*	Fair	Good	Very Good	Very Good	Very Good	Very Good	Very Good	Good	Good	Very Good
	Radish *p. 81*	Excellent	Very Good	Good	Very Good	Good	Very Good	Excellent	Fair	Fair	Very Good
	Rapeseed *p. 81*	Good	Fair	Good	Very Good	Good	Very Good	Good	Good	Fair	Good
LEGUMES	Berseem clover *p. 118*	Fair	Very Good	Very Good	Poor	Poor	Fair	Very Good	Good	Fair	Very Good
	Cowpeas *p. 125*	Good	Good	Very Good	Poor	Poor	Poor	Excellent	Very Good	Poor	Excellent
	Crimson clover *p. 130*	Fair	Good	Good	Fair	Good	Fair	Good	Very Good	Fair	Good
	Field peas *p. 135*	Fair	Fair	Very Good	Good	Very Good	Fair	Very Good	Very Good	Fair	Very Good
	Hairy vetch *p. 142*	Good	Good	Very Good	Fair	Good	Good	Very Good	Excellent	Poor	Poor
	Medics *p. 152*	Good	Fair	Fair	Good	Good	Fair	Very Good	Fair	Fair	Excellent
	Red clover *p. 159*	Very Good	Very Good	Good	Fair	Fair	Good	Good	Very Good	Good	Good
	Subterranean clover *p. 164*	Poor	Fair	Good	Fair	Fair	Very Good	Excellent	Very Good	Very Good	Very Good
	Sweetclovers *p. 171*	Excellent	Excellent	Excellent	Fair	Fair	Fair	Good	Very Good	Good	Poor
	White clover *p. 179*	Fair	Fair	Very Good	Poor	Poor	Good	Very Good	Good	Excellent	Good
	Woollypod vetch *p. 185*	Very Good	Good	Very Good	Fair	Fair	Good	Excellent	Very Good	Fair	Good

○=Poor; ◔=Fair; ◑=Good; ◕=Very Good; ●=Excellent

Chart 4B **POTENTIAL DISADVANTAGES**

Note change in symbols ○ = problem ● = not a problem

Group	Species	Increase Pest Risks: weed potential	Increase Pest Risks: insects/nematodes	Increase Pest Risks: crop disease	Management Challenges: hinder crops	Management Challenges: establish	Management Challenges: till-kill	Management Challenges: mow-kill	Management Challenges: mature incorp.	Comments Pro/Con
NONLEGUMES	Annual ryegrass	○[1]	◕	◕	◑	●	●	●	◑	If mowing, leave 3-4" to ensure regrowth.
NONLEGUMES	Barley	◔	◔	◔	◑	●	●	●	○	Can be harder than rye to incorporate when mature.
NONLEGUMES	Oats	●	◑	◕	◑	●	●	◔	◑	Cleaned, bin-run seed will suffice.
NONLEGUMES	Rye	◔	◑	◕	◔	◕	◕	●	○	Can become a weed if tilled at wrong stage.
NONLEGUMES	Wheat	◑	◔	◔	◑	●	●	◕	◑	Absorbs N and H_2O heavily during stem growth, so kill before then.
NONLEGUMES	Buckwheat	○	◕	●	●	●	●	●	●	Buckwheat sets seed quickly.
NONLEGUMES	Sorghum-sudangrass	◑	◕	●	◔	●	◕	◕	◔	Mature, frost-killed plants become quite woody.
BRASSICAS	Mustards	◔	◕	●	◑	●	●	◔	●	Great biofumigation potential; winterkills at 25° F.
BRASSICAS	Radish	◑	◕	●	●	●	●	●	●	Winter kills at 25° F; cultivars vary widely.
BRASSICAS	Rapeseed	◔	◔	●	◑	●	◑	◔	◕	Canola has less biotoxic activity than rape.
LEGUMES	Berseem clover	●	◑	◕	●	●	◔	◔	◕	Multiple cuttings needed to achieve maximum N.
LEGUMES	Cowpeas	●	◑	◕	●	●	●	●	●	Some cultivars, nematode resistant.
LEGUMES	Crimson clover	◑	○	◕	●	◕	◕	◕	◕	Good for underseeding, easy to kill by tillage or mowing.
LEGUMES	Field peas	●	◑	◔	●	●	●	●	◕	Susceptible to *sclerotinia* in East.
LEGUMES	Hairy vetch	◑	◑	●	●	◑	◔	●	◑	Tolerates low fertility, wide pH range, cold or fluctuating winters.
LEGUMES	Medics	◑	◔	●	◕	◕	◔	◕	◕	Perennials easily become weedy.
LEGUMES	Red clover	◑	◔	◕	●	◕	◑	◔	◕	Grows best where corn grows well.
LEGUMES	Subterranean clover	◑	○	◑	◑	●	◔	○	◕	Cultivars vary greatly.
LEGUMES	Sweetclovers	◔	◔	●	◑	●	◑	◕	◔	Hard seed possible problem; does not tolerate seeding year mowing
LEGUMES	White clover	◑	◑	◕	◔	◕	○	◑	◕	Can be invasive; survives tillage.
LEGUMES	Woollypod vetch	◑	◑	◔	◑	◑	◕	●	◔	Hard seed can be problematic; resident vegetation eventually displaces

[1]Note change in symbols, this page only: ○ = problem. ◔ = Could be a moderate problem. ◑ = Could be a minor problem. ◕ = Occasionally a minor problem. ● = not a problem

OVERVIEW OF NONLEGUME COVER CROPS

Commonly used nonlegume cover crops include:

- Annual cereals (rye, wheat, barley, oats)
- Annual or perennial forage grasses such as ryegrass
- Warm-season grasses like sorghum-sudangrass
- Brassicas and mustards

Nonlegume cover crops are most useful for:

- Scavenging nutrients—especially N—left over from a previous crop
- Reducing or preventing erosion
- Producing large amounts of residue and adding organic matter to the soil
- Suppressing weeds

Annual cereal grain crops have been used successfully in many different climates and cropping systems. Winter annuals usually are seeded in late summer or fall, establish and produce good root and topgrowth biomass before going dormant during the winter, then green up and produce significant biomass before maturing. Rye, wheat, and hardy triticale all follow this pattern, with some relatively small differences that will be addressed in the section for each cover crop.

There is growing interest in the use of brassica and mustard cover crops due to their "biofumigation" characteristics. They release biotoxic chemicals as they break down, and have been found to reduce disease, weed and nematode pressure in the subsequent crop. Brassicas and mustards provide most of the benefits of other nonlegume cover crops, while some (forage radish, for example) are thought to alleviate soil compaction. See the chapter, *Brassicas and Mustards* (pp. 81), for more information.

Perennial and warm-season forage grasses also can serve well as cover crops. Forage grasses, like sod crops, are excellent for nutrient scavenging, erosion control, biomass production and weed control. Perennials used as cover crops are usually grown for about one year. Summer-annual (warm-season) grasses may fill a niche for biomass production and weed or erosion control if the ground would otherwise be left fallow (between vegetable crops, for example). Buckwheat, while not a grass, is also a warm-season plant used in the same ways as summer-annual grasses.

Nonlegume cover crops are higher in carbon than legume cover crops. Because of their high carbon content, grasses break down more slowly than legumes, resulting in longer-lasting residue. As grasses mature, the carbon-to-nitrogen ratio (C:N) increases. This has two tangible results: The higher carbon residue is harder for soil microbes to break down, so the process takes longer, and the nutrients contained in the cover crop residue usually are less available to the next crop.

So although grass cover crops take up leftover N from the previous crop, as they mature the N is less likely to be released for use by a crop grown immediately after the grass cover crop. As an example of this, think of how long it takes for straw to decompose in the field. Over time, the residue does break down and nutrients are released. In general, this slower decomposition and the higher carbon content of grasses can lead to increased soil organic matter, compared to legumes.

The carbon content and breakdown rate of brassicas is usually intermediate to grasses and legumes, depending on maturity when terminated. Brassicas and mustards can take up as much N as grass cover crops, but may release that N more readily to the subsequent crop.

Nonlegume cover crops can produce a lot of residue, which contributes to their ability to prevent erosion and suppress weeds while they are growing or when left on the soil surface as a mulch.

Although grasses and other nonlegumes contain some nitrogen in their plant tissues, they generally are not significant sources of N for your cropping system. They do, however, keep excess soil N from leaching, and prevent the loss of soil organic matter through erosion.

Management of nonlegumes in your cropping system may involve balancing the amount of residue produced with the possibility of tying up N for more than one season. Mixtures of grass and legume cover crops can alleviate the N-immobilization effect, can produce as much or more dry matter as a pure grass stand and may provide better erosion control due to the differences in growth habit. Suggestions for cover crop mixtures are found in the individual cover crop sections.

In addition to grasses, another summer non-legume is buckwheat, which is described in detail in its own section (p. 90). Buckwheat is usually classed as a non-grass coarse grain. While it is managed like a quick-growing grain, it has a succulent stem, large leaves and white blossoms.

ANNUAL RYEGRASS

Lolium multiflorum

Also called: Italian ryegrass

Type: cool season annual grass

Roles: prevent erosion, improve soil structure and drainage, add organic matter, suppress weeds, scavenge nutrients

Mix with: legumes, grasses

See charts, pp. 66 to 72, for ranking and management summary.

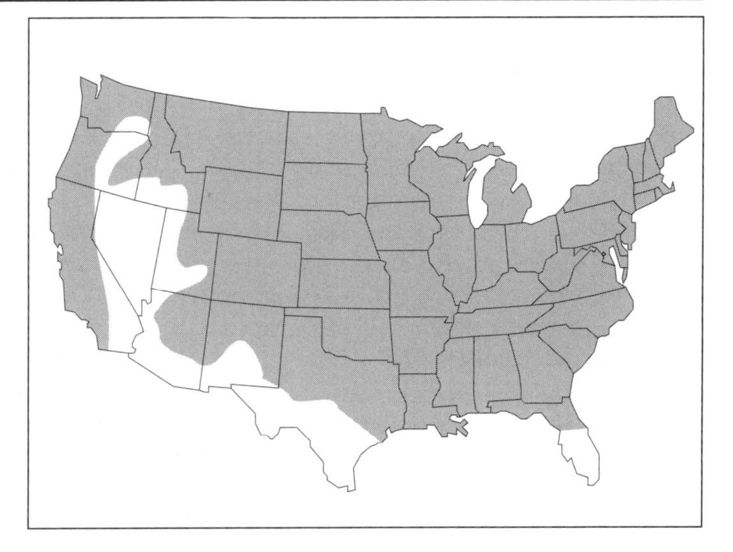

If you want to build soil without investing much in a cover crop, consider annual ryegrass. A quick-growing, nonspreading bunch grass, annual ryegrass is a reliable, versatile performer almost anywhere, assuming adequate moisture and fertility. It does a fine job of holding soil, taking up excess N and outcompeting weeds.

Ryegrass is an excellent choice for building soil structure in orchards, vineyards and other cropland to enhance water infiltration, water-holding capacity or irrigation efficiency. It can reduce soil splash on solanaceous crops and small fruit crops, decreasing disease and increasing forage quality. You also can overseed ryegrass readily into corn, soybeans and many high-value crops.

BENEFITS

Erosion fighter. Ryegrass has an extensive, soil-holding root system. The cover crop establishes quickly even in poor, rocky or wet soils and tolerates some flooding once established. It's well-suited for field strips, grass waterways or exposed areas.

Soil builder. Ryegrass's dense yet shallow root system improves water infiltration and enhances soil tilth. Rapid aboveground growth helps supply organic matter. Expect about 4,000 to 8,000 lb. dry matter/A on average with a multicut regimen, climbing as high as 9,000 lb. DM/A over a full field season with high moisture and fertility.

Weed suppressor. Mixed with legumes or grasses, annual ryegrass usually establishes first and improves early-season weed control. With adequate moisture, it serves well in Hardiness Zone 6 and warmer as a living mulch in high-value systems where you can mow it regularly. It may winterkill elsewhere, especially without protective snow cover during prolonged cold snaps. Even so, its quick establishment in fall still would provide an excellent, winterkilled mulch for early-spring weed suppression.

Nutrient catch crop. A high N user, ryegrass can capture leftover N and reduce nitrate leaching over winter. Provided it survives the winter, its extensive, fibrous root system can take up as much as 43 lb. N/A, a University of California study showed (445). It took up about 60 lb. N/A by mid-May following corn in a Maryland study. Cereal rye scavenged the same amount of N by mid-April on this silt loam soil (372). Ryegrass works well ahead of no-till corn or soybeans in the Corn Belt, sometimes winterkilling, or spray it for a weed-controlling mulch (302).

Nurse/companion crop. Ryegrass helps slow-growing, fall-seeded legumes establish and overwinter in the northern U.S., even if the ryegrass winterkills. It tends to outcompete legumes in the South, although low N fertility favors the legume.

Emergency forage. Ryegrass is a very palatable forage (132). You can extend the grazing period in late fall and early spring by letting livestock graze cover crops of ryegrass or a ryegrass-based mix. Annual ryegrass can be used as emergency forage if alfalfa winterkills. It establishes quickly and produces a lot of forage in a short amount of time.

MANAGEMENT

Ryegrass prefers fertile, well-drained loam or sandy loam soils, but establishes well on many soil types, including poor or rocky soils. It tolerates clay or poorly-drained soils in a range of climates and will outperform small grains on wet soils (132, 421).

Annual ryegrass has a biennial tendency in cool regions. If it overwinters, it will regrow quickly and produce seed in late spring. Although few

ANNUAL RYEGRASS (Lolium multiflorum)

plants survive more than a year, this reseeding characteristic can create a weed problem in some areas, such as the mid-Atlantic or other areas with mild winters. In the Midwest and Southern Plains, it can be a serious weed problem in oat and wheat crops. It has also been shown to develop herbicide resistance, compounding possible weed problems (161).

Establishment & Fieldwork

Annual ryegrass germinates and establishes well even in cool soil (421). Broadcast seed at 20 to 30 lb./A. You needn't incorporate seed when broadcasting onto freshly cultivated soil—the first good shower ensures seed coverage and good germination. Cultipacking can reduce soil heaving, however, especially with late-fall plantings. Drill 10 to 20 lb./A, 1/4 to 1/2 inch deep.

Noncertified seed will reduce seeding cost, although it can introduce weeds. Annual ryegrass also cross-pollinates with perennial ryegrass and turf-type annual ryegrass species, so don't expect a pure stand if seeding common annual ryegrass.

Winter annual use. Seed in fall in Zone 6 or warmer. In Zone 5 and colder, seed from midsummer to early fall—but at least 40 days before your area's first killing frost (194). Late seeding increases the probability of winterkill.

If aerially seeding, increase rates at least 30 percent compared to broadcast seeding rate (18). You can overseed into corn at last cultivation or later (consider adding 5 to 10 pounds of red or white clover with it) or plant right after corn silage harvest. Overseed into soybeans at leaf-yellowing or later (191, 194). When overseeding into solanaceous crops such as peppers, tomatoes and eggplant, wait until early to full bloom.

Spring seeding. Sow ryegrass right after small grains or an early-spring vegetable crop, for a four- to eight-week summer period before a fall vegetable crop (361).

Mixed seeding. Plant ryegrass at 8 to 15 lb./A with a legume or small grain, either in fall or early in spring. Ryegrass will dominate the mixture unless you plant at low rates or mow regularly. The legume will compete better in low-N conditions. Seed the legume at about two-thirds its normal rate. Adequate P and K levels are important when growing annual ryegrass with a legume.

In vineyards, a fall-seeded, 50:50 mix of ryegrass and crimson clover works well, some California growers have found (211).

Although not a frequent pairing, drilling ryegrass in early spring at 20 lb./A with an oats nurse crop or frost seeding 10 lb./A into overwintered small grains can provide some fine fall grazing. Frost seeding with red clover or other large-seeded, cool-season legumes also can work well, although the ryegrass could winterkill in some conditions.

Maintenance. Avoid overgrazing or mowing ryegrass closer than 3 to 4 inches. A stand can persist many years in orchards, vineyards, and other areas if allowed to reseed naturally and not subject to prolonged heat, cold or drought. That's rarely the case in Zone 5 and colder, however, where climate extremes take their toll. Perennial ryegrass may be a smarter choice if persistence is important. Otherwise, plan on incorporating the cover within a year of planting. Annual ryegrass is a relatively late maturing plant, so in vineyards it may use excessive water and N if left too long.

Killing & Controlling

You can kill annual ryegrass mechanically by disking or plowing, preferably during early bloom (usually in spring), before it sets seed (361, 422). Mowing may not kill ryegrass completely (103). You also can kill annual ryegrass with non-persistent contact herbicides, although some users report incomplete kill and/or resistance to glyphosate (161, 302).

To minimize N tie-up as the biomass decomposes, wait a few weeks after incorporation before you seed a subsequent crop. Growing ryegrass with a legume such as red clover would minimize the N concern. By letting the cover residue decompose a bit, you'll also have a seedbed that is easier to manage.

Pest Management

Weed potential. Ryegrass can become a weed if allowed to set seed (361). It often volunteers in vineyards or orchards if there is high fertility and may require regular mowing to reduce competition with vines (422). A local weed management specialist may be able to recommend a herbicide that can reduce ryegrass germination if the cover is becoming a weed in perennial grass stands. Chlorsulfuron is sometimes used for this purpose in California (422).

Insect and other pests. Ryegrass attracts few insect pests and generally can help reduce insect pest levels in legume stands and many vegetable crops, such as root crops and brassicas. Rodents are occasionally a problem when ryegrass is used as a living mulch.

Rust occasionally can be a problem with annual ryegrasses, especially crown and brown (stem) rust. Look for resistant, regionally adapted varieties. Annual ryegrass also can host high densities of pin nematodes (*Paratylenchus projectus*) and bromegrass mosaic virus, which plant-parasitic nematodes (*Xiphinema spp.*) transmit (422).

Other Options

Ryegrass provides a good grazing option that can extend the grazing season for almost any kind of livestock. Although very small-seeded, ryegrass does not tiller heavily, so seed at high rates if you expect a ryegrass cover crop also to serve as a pasture. Some varieties tolerate heat fairly well and can persist for several years under sound grazing practices that allow the grass to reseed. As a hay option, annual ryegrass can provide 2,000 to 6,000 pounds of dry forage per acre, depending on moisture and fertility levels (422). For highest quality hay, cut no later than the early bloom stage and consider growing it with a legume. When using ryegrass for grass waterways and conservation strips on highly erodible slopes, applying 3,000 to 4,000 pounds of straw per acre after seeding at medium to high rates can help keep soil and seed in place until the stand establishes (422).

Management Cautions

Ryegrass is a heavy user of moisture and N. It performs poorly during drought or long periods of high or low temperature, and in low-fertility soils. It can compete heavily for soil moisture when used as living mulch. It also can become a weed problem (361).

COMPARATIVE NOTES

- Establishes faster than perennial ryegrass but is less cold-hardy
- Less persistent but easier to incorporate than perennial ryegrass
- About half as expensive as perennial ryegrass
- In Southern USA, annual is more adapted and produces much greater biomass

Cultivars. Many varieties are widely available. Improved cultivars should be considered if growing for forage. There are diploid (2n = 14 chromosomes) and tetraploid (4n = 28 chromosomes) cultivars. Tetraploids produce larger plants with wider leaves and mature later.

Seed sources. See *Seed Suppliers* (p. 195). 

BARLEY

Hordeum vulgare

Type: cool season annual cereal grain

Roles: prevent erosion, suppress weeds, scavenge excess nutrients, add organic matter

Mix with: annual legumes, ryegrass or other small grains

See charts, pp. 66 to 72, for ranking and management summary.

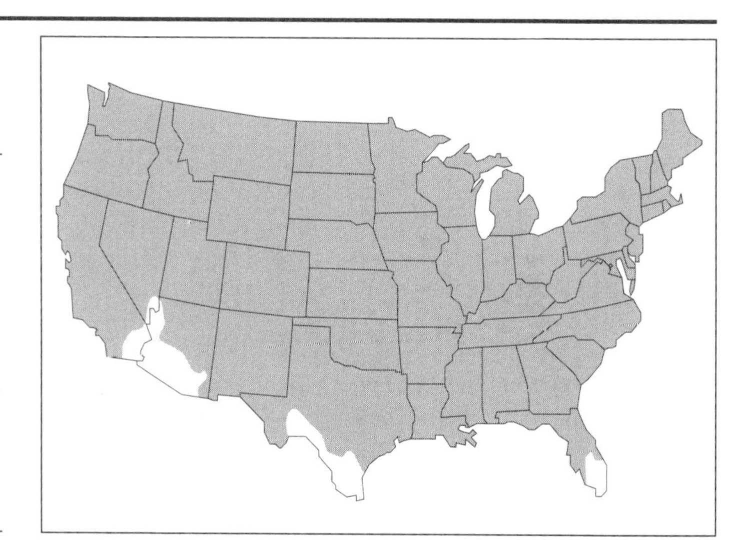

Inexpensive and easy to grow, barley provides exceptional erosion control and weed suppression in semi-arid regions and in light soils. It also can fill short rotation niches or serve as a topsoil-protecting crop during droughty conditions in any region. It is more salt tolerant than other small grains and can sop up excess subsoil moisture to help prevent saline seep formation (136).

It's a fine choice for reclaiming overworked, weedy or eroded fields, or as part of a cover crop mix for improving soil tilth and nutrient cycling in perennial cropping systems in Hardiness Zone 8 or warmer.

Barley prefers cool, dry growing areas. As a spring cover crop, it can be grown farther north than any other cereal grain, largely because of its short growing period. It also can produce more biomass in a shorter time than any other cereal crop (273).

BENEFITS

Erosion control. Use barley as an overwintering cover crop for erosion control in Zone 8 and warmer, including much of California, western Oregon and western Washington. It's well-suited for vineyards and orchards, or as part of a mixed seeding.

A fast-growing barley can be grown farther north and produce more biomass in a shorter time than any other cereal grain.

As a winter annual, barley develops a deep, fibrous root system. The roots can reach as deep as 6.5 feet. As a spring crop, barley has a comparatively shallow root system but holds soil strongly to minimize erosion during droughty conditions (71).

Nutrient recycler. Barley can scavenge significant amounts of nitrogen. It captured 32 lb. N/A as a winter cover crop following a stand of fava beans (*Vicia faba*) in a California study, compared with 20 lb./A for annual ryegrass. A barley cover crop reduced soil N an average of 64 percent at eight sites throughout North America that had received an average of 107 lb. N/A (265). Intercropping barley with field peas (*Pisum sativum*) can increase the amount of N absorbed by barley and returned to the soil in barley residue, other studies show (215, 218). Barley improves P and K cycling if the residue isn't removed.

Weed suppressor. Quick to establish, barley outcompetes weeds largely by absorbing soil moisture during its early growing stages. It also shades out weeds and releases allelopathic chemicals that help suppress them.

Tilth-improving organic matter. Barley is a quick source of abundant biomass that, along with its thick root system, can improve soil structure and water infiltration (273, 445). In California cropping systems, cultivars such as UC476 or COSINA can produce as much as 12,900 lb. biomass/A.

Nurse crop. Barley has an upright posture and relatively open canopy that makes it a fine nurse crop for establishing a forage or legume stand. Less competitive than other small grains, barley also uses less water than other covers crops. In weedy fields, wait to broadcast the forage or legume until after you've mechanically weeded barley at the four- or five-leaf stage to reduce weed competition.

As an inexpensive, easy-to-kill companion crop, barley can protect sugarbeet seedlings during their first two months while also serving as a soil protectant during droughty periods (details below).

Pest suppression. Barley can reduce incidence of leafhoppers, aphids, armyworms, root-knot nematodes and other pests, a number of studies suggest.

MANAGEMENT

Establishment & Fieldwork

Barley establishes readily in prepared seedbeds, and can also be successfully no-tilled. It prefers adequate but not excessive moisture and does poorly in waterlogged soils. It grows best in well-drained, fertile loams or light, clay soils in areas having cool, dry, mild winters. It also does well on light, droughty soils and tolerates somewhat alkaline soils better than other cereal crops.

With many varieties of barley to choose from, be sure to select a regionally adapted one. Many are well-adapted to high altitudes and cold, short growing seasons.

Spring annual use. Drill at 50 to 100 lb./A (1 to 2 bushels) from 3/4 to 2 inches deep into a prepared seedbed, or no-till using the same seeding rate.

If broadcasting, prepare the seedbed with at least a light field cultivation. Sow 80 to 125 lb./A (1.5 to 2.5 bushels) and harrow, cultipack or disk lightly to cover. Use a lower rate (25 to 50 pounds) if overseeding as a companion crop or a higher rate (140 pounds) for very weedy fields. When broadcasting, consider seeding half in one direction, then the rest in a perpendicular direction for better coverage (71).

Winter annual use. Barley can be used as a winter annual cover crop wherever it is grown as a winter grain crop. It is less winter-hardy than rye. In Zone 8 or warmer, it grows throughout the winter if planted from September through February. Plantings before November 1 generally fare best, largely due to warmer soil conditions.

Expect mixed results if trying to use barley as a self-reseeding cover crop.

Mixed seedings. Barley works well in mixtures with other grasses or legumes. In low-fertility soils or where you're trying to minimize tie-up of soil nitrogen, growing barley with one or more legumes can be helpful. Your seeding cost per pound will increase, but the reduced seeding rate can offset some of this. A short-season Canadian field pea would be a good companion, or try an oat/barley/pea mix, suggests organic farmer Jack Lazor, Westfield, Vt.

In northern California, Phil LaRocca (LaRocca Vineyards, Forest Ranch, Calif.) lightly disks his upper vineyard's soil before broadcasting a mix of barley, fescue, brome, LANA vetch, and crimson, red and subterranean clovers, usually during October. He seeds at 30 to 35 lb./A, with 10 to 20 percent being barley. "I've always added more barley to the seeding rate than recommended. More is better, especially with barley, if you want biomass and weed suppression," he says.

After broadcasting, LaRocca covers erosion-prone areas with 2 tons of rice straw per acre, which is "cheaper than oat straw here and has fewer weed seeds," he notes. "The straw decomposes quickly and holds seed and soil well." Besides contributing to soil humus (as the cover crop also does), the straw helps keep the seedbed warm and moist. That can be very helpful in LaRocca's upper vineyard, where it sometimes snows in winter.

In his other, less-erodible vineyard, LaRocca disks up the cover vegetation, then runs a harrow quickly on top of the disked alleyways to set a seedbed before broadcasting and cultipacking a similar mix of cover crops.

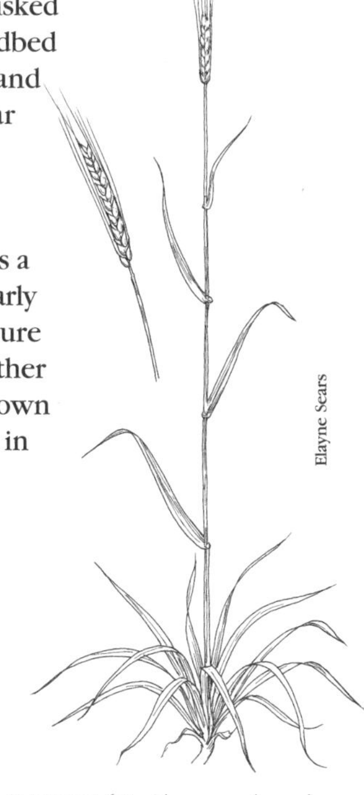

BARLEY (Hordeum vulgare)

Field Management

Although barley absorbs a lot of water in its early stages, it uses moisture more efficiently than other cereals and can be grown without irrigation in some situations. About half of the commercial barley acreage in dryland areas is irrigated, however. California cropping systems that include barley tend to be irrigated as well. Low seeding rates won't necessarily conserve moisture, as vegetative growth often increases.

LaRocca hasn't had any moisture problems or grape-yield concerns from growing barley or other cover crops, even in the 40 percent of his upper vineyard that isn't furrow-irrigated. "Once your vines are established, their root system is deeper and much more competitive than a typical cover crop's root system," he observes.

Mowing can postpone and prolong barley flowering, as with other cereal grains. As a spring cover, barley puts on biomass quickly, so you can kill it in plenty of time for seeding a following crop. If you want barley to reseed, don't mow until most of the stand has headed and seed is about to fall off.

To encourage reseeding of his cover mix, Phil LaRocca allows every other row in his upper vineyard to go to seed, then disks it down. That lets him skip reseeding some blocks.

If you're concerned about barley reseeding or crop competition when intercropped, however, plant a lighter stand, suggests Alan Brutlag, Wendell, Minn. During droughty conditions, he broadcasts

25 to 30 pounds of barley per acre as a soil-protective companion crop for sugarbeet seedlings. The low-density stand is easy to stunt or kill a month later with the combination of herbicides and crop oil that he uses for weed control in his sugarbeets. Another control option is a single application of an herbicide labeled for grass control.

Killing

Kill barley with a grass herbicide in late spring, or by rolling disking or mowing at the mid- to late-bloom stage but before it starts setting seed.

If plant-parasitic nematodes have been a problem, incorporate overwintered barley early in spring, before warm temperatures encourage nematode populations.

Pest Management

Annual weeds and lodging can occur when growing barley in high-fertility soils, although these wouldn't pose problems in a barley cover crop. Despite their less dense canopy, six-rowed varieties tend to be taller and more competitive against weeds than two-rowed varieties. If you're considering a grain option, harrowing or hoeing just before barley emergence could reduce weeds that already have sprouted.

Barley produces alkaloids that have been shown to inhibit germination and growth of white mustard (247). These exudates also protect barley plants from fungus, armyworm larvae, bacteria and aphids (248, 455).

Barley seems to reduce the incidence of grape leafhoppers in vineyards and increase levels of beneficial spiders, one California grower observed (211). Growing high-biomass cover crops such as barley or rye increased populations of centipedes, predator mites and other important predators, independent of tillage system used, a study in the Pacific Northwest found (444).

Cutworms and other small grain pests can be occasional problems. Some perennial crop growers in California report increased incidence of gophers when growing cover crop mixes and try to minimize this by encouraging owl populations.

Avoid seeding in cold, damp soils, which makes barley more prone to fungus and disease. Assuming adequate soil moisture, shallow seeding can hasten emergence and lessen incidence of root rot disease, if this has been a problem in your area (397). Varieties resistant to leaf diseases are available. Two-rowed varieties are more resistant to leaf rust and mildew. Also avoid planting barley after wheat.

If nematodes are likely to be a problem, plant late in fall or during winter to avoid warm-season growth and incorporate early in spring in Zone 8 and warmer. Barley can be a host for a nematode species (*Meloidogyne javanica*) that adversely affects Thompson seedless grapes.

Barley drastically reduced root-knot nematode (*Meloidogyne hapla* M. Chitwood) populations and increased marketable carrot yields by at least seventeen-fold in a Quebec study comparing three-year rotations (242).

Other Options

Barley can be grazed lightly in winter or spring or cut for hay/haylage (191). It has greater forage nutritive value than oats, wheat or triticale. It also can be grown as a specialty grain for malting, soups, bread and other uses. As a feed grain (in a hog ration, for example), it can replace some costlier corn.

COMPARATIVE NOTES

- Barley tillers more than oats and also is more drought-tolerant, but oats generally perform better as a companion crop or winterkilled nurse crop because they are less competitive than barley (397).
- Barley tolerates alkaline soils better than any other cereal.
- Winter cultivars are less winterhardy than winter wheat, triticale or cereal rye.

Cultivars. Many commercial varieties are available. Look for low-cost, regionally adapted cultivars with at least 95-percent germination.

Six-rowed cultivars are better for overseeding, and are more heat- and drought-tolerant. Two-rowed types have more symmetrical kernels and are more disease-resistant (e.g. leaf rust and mildew) than six-rowed types, in which two-thirds of the lateral rows of the spike are smaller and twisted.

Seed sources. See *Seed Suppliers* (p. 195). 

BRASSICAS AND MUSTARDS

Type: Annual (usually winter or spring; summer use possible)

Roles: Prevent erosion, suppress weeds and soilborne pests, alleviate soil compaction and scavenge nutrients

Mix with: Other brassicas or mustards, small grains or crimson clover

Species: *Brassica napus, Brassica rapa, Brassica juncea, Brassica hirta, Raphanus sativus, Sinapsis alba*

See charts, pp. 66 to 72, for ranking and management summary.

RAPE or CANOLA (Brassica rapa)

Nomenclature Note: The cover crops described in this chapter all belong to the **family** BRASSICACEAE. Most but not all of the species belong to the **genus** *Brassica*. In common usage, the various species are sometimes lumped together as "brassicas" and sometimes distinguished as "brassicas" vs. "mustards." In this book, we will use brassicas as an umbrella term for all species; mustards will be used to distinguish that subgroup, which has some unique characteristics.

Adaptation Note: This chapter addresses management of eight different cover crop species with varying degrees of winterhardiness. Some can be managed as winter or spring annuals. Others are best planted in late summer for cover crop use but will winterkill. Consult the information on management, winterhardiness and winter vs. spring use (pp. 87-88) and the examples throughout the chapter, then check with local experts for specific adaptation information for your brassica cover crop of choice.

Brassica and mustard cover crops are known for their rapid fall growth, great biomass production and nutrient scavenging ability. However, they are attracting renewed interest primarily because of their pest management characteristics. Most *Brassica* species release chemical compounds that may be toxic to soil borne pathogens and pests, such as nematodes, fungi and some weeds. The mustards usually have higher concentrations of these chemicals.

Brassicas are increasingly used as winter or rotational cover crops in vegetable and specialty crop production, such as potatoes and tree fruits. There is also growing interest in their use in row crop production, primarily for nutrient capture, nematode trapping, and biotoxic or biofumigation activity. Some brassicas have a large taproot that can break through plow pans better than the fibrous roots of cereal cover crops or the mustards. Those brassicas that winterkill decompose very quickly and leave a seedbed that is mellow and easy to plant in.

With a number of different species to consider, you will likely find one or more that can fit your farming system. Don't expect brassicas to eliminate your pest problems, however. They are a good tool and an excellent rotation crop, but pest management results are inconsistent. More research is needed to further clarify the variables affecting the release and toxicity of the chemical compounds involved (see p. 82).

BENEFITS

Erosion control and nutrient scavenging. Brassicas can provide greater than 80% soil coverage when used as a winter cover crop (176). Depending on location, planting date and soil fertility, they produce up to 8,000 lb. biomass/A. Because of their fast fall growth, brassicas are well-suited to capture soil nitrogen (N) remaining after crop harvest. The amount of nitrogen captured is mainly related to biomass accumulation and the amount of N available in the soil profile.

Because they immobilize less nitrogen than some cereal cover crops, much of the N taken up can become available for uptake by main crops in early to late spring (see also *Building Soil Fertility and Tilth with Cover Crops*, pp. 16–24). Brassicas can root to depths of six feet or more, scavenging nutrients from below the rooting depth of most crops. To maximize biomass production and nutrient scavenging in the fall, brassicas must be planted earlier than winter cereal cover crops in most regions, making them more difficult to fit into grain production rotations.

Brassicas must be planted earlier than winter cereal cover crops in most regions.

Pest management. All brassicas have been shown to release bio-toxic compounds or metabolic by-products that exhibit broad activity against bacteria, fungi, insects, nematodes, and weeds. Brassica cover crops are often mowed and incorporated to maximize their natural fumigant potential. This is because the fumigant chemicals are produced only when individual plant cells are ruptured.

Pest suppression is believed to be the result of glucosinolate degradation into biologically active sulfur containing compounds call thiocyanates (152, 320). To maximize pest suppression, incorporation should occur during vulnerable life-stages of the pest (446).

The biotoxic activity of brassica and mustard cover crops is low compared to the activity of commercial fumigants (388). It varies depending on species, planting date, growth stage when killed, climate and tillage system. Be sure to consult local expertise for best results.

▼ **Precaution.** The use of brassicas for pest management is in its infancy. Results are inconsistent from year to year and in different geographic regions. Different species and varieties contain different amounts of bioactive chemicals. Be sure to consult local expertise and begin with small test plots on your farm.

Disease

In Washington, a SARE-funded study of brassica green manures in potato cropping systems compared winter rape (*Brassica napus*) and white mustard (*Sinapis alba*) to no green manure, with and without herbicides and fungicides. The winter rape system had a greater proportion of *Rhizoctonia*-free tubers (64%) than the white mustard (27%) and no green manure (28%) treatments in the non-fumigated plots. There was less *Verticillium* wilt incidence with winter rape incorporation (7%) than with white mustard (21%) or no green manure incorporation (22%) in non-fumigated plots (88).

In Maine, researchers have documented consistent reductions in *Rhizoctonia* (canker and black scurf) on potato following either rapeseed green manure or canola grown for grain (459, 460). They have also observed significant reductions in powdery scab (caused by *Spongospora subterranea*) and common scab (*Streptomyces scabiei*) following brassica green manures, especially an Indian mustard (*B. juncea*) green manure (458, 459).

Nematodes

In Washington state, a series of studies addressed the effect of various brassica and mustard cover crops on nematodes in potato systems (260, 266, 353, 283, 284, 285).

The Columbia root-knot nematode (*Meloidogyne chitwoodi*) is a major pest in the Pacific Northwest. It is usually treated with soil fumigants costing $20 million in Washington alone.

Rapeseed, arugula and mustard were studied as alternatives to fumigation. The brassica cover crops are usually planted in late summer (August) or early fall and incorporated in spring before planting mustard.

Results are promising, with nematodes reduced up to 80%, but—because of the very low damage threshold—*green manures alone cannot be recommended for adequate control of* Meloidogyne chitwoodi *in potatoes.* The current recommended alternative to fumigation is the use of rapeseed or mustard cover crop plus the application of MOCAP. This regimen costs about the same as fumigation (2006 prices).

Several brassicas are hosts for plant parasitic nematodes and can be used as trap crops followed by an application of a synthetic nematicide. Washington State University nematologist Ekaterini Riga has been planting arugula in the end of August and incorporating it in the end of October.

Nematicides are applied two weeks after incorporation, either at a reduced rate using Telone or the full rate of Mocap and Temik. Two years of field trials have shown that arugula in combination with synthetic nematicides reduced *M. chitwoodi* to economic thresholds.

Longer crop rotations that include mustards and non-host crops are also effective for nematode management. For example, a 3-year rotation of potatoes>corn>wheat provides nearly complete control of the northern root-knot nematode (*Meloidogyne hapla*) compared to methyl bromide and other broad-spectrum nematicides.

However, because the rotation crops are less profitable than potatoes, they are less commonly used. Not until growers better appreciate the less tangible long-term cover crop benefits of soil improvement, nutrient management and pest suppression will such practices be more widely adopted.

In Wyoming, oilseed radish (*Raphanus sativus*) and yellow mustard (*Sinapsis alba*) reduced the sugar beet cyst nematode populations by 19-75%, with greater suppression related to greater amount of cover crop biomass (231).

In Maryland, rapeseed, forage radish and a mustard blend did not significantly reduce incidence of soybean cyst nematode (which is closely related to the sugar beet cyst nematode). The same species, when grown with rye or clover, did reduce incidence of stubby root nematode (432).

Also in Maryland, in no-till corn on a sandy soil, winterkilled forage radish increased bacteria-eating nematodes, rye and rapeseed increased the proportion of fungal feeding nematodes, while nematode communities without cover crops were intermediate. The Enrichment Index, which indicates a greater abundance of opportunistic bacteria-eating nematodes, was 23% higher in soils that had brassica cover crops than the unweeded control plots.

These samples, taken in November, June (a month after spring cover crop kill), and August (under no-till corn), suggest that the cover crops, living or dead, increased bacterial activity and may have enhanced nitrogen cycling through the food web (432).

Weeds

Like most green manures, brassica cover crops suppress weeds in the fall with their rapid growth and canopy closure. In spring, brassica residues can inhibit small seeded annual weeds such as, pigweed, shepherds purse, green foxtail, kochia, hairy nightshade, puncturevine, longspine sandbur, and barnyardgrass (293), although pigweed was not inhibited by yellow mustard (178).

In most cases, early season weed suppression obtained with brassica cover crops must be supplemented with herbicides or cultivation to avoid crop yield losses from weed competition later in the season. As a component of integrated weed management, using brassica cover crops in vegetable rotations could improve weed control and reduce reliance on herbicides (39).

In Maine, the density of sixteen weed and crop species was reduced 23 to 34% following incorporation of brassica green manures, and weed establishment was delayed by 2 days, compared to a fallow treatment. However, other short-season green manure crops including oat, crimson clover and buckwheat similarly affected establishment (176).

In Maryland and Pennsylvania, forage radish is planted in late August and dies with the first hard frost (usually December). The living cover crop and the decomposing residues suppress winter annual weeds until April and result in a mellow, weed-free seedbed into which corn can be no-tilled without any preplant herbicides. Preliminary data show summer suppression of horseweed but not lambsquarters, pigweed, or green foxtail (432).

Mustard cover crops have been extremely effective at suppressing winter weeds in tillage intensive, high value vegetable production systems in Salinas, California. Mustards work well in tillage intensive systems because they are relatively easy to incorporate into the soil prior to planting vegetables. However, the growth and biomass production by mustards in the winter is not usually as reliable as that of other cover crops such as cereal rye and legume/cereal mixtures (45).

Deep tillage. Some brassicas (forage radish, rapeseed, turnip) produce large taproots that can penetrate up to six feet to alleviate soil compaction (432). This so-called "biodrilling" is most effective when the plants are growing at a time of year when the soil is moist and easier to penetrate.

Their deep rooting also allows these crops to scavenge nutrients from deep in the soil profile. As the large tap roots decompose, they leave channels open to the surface that increase water infiltration and improve the subsequent growth and soil penetration of crop roots. Smaller roots decompose and leave channels through the plow plan and improve the soil penetration by the roots of subsequent crops (446).

Most mustards have a fibrous root system, and rooting effects are similar to small grain cover crops in that they do not root so deeply but develop a large root mass more confined to the soil surface profile.

SPECIES

Rapeseed (or Canola). Two *Brassica* species are commonly grown as rapeseed, *Brassica napus* and *Brassica rapa*. Rapeseed that has been bred to have low concentrations of both erucic acid and glucosinolates in the seed is called canola, which is a word derived from Canadian Oil.

Annual or spring type rapeseed belongs to the species *B. napus*, whereas winter-type or biennial rapeseed cultivars belong to the species *B. rapa*. Rapeseed is used as industrial oil while canola is used for a wider range of products including cooking oils and biodiesel.

Besides their use as an oil crop, these species are also used for forage. If pest suppression is an objective, rapeseed should be used rather than canola since the breakdown products of glucosinolates are thought to be a principal mechanism for pest control with these cover crops.

Rapeseed has been shown to have biological activity against plant parasitic nematodes as well as weeds (176, 365).

Due to its rapid fall growth, rapeseed captured as much as 120 lb. of residual nitrogen per acre in Maryland (6). In Oregon, aboveground biomass accumulation reached 6,000 lb./A and N accumulation was 80 lb./A.

Some winter-type cultivars are able to withstand quite low temperatures (10° F) (352). This makes rapeseed one of the most versatile cruciferous cover crops, because it can be used either as a spring- or summer-seeded cover crop or a fall-seeded winter cover crop. Rapeseed grows 3 to 5 feet tall.

Mustard. Mustard is a name that is applied to many different botanical species, including white or yellow mustard (*Sinapis alba*, sometimes referred to as *Brassica hirta*), brown or Indian mustard (*Brassica juncea*)—sometimes erroneously referred to as canola—and black mustard (*B. nigra* (L.) (231).

The glucosinolate content of most mustards is very high compared to the true *Brassicas*.

In the Salinas Valley, California, mustard biomass reached 8,500 lb./A. Nitrogen content on high residual N vegetable ground reached 328 lb. N/A (388, 422).

Because mustards are sensitive to freezing, winterkilling at about 25° F, they are used either as a spring/summer crop or they winter kill except in areas with little freeze danger. Brown and field mustard both can grow to 6 feet tall.

In Washington, a wheat/mustard-potato system shows promise for reducing or eliminating the soil fumigant metam sodium. White mustard and oriental mustard both suppressed potato early dying (*Verticillium dahliae*) and resulted in tuber yields equivalent to fumigated soils, while also improving infiltration, all at a cost savings of about $66/acre (see www.plantmanagementnetwork.org/pub/cm/research/2003/mustard/).

Mustards have also been shown to suppress growth of weeds (See "Weeds" p. 99 and 39, 176, 365).

Radish. The true radish or forage radish (*Raphanus sativus*) does not exist in the wild and has only been known as a cultivated species since ancient times. Cultivars developed for high forage biomass or high oilseed yield are also useful for cover crop purposes. Common types include oilseed and forage radish.

Their rapid fall growth has the potential to capture nitrogen in large amounts and from deep in the soil profile (170 lb./acre in Maryland (234). Above ground dry biomass accumulation reached 8,000 lb./acre and N accumulation reached 140 lb./acre in Michigan (304). Below ground biomass of radishes can be as high as 3,700 lb./acre.

Oilseed radish is less affected by frost than forage radish, but may be killed by heavy frost below 25° F. Radish grows about 2-3 feet tall.

Radishes have been shown to alleviate soil compaction and suppress weeds (177, 446).

Turnips. Turnips (*B. rapa L. var. rapa (L.) Thell*) are used for human and animal food because of their edible root. Turnip has been shown to alleviate soil compaction. While they usually do not produce as much biomass as other brassicas, they provide many macrochannels that facilitate water infiltration (359). Similar to radish, turnip is unaffected by early frost but will likely be killed by temperatures below 25° F.

In an Alabama study of 50 cultivars belonging to the genera *Brassica, Raphanus*, and *Sinapis*, forage and oilseed radish cultivars produced the largest amount of biomass in central and south Alabama, whereas winter-type rapeseed cultivars had the highest production in North Alabama (425).

Some brassicas are also used as vegetables (greens). Cultivated varieties of *Brassica rapa* include bok choy (*chinensis group*), mizuna (*nipposinica group*), flowering cabbage (*parachinensis group*), chinese cabbage (*pekinensis group*) and turnip (*rapa group*). Varieties of *Brassica napus* include Canadian turnip, kale, rutabaga, rape, swede, swedish turnip and yellow turnip. Collard, another vegetable, is a cabbage, *B. oleracea* var. *acephala*, and *B. juncea* is consumed as mustard greens.

A grower in Maryland reported harvesting the larger roots of forage radish (cultivar DAIKON) cover crop to sell as a vegetable. In California, broccoli reduced the incidence of lettuce drop caused by *Sclerotinia* minor (175).

AGRONOMIC SYSTEMS

Brassicas must be planted earlier than small grain cover crops for maximum benefits, making it difficult to integrate them into cash grain rotations.

Broadcasting seeding (including aerial seeding) into standing crops of corn or soybean has been successful in some regions (235). See also *After 25 Years, Improvements Keep Coming*, (p. 52). Brassica growth does not normally interfere with soybean harvest, although could be a problem if soybean harvest is delayed. The shading by the crop canopy results in less cover crop biomass and especially less root growth, so this option is not recommended where the brassica cover crop is intended for compaction alleviation.

In a Maryland SARE-funded project, dairy farmers planted forage radish immediately after corn silage harvest. With a good stand of forage radish, which winterkills, corn can be planted in early spring without tillage or herbicides, resulting in considerable savings. The N released by the decomposed forage radish residues increased corn yield boost in most years. This practice is particularly useful when manure is fall-applied to corn silage fields. (For more information see SARE project report LNE03-192 http://www.sare.org/reporting/report_viewer.asp?pn=LNE03-192).

▼ **Precaution:** Brassica cover crops may be susceptible to carry-over from broadleaf herbicides applied to the previous grain crop.

Mustard Mix Manages Nematodes in Potato/Wheat System

Looking for a green manure crop to maintain soil quality in his intensive potato/wheat rotation, Dale Gies not only improved infiltration and irrigation efficiency, he also found biofumigation, a new concept in pest management.

Farming 750 irrigated acres with two sons and a son-in-law in the Columbia basin of Grant County, Wash., Gies started growing green manure crops in 1990 because he wanted to improve his soils for future generations. Since then, he has reduced his use of soil fumigants thanks to the biocidal properties of *Brassica* cover crops. In particular, Gies is most excited about results using a mixture of white or oriental mustard and arugula (*Eruca sativa*), also a brassica, to manage nematodes and potato early dying disease.

"We use the mustards to augment other good management practices," Gies cautions. "Don't expect a silver bullet that will solve your pest problems with one use."

Controlling nematodes is essential to quality potato production, both for the domestic and the international market. Farmers typically manage root knot nematodes (*Meloidogyne chitwoodi*) and fungal diseases with pesticides, such as Metam sodium, a fumigant used routinely to control early dying disease (*Verticillium dahliae*), that cost that up to $500 per acre. Farmers are especially vulnerable to early dying disease if their rotations contain fewer than three years between potato crops.

However, with potato prices dropping, potato farmers in Washington and elsewhere started looking for ways to reduce costs. Gies contacted Andy McGuire at Washington State University Extension for help documenting the results he was seeing with brassicas. With research funding from SARE, McGuire confirmed that the mustards improved infiltration. He also showed that white mustard was as effective as metam sodium in controlling potato early dying disease.

"The findings suggest that mustard green manures may be a viable alternative to the fumigant metam sodium in some potato cropping systems," says McGuire. "The practice can also improve water infiltration rates and provide substantial savings to farmers. Until more research is done, however, mustard cover crops should be used to *enhance*, not eliminate, chemical control of nematodes."

Researchers have found that mustards can also suppress common root rot (*Aphanomyces euteiches*) and the northern root-knot nematode (*Meloidogyne hapla*).

Vegetable Systems. Fall-planted brassica cover crops fit well into vegetable cropping systems following early harvested crops. White mustard and brown mustard have become popular fall-planted cover crops in the potato producing regions of the Columbia Basin of eastern Washington.

Planted in mid to late August, white mustard emerges quickly and produces a large amount of biomass before succumbing to freezing temperatures. As a component of integrated weed management, using brassica cover crops in vegetable rotations could improve weed control and reduce reliance on herbicides (39).

Winter-killed forage radish leaves a nearly weed- and residue-free seedbed, excellent for early spring "no-till" seeding of crops such as carrots, lettuce, peas and sweet corn. This approach can save several tillage passes or herbicide applications for weed control in early spring and can take advantage of the early nitrogen release by the forage radish. Soils warm up faster than under heavy residue, and because no seedbed preparation or weed control is needed, the cash crop can be seeded earlier than normal.

Two types of mustard commonly used in the Columbia Basin are white mustard (*Sinapis alba*, also called *Brassica hirta* or yellow mustard), and Oriental mustard (*Brassica juncea*, also called Indian or brown mustard). Blends of the two are often planted as green manures. Fall incorporation seems to be best to control nematodes and soil-borne diseases, and Oriental mustard may be better at it than white mustard.

Gies plants a mix of mustards and NEMAT, an arugula variety developed in Italy for nematode suppression. The arugula attracts nematodes but they cannot reproduce on its roots, so nematode populations reduce, according to Washington State University researcher Ekaterini Riga.

Riga's greenhouse studies showed that arugula reduced Columbia root knot nematode (*Meloidogyne chitwoodi*) populations compared to the control or other green manure treatments. Subsequent field trial in 2005 and 2006 showed that arugula in combination with half the recommended rate of Telone (another fumigant) or full rates of Mocap and Temik reduced root knot nematode populations from 700 nematodes per gram of soil to zero. The combination also improved potato yield and tuber quality and it is still affordable by the growers.

"Arugula acts both as a green manure and a nematode trap crop," says Riga.

"It contains chemicals with high biocidal activity that mimic synthetic fumigants. Since nematodes are attracted to the roots of Arugula, it can be managed as a trap crop."

What causes brassicas to have biocidal properties? Researchers are keying in on the presence of glucosinolates in mustards. When the crop is incorporated into the soil, the breakdown of glucosinolates produces other chemicals that act against pests. Those secondary chemicals behave like the active chemical in commercial fumigants like metam sodium.

More research is needed to better determine site- and species-specific brassica cover crop effects on pests. It seems to be working for Dale Gies, however, "whose short season fresh market potato system probably functions differently than processing potatoes" according to WSU's Andy McGuire. To stay updated on cover crop work in Washington State, see www.grant-adams.wsu.edu/ agriculture/ covercrops/green_manures/.

For Gies, however, "Tying the whole system together makes it work economically, *and* it improves the soil."

—Andy Clark

MANAGEMENT

Establishment

Most *Brassica* species grow best on well drained soils with a pH range of 5.5–8.5. Brassicas do not grow well on poorly drained soils, especially during establishment. Winter cover crops should be established as early as possible. A good rule of thumb is to establish brassicas about 4 weeks prior to the average date of the first 28° F freeze. The minimum soil temperature for planting is 45° F; the maximum is 85° F.

Winter hardiness

Some brassicas and most mustards may winterkill, depending on climate and species. Forage radish normally winter kills when air temperatures drop below 23° F for several nights in a row. Winter hardiness is higher for most brassicas if plants reach a rosette stage between six to eight leaves before the first killing frost.

Some winter-type cultivars of rapeseed are able to withstand quite low temperatures (10° F) (352).

Late planting will likely result in stand failure and will certainly reduce biomass production and nutrient scavenging. Planting too early, however, may increase winterkill in northern zones (166).

In Washington (Zone 6), canola and rapeseed usually overwinter, mustards do not. Recent work with arugula (*Eruca sativa*) shows that it does overwinter and may provide similar benefits as the mustards (430).

In Michigan, mustards are planted in mid-August, and winterkill with the first hard frost, usually in October. When possible, plant another winter cover crop such as rye or leave strips of untilled brassica cover crop rather than leave the soil without growing cover over the winter (391).

In Maine, all brassica and mustards used as cover crops winterkill (166).

Winter vs. spring annual use

Brassica and mustard cover crops can be planted in spring or fall. Some species can be managed to winterkill, leaving a mellow seedbed requiring little or no seedbed preparation. For the maximum benefits offered by brassicas as cover crops, fall-planting is usually preferable because planting conditions (soil temperature and moisture) are more reliable and the cover crops produce more dry matter.

In Maryland, rapeseed and forage radish were more successful as winter- rather than spring-annual cover crops. The early spring planted brassicas achieved about half the quantities of biomass and did not root as deeply, before bolting in spring (432).

In Michigan, mustards can be planted in spring following corn or potatoes or in fall into wheat residue or after snap beans. Fall seedings need about 900 growing-degree-days to produce acceptable biomass, which is usually incorporated at first frost (usually October). Spring seeding is less reliable due to cool soil temperatures, and its use is limited mostly to late-planted vegetable crops (391).

In Maine, brassicas are either planted in late summer after the cash crop and winterkill, or they are spring-seeded for a summer cover crop (166).

Rapeseed planted in late spring to summer has been used with some success in the mid-Atlantic region to produce high biomass for incorporation to biofumigate soil for nematodes and diseases prior to planting strawberries and fruit trees.

Mixtures. Mix with small grains (oats, rye), other brassicas or legumes (e.g. clover). Brassicas are very competitive and can overwhelm the other species in the mixture. The seeding rate must be adjusted so ensure adequate growth of the companion species. Consult local expertise and start with small plots or experiment with several seeding rates.

Washington farmers use mixtures of white and brown mustard, usually with a greater proportion of brown mustard.

In Maryland and Pennsylvania, farmers and researchers seed the small grain and forage radish in separate drill rows rather than mixing the seed. This is done by taping closed alternate holes in the two seeding boxes of a grain drill with both small seed and large seed boxes. Two rows of oats between each row of forage radish has also proven successful (432). Rye (sown at 48 lb./A) can be grown successfully as a mixture with winter-killing forage radish (13 lb./A).

Killing

Brassica cover crops that do not winterkill can be terminated in spring by spraying with an appropriate herbicide, mowing, and/or incorporating above-ground biomass by tillage before the cover crop has reached full flower. Rolling may also be used to kill these covers if they are in flower.

Rapeseed has proved difficult to kill with glyphosate, requiring a higher than normal rate of application—at least 1 quart/acre of glyphosate—and possibly multiple applications. Radish, mustard, and turnip can be killed using a full rate of paraquat, multiple applications of glyphosate, or glyphosate plus 1pt/acre 2,4-D.

In Alabama and Georgia, brassica cover crops were reportedly harder to chemically kill than winter cereals. Timely management and multiple herbicide applications may be necessary for successful termination. If not completely killed, rapeseed volunteers can be a problem in the subsequent crop. Always check herbicide rotation restrictions before applying.

Another no-till method for terminating mature brassicas is flail mowing. Be sure to evenly distribute residue to facilitate planting operations and reduce allelopathic risk for cash crop. As mentioned above, many producers incorporate brassica residues using conventional tillage methods to enhance soil biotoxic activity especially in plasticulture systems.

Brassica pest suppression may be more effective if the cover crop is incorporated.

Seed and Planting

Because *Brassica* spp. seed may be scarce, it is best to call seed suppliers a few months prior to planting to check on availability. *Brassica* seeds in general are relatively small; a small volume of seed goes a long way.

- Rapeseed (Canola). Drill 5-10 lb./A no deeper than ¾ in. or broadcast 8–14 lb./A.
- Mustard. Drill 5-12 lb./A ¼–¾ in. deep or broadcast 10-15 lb./A.
- Radish. Drill 8 to 12 lb/A. ¼–½ in. deep, or broadcast 12-20 lb./A. Plant in late summer or early fall after the daytime average temperature is below 80° F.
- Turnip. Drill 4-7 lb./A about ½ in. deep or broadcast 10-12 lb./A. Plant in the fall after the daytime average temperature is below 80° F.

Nutrient Management

Brassicas and mustards need adequate nitrogen and sulfur fertility. Brassica sulfur (S) nutrition needs and S uptake capacity exceed those of many other plant species, because S is required for oil and glucosinolate production. A 7:1 N/S ratio in soils is optimum for growing rape, while N/S ratios ranging from 4:1 to 8:1 work well for brassica species in general.

Ensuring sufficient N supply to brassicas during establishment will enhance their N uptake and early growth. Some brassicas, notably rape, can scavenge P by making insoluble P more available to them via the excretion of organic acids in their root zone (168).

Brassicas decompose quickly. Decomposition and nutrient turnover from roots (C:N ratios 20-30) is expected to be slower than that from shoots (C:N ratios 10-20), but overall faster than that of winter rye. A winter-killed radish cover crop releases plant available nitrogen especially early in spring, so it should be followed by an early planted nitrogen demanding crop to avoid leaching losses (432).

COMPARATIVE NOTES

Canola is more prone to insect problems than mustards, probably because of its lower concentration of glucosinolates.

In the Salinas Valley, which has much milder summer and winter temperatures than the Central Valley of California, brassica cover crops are generally less tolerant of suboptimal conditions (i.e. abnormally low winter temperatures, low soil nitrogen, and waterlogging), and hence are more likely to produce a nonuniform stand than other common cover crops (45).

▼ **Precautions.** The use of brassicas for pest management is in its infancy. Results are inconsistent from year to year and in different geographic regions. Be sure to consult local expertise and begin with small test plots on your farm.

Bio-toxic activity can stunt cash crop growth, thus avoid direct planting into just-killed green residue.

Brassica cover crops should NOT be planted in rotation with other brassica crops such as cabbage, broccoli, and radish because the latter are susceptible to similar diseases. Also, scattered volunteer brassica may appear in subsequent crops. Controlling brassica cover crop volunteers that come up in brassica cash crops would be challenging if not impossible.

Black mustard (*Brassica nigra*) is hardseeded and could cause weed problems in subsequent crops (39).

Rapeseed contains erucic acid and glucosinolates, naturally occurring internal toxicants. These compounds are anti-nutritional and are a concern when feeding to livestock. Human consumption of brassicas has been linked to reducing incidence of cancer. All canola cultivars have been improved through plant breeding to contain less than 2% erucic acid.

Winter rape is a host for root lesion nematode. In a SARE funded study in Washington, root lesion nematode populations were 3.8 times higher in the winter rape treatment than in the white mustard and no green manure treatments after green manure incorporation in unfumigated plots. However, populations in the unfumigated winter rape treatment were below the economic threshold both years of the study. For more information, go to www.sare.org/projects/ and search for SW95-021. See also SW02-037).

Rapeseed may provide overwintering sites for harlequin bug in Maryland (432).

Contributors: Guihua Chen, Andy Clark, Amy Kremen, Yvonne Lawley, Andrew Price, Lisa Stocking, Ray Weil

BUCKWHEAT

Fagopyrum esculentum

Type: summer or cool-season annual broadleaf grain

Roles: quick soil cover, weed suppressor, nectar for pollinators and beneficial insects, topsoil loosener, rejuvenator for low-fertility soils

Mix with: sorghum-sudangrass hybrids, sunn hemp

See charts, pp. 66 to 72, for ranking and management summary.

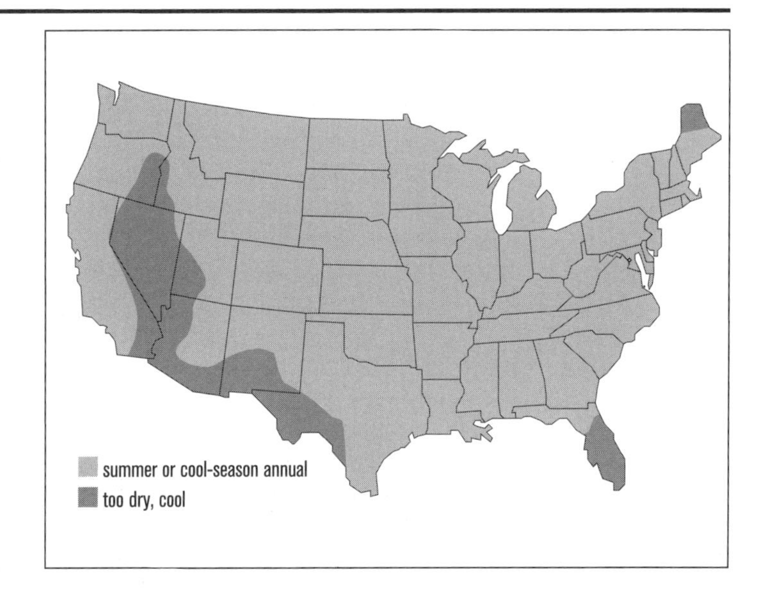

Buckwheat is the speedy short-season cover crop. It establishes, blooms and reaches maturity in just 70 to 90 days and its residue breaks down quickly. Buckwheat suppresses weeds and attracts beneficial insects and pollinators with its abundant blossoms. It is easy to kill, and reportedly extracts soil phosphorus from soil better than most grain-type cover crops.

Buckwheat thrives in cool, moist conditions but it is not frost tolerant. Even in the South, it is not grown as a winter annual. Buckwheat is not particularly drought tolerant, and readily wilts under hot, dry conditions. Its short growing season may allow it to avoid droughts, however.

BENEFITS

Quick cover. Few cover crops establish as rapidly and as easily as buckwheat. Its rounded pyramid-shaped seeds germinate in just three to five days. Leaves up to 3 inches wide can develop within two weeks to create a relatively dense, soil shading canopy. Buckwheat typically produces only 2 to 3 tons of dry matter per acre, but it does so quickly—in just six to eight weeks (257). Buckwheat residue also decomposes quickly, releasing nutrients to the next crop.

Weed suppressor. Buckwheat's strong weed-suppressing ability makes it ideal for smothering warm-season annual weeds. It's also planted after intensive, weed-weakening tillage to crowd out perennials. A mix of tillage and successive dense seedings of buckwheat can effectively suppress Canada thistle, sowthistle, creeping jenny, leafy spurge, Russian knapweed and perennial peppergrass (257). While living buckwheat may have an allelopathic weed-suppressing effect (351), its primary impact on weeds is through shading and competition.

Phosphorus scavenger. Buckwheat takes up phosphorus and some minor nutrients (possibly including calcium) that are otherwise unavailable to crops, then releasing these nutrients to later crops as the residue breaks down. The roots of the plants produce mild acids that release nutrients from the soil. These acids also activate slow-releasing organic fertilizers, such as rock phosphate. Buckwheat's dense, fibrous roots cluster in the top 10 inches of soil, providing an extensive root surface area for nutrient uptake.

Thrives in poor soils. Buckwheat performs better than cereal grains on low-fertility soils and soils with high levels of decaying organic matter. That's why it was often the first crop planted on cleared land during the settlement of woodland areas and is still a good first crop for rejuvenating over-farmed soils. However, buckwheat does not do well in compacted, droughty or excessively wet soils.

Quick regrowth. Buckwheat will regrow after mowing if cut before it reaches 25 percent bloom. It also can be lightly tilled after the midpoint of its long flowering period to reseed a second crop. Some growers bring new land into production by raising three successive buckwheat crops this way.

Soil conditioner. Buckwheat's abundant, fine roots leave topsoil loose and friable after only minimal tillage, making it a great mid-summer soil conditioner preceding fall crops in temperate areas.

Nectar source. Buckwheat's shallow white blossoms attract beneficial insects that attack or parasitize aphids, mites and other pests. These beneficials include hover flies (*Syrphidae*), predatory wasps, minute pirate bugs, insidious flower bugs, tachinid flies and lady beetles. Flowering may start within three weeks of planting and continue for up to 10 weeks.

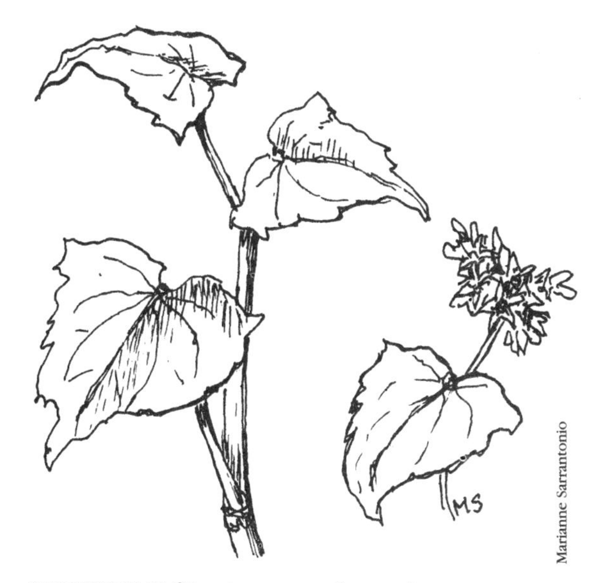

BUCKWHEAT (Fagopyrum esculentum)

Nurse crop. Due to its quick, aggressive start, buckwheat is rarely used as a nurse crop, although it can be used anytime you want quick cover. It is sometimes used to protect late-fall plantings of slow-starting, winter-hardy legumes wherever freezing temperatures are sure to kill the buckwheat.

MANAGEMENT

Buckwheat prefers light to medium, well-drained soils—sandy loams, loams, and silt loams. It performs poorly on heavy, wet soils or soils with high levels of limestone. Buckwheat grows best in cool, moist conditions, but is not frost-tolerant. It is also not drought tolerant. Extreme afternoon heat will cause wilting, but plants bounce back overnight.

Establishment

Plant buckwheat after all danger of frost. In untilled, minimally tilled or clean-tilled soils, drill 50 to 60 lb./A at $^1/_2$ to $1^1/_2$ inches deep in 6 to 8 inch rows. Use heavier rates for quicker canopy development. For a fast smother crop, broadcast up to 96 lb./A (2 bu./A) onto a firm seedbed and incorporate with a harrow, tine weeder, disk or field cultivator. Overall vigor is usually better in drilled seedings. As a nurse-crop for slow-growing, winter annual legumes planted in late summer or fall, seed at one-quarter to one-third of the normal rate.

Buckwheat germinates and grows quickly, producing 2 to 3 tons of dry matter in just 6 to 8 weeks.

Buckwheat compensates for lower seeding rates by developing more branches per plant and more seeds per blossom. However, skimping too much on seed makes stands more vulnerable to early weed competition until the canopy fills in. Using cleaned, bin-run or even birdseed-grade seed can lower establishment costs, but increases the risk of weeds. As denser stands mature, stalks become spindly and are more likely to lodge from wind or heavy rain.

Rotations

Buckwheat is used most commonly as a mid-summer cover crop to suppress weeds and replace bare fallow. In the Northeast and Midwest, it is often planted after harvest of early vegetable crops, then followed by a fall vegetable, winter grain, or cool-season cover crop. Planted later, winterkilled residue provides decent soil cover and is easy to no-till into. In many areas, it can be planted following harvest of winter wheat or canola.

In parts of California, buckwheat grows and flowers between the killing of winter annual legume cover crops in spring and their re-establishment in fall. Some California vineyard managers seed 3-foot strips of buckwheat in row middles, alternating it and another summer cover crop, such as sorghum-sudangrass.

Buckwheat is sensitive to herbicide residues from previous crops, especially in no-till seedbeds. Residue from trifluralin and from triazine and sulfonylurea herbicides have damaged or killed buckwheat seedlings (79). When in doubt, sow and water a small test plot of the fast-germinating seed to detect stunting or mortality.

Pest Management

Few pests or diseases bother buckwheat. Its most serious weed competitors are often small grains from preceding crops, which only add to the cover crop biomass. Other grass weeds can be a problem, especially in thin stands. Weeds also can increase after seed set and leaf drop. Diseases include a leaf spot caused by the fungus *Ramularia* and *Rhizoctonia* root rot.

Other Options

Plant buckwheat as an emergency cover crop to protect soil and suppress weeds when your main crop fails or cannot be planted in time due to unfavorable conditions.

To assure its role as habitat for beneficial insects, allow buckwheat to flower for at least 20 days—the time needed for minute pirate bugs to produce another generation.

Buckwheat can be double cropped for grain after harvesting early crops if planted by mid-July in northern states or by early August in the South. It requires a two-month period of relatively cool, moist conditions to prevent blasting of the blossoms. There is modest demand for organic and specially raised food-grade buckwheat in domestic and overseas markets. Exporters usually specify variety, so investigate before planting buckwheat for grain.

Management Cautions

Buckwheat can become a weed. Kill within 7 to 10 days after flowering begins, before the first seeds begin to harden and turn brown. Earliest maturing seed can shatter before plants finish blooming. Some seed may overwinter in milder regions.

Buckwheat can harbor insect pests including Lygus bugs, tarnished plant bugs and *Pratylynchus penetrans* root lesion nematodes (256).

COMPARATIVE NOTES

- Buckwheat has only about half the root mass as a percent of total biomass as small grains (355). Its succulent stems break down quickly, leaving soils loose and vulnerable to erosion, particularly after tillage. Plant a soil-holding crop as soon as possible.
- Buckwheat is nearly three times as effective as barley in extracting phosphorus, and more than 10 times more effective than rye—the poorest P scavenger of the cereal grains (355).
- As a cash crop, buckwheat uses only half as much soil moisture as soybeans (299).

Seed sources. See *Seed Suppliers* (p. 195). 

OATS

Avena sativa

Also called: spring oats

Type: cool season annual cereal

Roles: suppress weeds, prevent erosion, scavenge excess nutrients, add biomass, nurse crop

Mix with: clover, pea, vetch, other legumes or other small grains

See charts, pp. 66 to 72, for ranking and management summary.

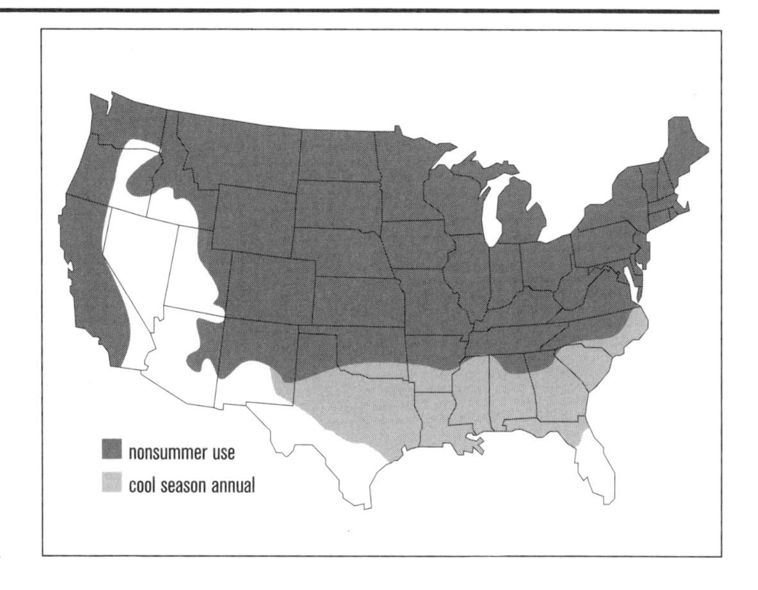

If you need a low-cost, reliable fall cover that winterkills in Hardiness Zone 6 and colder and much of Zone 7, look no further. Oats provide quick, weed-suppressing biomass, take up excess soil nutrients and can improve the productivity of legumes when planted in mixtures. The cover's fibrous root system also holds soil during cool-weather gaps in rotations, and the ground cover provides a mellow mulch before low-till or no-till crops.

An upright, annual grass, oats thrive under cool, moist conditions on well-drained soil. Plants can reach heights in excess of 4 feet. Stands generally fare poorly in hot, dry weather.

BENEFITS

You can depend on oats as a versatile, quick-growing cover for many benefits:

Affordable biomass. With good growing conditions and sound management (including timely planting), expect 2,000 to 4,000 pounds of dry matter per acre from late-summer/early fall-seeded oats and up to 8,000 pounds per acre from spring stands.

Nutrient catch crop. Oats take up excess N and small amounts of P and K when planted early enough. Late-summer plantings can absorb as much as 77 lb. N/A in an eight- to ten-week period, studies in the Northeast and Midwest have shown (313, 329).

Where the plant winterkills, some farmers use oats as a nitrogen catch crop after summer legume plowdowns, to hold some N over winter without needing to kill the cover in spring. Some of the N in the winterkilled oats may still be lost by spring, either through denitrification into the atmosphere or by leaching from the soil profile. Consider mixing oats with an overwintering legume if your objective is to maximize N contribution to the next crop.

Oats are a reliable, low-cost cover that winterkill in Zone 6 and much of Zone 7.

Smother crop. Quick to germinate, oats are a great smother crop that outcompetes weeds and also provides allelopathic residue that can hinder germination of many weeds—and some crops (see below)—for a few weeks. Reduce crop suppression concerns by waiting two- to three weeks after killing oats before planting a subsequent crop.

Fall legume nurse crop. Oats have few equals as a legume nurse crop or companion crop. They can increase the fertilizer replacement value of legumes. Adding about 35 to 75 lb. oats/A to the seeding mix helps slow-establishing legumes such as hairy vetch, clovers or winter peas, while increasing biomass. It also helps reduce fall weeds. The oats will winterkill in many areas while improving the legume's winter survival.

Spring green manure or companion crop. Spring-seeded with a legume, oats can provide hay or grain and excellent straw in the Northern U.S., while the legume remains as a summer—or even later—cover. There's also a haylage option with a fast-growing legume if you harvest when oats are in the dough stage. The oats will increase the dry matter yield and boost the total protein, but, because of its relatively high nitrogen content, could pose a nitrate-poisoning threat to livestock, especially if you delay harvesting until oats are nearing the flowering stage.

The climbing growth habit of some viny legumes such as vetch can contribute to lodging and make oat grain harvest difficult. If you're growing the legume for seed, the oats can serve as a natural trellis that eases combining.

MANAGEMENT

Establishment & Fieldwork

Time seeding to allow at least six to 10 weeks of cool-season growth. Moderately fertile soil gives the best stands.

Late-summer/early-fall planting. For a winterkilled cover, spring oats usually are seeded in late summer or early fall in Zone 7 or colder. Broadcasting or overseeding will give the best results for the least cost, unless seeding into heavy residue. Cleaned, bin-run seed will suffice.

If broadcasting and you want a thick winterkilled mulch, seed at the highest locally recommended rate (probably 3 to 4 bushels per acre) at least 40 to 60 days before your area's first killing frost. Assuming adequate moisture for quick germination, the stand should provide some soil-protecting, weed-suppressing mulch.

Disk lightly to incorporate. In many regions, you'll have the option of letting it winterkill or sending in cattle for some fall grazing.

If seeding oats as a fall nurse crop for a legume, a low rate (1 to 2 bushels per acre) works well.

If drilling oats, seed at 2 to 3 bushels per acre 1/2 to 1 inch deep, or 1 1/2 inches when growing grain you plan to harrow for weed control.

Shallow seeding in moist soil provides rapid emergence and reduces incidence of root rot disease.

Timing is critical when you want plenty of biomass or a thick ground cover. As a winter cover following soybeans in the Northeast or Midwest, overseeding spring oats at the leaf-yellowing or early leaf-drop stage (and with little residue present) can give a combined ground cover as high as 80 percent through early winter (200). If you wait until closer to or after soybean harvest, however, you'll obtain much less oat biomass to help retain bean residue, Iowa and Pennsylvania studies have shown.

Delaying planting by as little as two weeks in late summer also can reduce the cover's effectiveness as a spring weed fighter, a study in upstate New York showed. By spring, oat plots that had been planted on August 25 had 39 percent fewer weed plants and one-seventh the weed biomass of control plots with no oat cover, while oats planted two weeks later had just 10 percent fewer weed plants in spring and 81 percent of the weed biomass of control plots (329, 330).

No-hassle fieldwork. As a winterkilled cover, just light disking in spring will break up the brittle oat residue. That exposes enough soil for warming and timely planting. Or, no-till directly into the mulch, as the residue will decompose readily early in the season.

Winter planting. As a fall or winter cover crop in Zone 8 or warmer, seed oats at low to medium rates. You can kill winter-planted oats with spring plowing, or with herbicides in reduced-tillage systems.

Spring planting. Seeding rate depends on your intended use: medium to high rates for a spring green manure and weed suppressor, low rates for mixtures or as a legume companion crop. Higher rates may be needed for wet soils or thicker ground cover. Excessive fertility can encourage lodging, but if you're growing oats just for its cover value, that can be an added benefit for weed suppression and moisture conservation.

Easy to kill. Oats will winterkill in most of zone 7 or colder. Otherwise, kill by mowing or spraying soon after the vegetative stage, such as the milk or soft dough stage. In no-till systems, rolling/crimping will also work (best at dough stage or later). See *Cover Crop Roller Design Holds Promise For No-Tillers,* p. 146. If speed of spring soil-warming is not an issue, you can spray or mow the oats and leave on the soil surface for mulch.

OATS (Avena sativa)

If you want to incorporate the stand, allow at least two to three weeks before planting the next crop.

Killing too early reduces the biomass potential and you could see some regrowth if killing mechanically. But waiting too long could make tillage of the heavier growth more difficult in a conventional tillage system and could deplete soil moisture needed for the next crop. Timely killing also is important because mature oat stands can tie up nitrogen.

Pest Management

Allelopathic (naturally occurring herbicidal) compounds in oat roots and residue can hinder weed growth for a few weeks. These compounds also can slow germination or root growth of some subsequent crops, such as lettuce, cress, timothy, rice, wheat and peas. Minimize this effect by waiting three weeks after oat killing before seeding a susceptible crop, or by following with an alternate crop. Rotary hoeing or other pre-emerge mechanical weeding of solo-seeded oats can improve annual broadleaf control.

Oats are **less prone to insect problems** than wheat or barley. If you're growing oats for grain or forage, armyworms, various grain aphids and mites, wireworms, cutworms, thrips, leafhoppers, grubs and billbugs could present occasional problems.

Oats, Rye Feed Soil in Corn/Bean Rotation

Bryan and Donna Davis like what cover crops have done for their corn/soybean rotation. They use less grass herbicide, have applied insecticides only once in the last six years, and they have seen organic matter content almost double from less than 2% to almost 4%.

Rye and oats are the cover crop mainstays on the nearly 1,000 acres they farm near Grinnell, Iowa. Bryan and Donna purchased the farm—in the family since 1929—in 1987 and almost immediately put most of the operation under 100% no-till, a system they had experimented with over the years. They now till some acres and are also in the process of transitioning 300 acres to organic.

Moving ⅓ of their acreage toward organic seems the logical culmination of the Davis' makeover of their farm that started with a desire to "get away from the chemicals." That was what motivated them to start using cover crops to feed the soil and help manage pests.

"We were trying to get away from the idea that every bug and weed must be exterminated. Rather, we need to 'manage' the system and tolerate some weed and insect pressure. It should be more of a balance," says Bryan.

Bryan and Donna are practitioners and proponents of "biological farming," a systems approach based on such principles as feeding the soil to keep it biologically active, reducing chemical inputs and paying attention to trace elements or micronutrients in order to maintain balance in the system. Cover crops play an integral role in this system.

They seed oats at 2-3 bu/A in spring or fall, depending on time and labor availability. Donna does most of the combining and planting, but even with a lot of acres for two people to manage, cover crops are a high priority on their schedule. Fall-seeded oats are planted after soybean harvest and "need rain on them soon after planting to get them started." They'll put on about a foot of growth before winterkilling, usually in December in their south-central Iowa conditions.

Spring oats are broadcast in mid or late March with a fertilizer cart and then rotary harrowed. If going back to corn, they seed at a heavier, 3.5 bu rate, expecting only about 5 or 6 weeks of growth before they work down the cover crop with a soil finisher and plant corn in early May. For soybeans, they either kill chemically and no-till the beans, or work down and seed conventionally.

They have managed rye in different ways over the years depending on its place in the rotation, but prefer to seed into killed or tilled rye rather than a living cover crop. They figure that they get about 35 lb. N from oats and up to 60 lb. from rye.

On their organic transition acres they are applying chicken manure (2 tons/A), and cover crops are crucial to sopping up excess nutrients and crowding out the weeds that crop up in response to the extra nutrients. They feel that their efforts to balance nutrients are also helping with weed control, because weeds feed on nutrient imbalances.

In addition to the increase in soil organic matter, attributed to cover crops and no-tillage, they've also seen improvements in soil moisture and infiltration. Fields that used to pond after heavy rains no longer do. Soybeans are weathering drought better, and corn stays green longer during a "more natural" drying down process.

"Our system takes more time and is more labor intensive, but if you look at the whole budget, we are doing much better now. We have cut our chemical costs dramatically, and have reduced fertility costs—in some fields—by ⅓ to ½," says Bryan. "With energy costs these days, you can't afford not to do this."

Davis is careful to note that this is not just about adding one component such as cover crops. "You need to address the whole system, not just one piece of the pie. To be able to have a sustaining system, you must work with the living system. Feed the soil and give it a roof over its head." Cover crops play a crucial role in that system.

—Andy Clark

Resistant oat varieties can minimize rusts, smuts and blights if they are a concern in your area or for your cropping system. Cover crops such as oats help reduce root-knot nematodes and vegetable crop diseases caused by *Rhizoctonia*, results of a producer study in South Carolina show (448), although brassicas are better. To reduce harmful nematodes that oats could encourage, avoid planting oats two years in a row or after nematode-susceptible small grains such as wheat, rye or triticale (71).

Other Options

There are many low-cost, regionally adapted and widely available oat varieties, so you have **hay, straw, forage or grain options**. Select for cultural and local considerations that best fit your intended uses. Day-length, stalk height, resistance to disease, dry matter yield, grain test weight and other traits may be important considerations. In the Deep South, fast-growing black oats (*Avena strigosa*) look promising as a weed-suppressive cover for soybeans. See *Up-and-Coming Cover Crops* (p. 191).

Aside from their value as a cover crop, oats are a great feed supplement, says grain and hog farmer Carmen Fernholz, Madison, Minn. A niche market for organic oats also could exist in your area, he observes.

Oats are more palatable than rye and easily overgrazed. If using controlled grazing in oat stands, watch for high protein levels, which can vary from 12 to 25 percent (434). The potassium level of oat hay also is sometimes very high and could cause metabolic problems in milking cows if it's the primary forage. Underseeding a legume enhances the forage option for oats by increasing the biomass (compared with solo-cropped oats) and providing nitrogen for a subsequent crop.

COMPARATIVE NOTES

- Fall brassicas grow faster, accumulate more N and may suppress weeds, nematodes and disease better than oats.
- Rye grows more in fall and early spring, absorbs more N and matures faster, but is harder to establish, to kill and to till than oats.
- As a legume companion/nurse crop, oats outperform most varieties of other cereal grains.
- Oats are more tolerant of wet soil than is barley, but require more moisture.

Seed sources. See *Seed Suppliers* (p. 195). 

RYE

Secale cereale

Also called: cereal rye, winter rye, grain rye

Type: cool season annual cereal grain

Roles: scavenge excess N, prevent erosion, add organic matter, suppress weeds

Mix with: legumes, grasses or other cereal grains

See charts, pp. 66 to 72, for ranking and management summary.

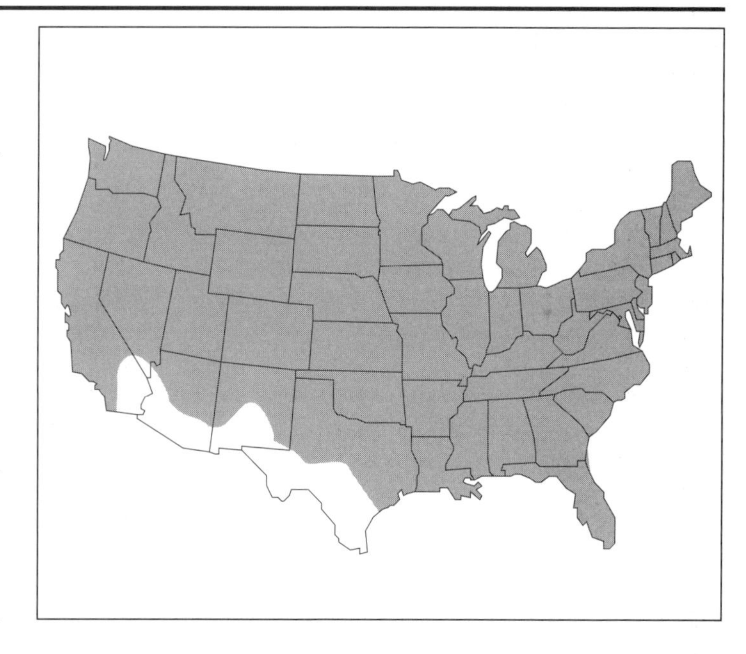

The hardiest of cereals, rye can be seeded later in fall than other cover crops and still provide considerable dry matter, an extensive soil-holding root system, significant reduction of nitrate leaching and exceptional weed suppression. Inexpensive and easy to establish, rye outperforms all other cover crops on infertile, sandy or acidic soil or on poorly prepared land. It is widely adapted, but grows best in cool, temperate zones.

Taller and quicker-growing than wheat, rye can serve as a windbreak and trap snow or hold rainfall over winter. It overseeds readily into many high-value and agronomic crops and resumes growth quickly in spring, allowing timely killing by rolling, mowing or herbicides. Pair rye with a winter annual legume such as hairy vetch to offset rye's tendency to tie up soil nitrogen in spring.

BENEFITS

Nutrient catch crop. Rye is the best cool-season cereal cover for absorbing unused soil N. It has no taproot, but rye's quick-growing, fibrous root system can take up and hold as much as 100 lb. N/A until spring, with 25 to 50 lb. N/A more typical (422). Early seeding is better than late seeding for scavenging N (46).

- A Maryland study credited rye with holding 60 percent of the residual N that could have leached from a silt loam soil following intentionally over-fertilized corn (372).
- A Georgia study estimated rye captured from 69 to 100 percent of the residual N after a corn crop (220).
- In an Iowa study, overseeding rye or a rye/oats mix into soybeans in August limited leaching loss from September to May to less than 5 lb. N/A (313).

Rye increases the concentration of exchangeable potassium (K) near the soil surface, by bringing it up from lower in the soil profile (123).

Rye's rapid growth (even in cool fall weather) helps trap snow in winter, further boosting winterhardiness. The root system promotes better drainage, while rye's quick maturity in spring—compared with other cover crops—can help conserve late-spring soil moisture.

Reduces erosion. Along with conservation tillage practices, rye provides soil protection on sloping fields and holds soil loss to a tolerable level (124).

Fits many rotations. In most regions, rye can serve as an overwintering cover crop after corn or before or after soybeans, fruits or vegetables. It's not the best choice before a small grain crop such as wheat or barley unless you can kill rye reliably and completely, as volunteer rye seed would lower the value of other grains.

Rye also works well as a strip cover crop and windbreak within vegetables or fruit crops and as a quick cover for rotation gaps or if another crop fails.

You can overseed rye into vegetables and tasseling or silking corn with consistently good results. You also can overseed rye into brassicas (369, 422), into soybeans just before leaf drop or between pecan tree rows (61).

Plentiful organic matter. An excellent source of residue in no-till and minimum-tillage systems and as a straw source, rye provides up to 10,000 pounds of dry matter per acre, with 3,000 to 4,000 pounds typical in the Northeast (118, 361). A rye cover crop might yield too much residue, depending on your tillage system, so be sure your planting regime for subsequent crops can handle this. Rye overseeded into cabbage August 26 covered nearly 80 percent of the between-row plots by mid-October and, despite some summer heat, already had accumulated nearly half a ton of biomass per acre in a New York study. By the May 19 plowdown, rye provided 2.5 tons of dry matter per acre and had accumulated 80 lb. N/A. Cabbage yields weren't affected, so competition wasn't a problem (329).

Weed suppressor. Rye is one of the best cool season cover crops for outcompeting weeds, especially small-seeded, light-sensitive annuals such as lambsquarters, redroot pigweed, velvetleaf, chickweed and foxtail. Rye also suppresses many weeds allelopathically (as a natural herbicide), including dandelions and Canada thistle and has been shown to inhibit germination of some triazine-resistant weeds (336).

Rye reduced total weed density an average of 78 percent when rye residue covered more than 90 percent of soil in a Maryland no-till study (410), and by 99 percent in a California study (422). You can increase rye's weed-suppressing effect before no-till corn by planting rye with an annual legume such as hairy vetch. Don't expect complete weed control, however. You'll probably need complementary weed management measures.

Pest suppressor. While rye is susceptible to the same insects that attack other cereals, serious infestations are rare. Rye reduces insect pest problems in rotations (448) and attracts significant numbers of beneficials such as lady beetles (56).

Rye can be planted later in fall than other cover crops.

Fewer diseases affect rye than other cereals. Rye can help reduce root-knot nematodes and other harmful nematodes, research in the South suggests (20, 448).

Companion crop/legume mixtures. Sow rye with legumes or other grasses in fall or overseed a legume in spring. A legume helps offset rye's tendency to tie up N. A legume/rye mixture adjusts to residual soil N levels. If there's plenty of N, rye tends to do better; if there is insufficient N, the legume component grows better, Maryland research shows (86). Hairy vetch and rye are a popular mix, allowing an N credit before corn of 50 to 100 lb. N/A. Rye also helps protect the less hardy vetch seedlings through winter.

MANAGEMENT

Establishment & Fieldwork

Rye prefers light loams or sandy soils and will germinate even in fairly dry soil. It also will grow in heavy clays and poorly drained soils, and many cultivars tolerate waterlogging (63).

Rye can establish in very cool weather. It will germinate at temperatures as low as 34° F. Vegetative growth requires 38° F or higher (361).

Winter annual use. Seed from late summer to midfall in Hardiness Zones 3 to 7 and from fall to midwinter in Zones 8 and warmer. In the Upper

Midwest and cool New England states, seed two to eight weeks earlier than a wheat or rye grain crop to ensure maximum fall, winter and spring growth. Elsewhere, your tillage system and the amount of fall growth you prefer will help determine planting date. Early planting increases the amount of N taken up before winter, but can make field management (especially killing the cover crop and tillage) more difficult in spring. See *Rye Smothers Weeds Before Soybeans* (p. 104).

Rye is more sensitive to seeding depth than other cereals, so plant no deeper than 2 inches (71). Drill 60 to 120 lb./A (1 to 2 bushels) into a prepared seedbed or broadcast 90 to 160 lb./A (1.5 to 3 bushels) and disk lightly or cultipack (361, 422).

If broadcasting late in fall and your scale and budget allow, you can increase the seeding rate to as high as 300 or 350 lb./A (about 6 bushels) to ensure an adequate stand. Rye can be overseed by air more consistently than many other cover crops.

"I use a Buffalo Rolling Stalk Chopper to help shake rye seeds down to the soil surface," says Steve Groff, a Holtwood, Pa., vegetable grower. "It's a very consistent, fast and economical way to establish rye in fall." (Groff's farming system is described in detail at www.cedarmeadow farm.com).

Mixed seeding. Plant rye at the lowest locally recommended rate when seeding with a legume (361), and at low to medium rates with other grasses. In a Maryland study, a mix of 42 pounds of rye and 19 pounds of hairy vetch per acre was the optimum fall seeding rate before no-till corn on a silt loam soil (81). If planting with clovers, seed rye at a slightly higher rate, about 56 lb. per acre.

For transplanting tomatoes into hilly, erosion-prone soil, Steve Groff fall-seeds a per-acre mix of 30 pounds rye, 25 pounds hairy vetch and 10 pounds crimson clover. He likes how the three-way mix guarantees biomass, builds soil and provides N.

Spring seeding. Although it's not a common practice, you can spring seed cereals such as rye as a weed-suppressing companion, relay crop or early forage. Because it won't have a chance to vernalize (be exposed to extended cold after ger-

CEREAL RYE (Secale cereale)

mination), the rye can't set seed and dies on its own within a few months in many areas. This provides good weed control in asparagus, says Rich de Wilde, Viroqua, Wis.

After drilling a large-seeded summer crop such as soybeans, try broadcasting rye. The cover grows well if it's a cool spring, and the summer crop takes off as the temperature warms up. Secondary tillage or herbicides would be necessary to keep the rye in check and to limit the cover crop's use of soil moisture.

Killing & Controlling

Nutrient availability concern. Rye grows and matures rapidly in spring, but its maturity date varies depending on soil moisture and temperature. Tall and stemmy, rye immobilizes N as it decomposes. The N tie-up varies directly with the maturity of the rye. Mineralization of N is very slow, so don't count on rye's overwintered N becoming available quickly.

Killing rye early, while it's still succulent, is one way to minimize N tie-up and conserve soil moisture. But spring rains can be problematic with rye, especially before an N-demanding crop, such as corn. Even if plentiful moisture hastens the optimal kill period, you still might get too much rain in the following weeks and have significant nitrate leaching, a Maryland study showed (109). Soil compaction also could be a problem if you're mowing rye with heavy equipment.

Late killing of rye can deplete soil moisture and could produce more residue than your tillage system can handle. For no-till corn in humid climates, however, summer soil-water conservation by cover crop residues often was more important than spring moisture depletion by growing cover crops, Maryland studies showed (82, 84, 85).

Legume combo maintains yield. One way to offset yield reductions from rye's immobilization of N would be to increase your N application. Here's another option: Growing rye with a legume allows you to delay killing the covers by a few weeks and sustain yields, especially if the legume is at least half the mix. This gives the legume more time to fix N (in some cases doubling the N contribution) and rye more time to scavenge a little more leachable N. Base the kill date on your area's normal kill date for a pure stand of the legume (109).

A legume/rye mix generally increases total dry matter, compared with a pure rye stand. The higher residue level can conserve soil moisture. For best results, wait about 10 days after killing the covers before planting a crop. This ensures adequate soil warming, dry enough conditions for planter coulters to cut cleanly and minimizes allelopathic effects from rye residue (84, 109). If using a herbicide, you might need a higher spray volume or added pressure for adequate coverage. Legume/rye mixes can be rolled once the legume is at full bloom (303).

Kill before it matures. Tilling under rye usually eliminates regrowth, unless the rye is less than 12 inches tall (361, 422). Rye often is plowed or disked in the Midwest when it's about 20 inches tall (307). Incorporating the rye before it's 18 in. high could decrease tie-up of soil N (361, 422). In Pennsylvania (118) and elsewhere, kill at least 10 days before planting corn.

For best results when mow-killing rye, wait until it has begun flowering. A long-day plant, rye is encouraged to flower by 14 hours of daylight and a temperature of at least 40° F. A sickle bar mower can give better results than a flail mower, which causes matting that can hinder emergence of subsequent crops (116).

Mow-kill works well in the South after rye sheds pollen in late April (101). If soil moisture is adequate, you can plant cotton three to five days after mowing rye when row cleaners are used in reduced-tillage systems.

Some farmers prefer to chop or mow rye by late boot stage, before it heads or flowers. "If rye gets away from you, you'd be better off baling it or harvesting it for seed," cautions southern Illinois organic grain farmer Jack Erisman (38). He often overwinters cattle in rye fields that precede soybeans. But he prefers that soil temperature be at least 60° F before planting beans, which is too late for him to no-till beans into standing rye.

Rye is the best cool-season cover crop for scavenging N, typically carrying 25 to 50 lb. N/A over to spring.

"If rye is at least 24 inches tall, I control it with a rolling stalk chopper that thoroughly flattens and crimps the rye stems," says Pennsylvania vegetable grower Steve Groff. "That can sometimes eliminate a burndown herbicide, depending on the rye growth stage and next crop."

A heavy duty rotavator set to only 2 inches deep does a good job of tilling rye, says Rich de Wilde, Viroqua, Wis.

Can't delay a summer planting by a few weeks while waiting for rye to flower? If early rye cultivars aren't available in your area and you're in Zone 5 or colder, you could plow the rye and use secondary tillage. Alternately, try a knockdown herbicide and post-emergent herbicide or spot-spraying for residual weed control.

For quicker growth of a subsequent crop such as corn or soybeans, leave the residue upright after killing (rather than flat). That hastens crop development—unless it's a dry year—via warmer soil temperatures and a warmer seed zone, according to a three-year Ontario study (146). This rarely influences overall crop yield, however, unless you plant too early and rye residue or low soil temperature inhibits crop germination.

Cereal Rye: Cover Crop Workhorse

Talk to farmers across America about cover crops and you'll find that most of them have planted a cereal rye cover crop. Almost certainly the most commonly planted cover crop, cereal rye can now be seen growing on millions of acres of farmland each year.

There are almost as many ways to manage cover crop rye as there are farmers using it. Climate, production system, soil type, equipment and labor are the principal factors that will determine how you manage rye. Your own practical experience will ultimately determine what works best for you.

Check out how others are managing rye in this book, on the Web and around your region. Test alternatives management practices that allow you to seed earlier or manage cover crop residue differently. Add a legume, a brassica or another grass to increase diversity on your farm.

Reasons for rye's widespread use include:

- It is winter-hardy, allowing it to grow longer into fall and resume growth earlier in the spring than most other cover crops.
- It produces a lot of biomass, which translates into a long-lasting residue cover in conservation tillage systems.
- It crowds out and out-competes winter annual weeds, while rye residue helps suppress summer weeds.
- It scavenges nutrients—particularly nitrogen—very effectively, helping keep nutrients on the farm and out of surface and ground water.
- It is relatively inexpensive and easy to seed.
- It works well in mixtures with legumes, resulting in greater biomass production and more complete fall/winter ground cover.
- It can be used as high-quality forage, either grazed or harvest as ryelage.
- It can fit into many different crop and livestock systems, including corn/soybean rotations, early or late vegetable crops, and dairy or beef operations.

Fall management (planting):

- While results are best if you plant rye by early fall, it also can be planted in November or December in much of the country—even into January in the deep South—and still provide tangible benefits.
- It can be drilled or broadcast after main crop harvest, with or without cultivation.
- It can be seeded before main crop harvest, usually by broadcasting, sometimes by plane or helicopter, and in northern climates, at last cultivation of the cash crop. Soil moisture availability is crucial to many of these pre-harvest seeding methods, either for germination of the cover crop or to avoid competition for water with the main crop.

Spring management (termination) is even more diverse:

- Rye can be killed with tillage, mowing, rolling or spraying.
- It can be killed before or after planting the cash crop, which can be drilled into standing cover crops in conservation tillage systems.
- Some want to leave rye growing as long as possible; others insist on terminating it as soon as possible in spring.
- Vegetable growers may leave walls of standing rye all season long between crop rows, usually to alleviate wind erosion.

Some examples of rye management wisdom from practitioners around the country:

- Pat Sheridan Jr., Fairgrove, Mich. Continuous no-till corn, sugar beets, soybeans, dry beans: "In late August, we fly rye into standing corn (or soybeans if we're coming back with soybeans the following year). We learned that rye is easier to burn down when it's more than two feet tall than when it has grown only a foot or less."
- Barry Martin, Hawksville, Ga. Peanuts and cotton. "After cotton, in late October or

November, we use a broadcast spreader (two bushels of rye per acre), then shred or mow to cover the seed. We usually get enough moisture in November and December for germination. After peanuts, we use a double disc grain drill (1.5 bushels of rye per acre) in mid-September to mid-October."

- Bryan and Donna Davis, Grinell, Iowa. Corn, soybeans, hay. "We tried to no-till corn and beans into rye three feet tall, but failed. The C:N ratio was way out of whack. The corn looked like it had been sprayed. If you don't kill before planting, you will invite insects." See also *Oats, Rye Feed Soil in Corn/Bean Rotation*, p. 96.
- Ed Quigley, Spruce Creek, PA. Dairy. "We seed cereal rye (two bushels per acre) immediately after corn silage. We allow as much spring growth as possible up to about 10 inches, at which point it becomes more difficult to kill, especially with cool/overcast conditions. We will also wait to make rylage in spring if we need feed, and then plant corn a bit later."

In some areas, farmers substitute other small grain cover crops for rye. They are doing so to better fit their particular niches, better manage their systems, or to cut costs by saving small grain seeds. Wheat is a popular alternative to rye. Look around and experiment!
—Andy Clark

Pest Management

Thick stands ensure excellent weed suppression. To extend rye's weed-management benefits, you can allow its allelopathic effects to persist longer by leaving killed residue on the surface rather than incorporating it. Allelopathic effects usually taper off after about 30 days. After killing rye, it's best to wait three to four weeks before planting small-seeded crops such as carrots or onions. If strip tilling vegetables into rye, be aware that rye seedlings have more allelopathic compounds than more mature rye residue. Transplanted vegetables, such as tomatoes, and larger-seeded species, especially legumes, are less susceptible to rye's allelopathic effects (117).

In an Ohio study, use of a mechanical under-cutter to sever roots when rye was at mid- to late bloom—and leaving residue intact on the soil surface (as whole plants)—increased weed suppression, compared with incorporation or mowing. The broadleaf weed reduction was comparable to that seen when sickle-bar mowing, and better than flail-mowing or conventional tillage (96).

If weed suppression is an important objective when planting a rye/legume mixture, plant early enough for the legume to establish well. Otherwise, you're probably better off with a pure stand. Overseeding may not be cost-effective before a crop such as field corn, however. A mix of rye and bigflower vetch (a quick-establishing, self-seeding, winter-annual legume that flowers and matures weeks ahead of hairy vetch) can suppress weeds significantly more than rye alone, while also allowing higher N accumulations (110).

Rye can effectively suppress weeds by shading, competition and allelopathy.

"Rye can provide the best and cleanest mulch you could want if it's cut or baled in spring before producing viable seed," says Rich de Wilde. Rye can become a **volunteer weed** if tilled before it's 8 inches high, however, or if seedheads start maturing before you kill it. Minimize regrowth by waiting until rye is at least 12 inches high before incorporating or by mow-killing after flowering but before grain fill begins.

Insect pests rarely a problem. Rye can reduce insect pest problems in crop rotations, southern research suggests (448). In a number of mid-Atlantic locations, Colorado potato beetles have been virtually absent in tomatoes no-till transplanted into a mix of rye/vetch/crimson clover, perhaps because the beetles can't navigate through the residue.

Rye Smothers Weeds Before Soybeans

An easy-to-establish rye cover crop helps Napoleon, Ohio, farmer Rich Bennett enrich his sandy soil while trimming input costs in no-till soybeans. Bennett broadcasts rye at 2 bushels per acre on corn stubble in late October. He incorporates the seed with a disc and roller.

The rye usually breaks through the ground but shows little growth before winter dormancy. Seeded earlier in fall, rye would provide more residue than Bennett prefers by bean planting—and more effort to kill the cover. "Even if I don't see any rye in fall, I know it'll be there in spring, even if it's a cold or wet one," he says.

By early May, the rye is usually at least 1.5 feet tall and hasn't started heading. He no-tills soybeans at 70 pounds per acre on 30-inch rows directly into standing rye cover crop. Then, depending on the amount of rye growth, he kills the rye with herbicide immediately after planting, or waits for more rye growth.

"If it's shorter than 15 to 18 inches, rye won't do a good enough job of shading out broadleaf weeds," notes Bennett, who likes how rye suppresses foxtail, pigweed and lambsquarters. "I sometimes wait up to two weeks to get more rye residue," he says.

"I kill the rye with 1.5 pints of Roundup per acre—about half the recommended rate. Adding 1.7 pounds of ammonium sulfate and 13 ounces of surfactant per acre makes it easier for Roundup to penetrate rye leaves," he explains.

The cover dies in about two weeks. The slow kill helps rye suppress weeds while soybeans establish. In this system, Bennett doesn't have to worry about rye regrowing.

Roundup Ready® beans have given him greater flexibility in this system. He used to cultivate beans twice using a Buffalo no-till cultivator. Now, depending on weed pressure (often giant ragweed and velvetleaf) he will spot treat or spray the whole field once with Roundup. Bennett figures the rye saves him $15 to $30 per acre in material costs and fieldwork, compared with conventional no-till systems for soybeans.

Rye doesn't hurt his bean yields, either. Usually at or above county average, his yields range from 45 to 63 bushels per acre, depending on rainfall, says Bennett.

"I really like rye's soil-saving benefits," he says. "Rye reduces our winter wind erosion, improves soil structure, conserves soil moisture and reduces runoff." Although he figures the rye's restrained growth (from the late fall seeding) provides only limited scavenging of leftover N, any that it does absorb and hold overwinter is a bonus.

Updated in 2007 by Andy Clark

While insect infestations are rarely serious with rye, as with any cereal grain crop occasional problems occur. If armyworms have been a problem, for example, burning down rye before a corn crop could move the pests into the corn. Purdue Extension entomologists note many northeastern Indiana corn farmers reported this in 1997. Crop rotations and IPM can resolve most pest problems you might encounter with rye.

Few diseases. Expect very few diseases when growing rye as a cover crop. A rye-based mulch can reduce diseases in some cropping systems. No-till transplanting tomatoes into a mix of rye/vetch/crimson clover, for example, consistently has been shown to delay the onset of early blight in several locations in the Northeast. The mulch presumably reduces soil splashing onto the leaves of the tomato plants.

If you want the option of harvesting rye as a grain crop, use of resistant varieties, crop rotation and plowing under crop residues can minimize rust, stem smut and anthracnose.

Other Options

Quick to establish and easy to incorporate when succulent, rye can fill rotation gaps in reduced-tillage, semi-permanent bed systems without increasing pest concerns or delaying crop plantings, a California study showed (216).

Erol Maddox, a Hebron, Md. grower, takes advantage of rye's relatively slow decomposition when double cropping. He likes transplanting spring cole crops into rye/vetch sod, chopping the cover mix at bloom stage and letting it lay until August, when he plants fall cole crops.

Mature rye isn't very palatable and provides poor-quality forage. It makes high quality hay or balage at boot stage, however, or grain can be ground and fed with other grains. Avoid feeding ergot-infected grain because it may cause abortions.

Rye can extend the grazing season in late fall and early spring. It tolerates fall grazing or mowing with little effect on spring regrowth in many areas (210). Growing a mixture of more palatable cover crops (clovers, vetch or ryegrass) can encourage regrowth even further by discouraging overgrazing (329).

Management Cautions

Although rye's extensive root system provides quick weed suppression and helps soil structure, don't expect dramatic soil improvement from a single stand's growth. Left in a poorly draining field too long, a rye cover could slow soil and warming even further, delaying crop planting. It's also not a silver bullet for eliminating herbicides. Expect to deal with some late-season weeds in subsequent crops (410).

COMPARATIVE NOTES

- Rye is more cold- and drought-tolerant than wheat.
- Oats and barley do better than rye in hot weather.
- Rye is taller than wheat and tillers less. It can produce more dry matter than wheat and a few other cereals on poor, droughty soils but is harder to burn down than wheat or triticale (241, 361).
- Rye is a better soil renovator than oats (422), but brassicas and sudangrass provide deeper soil penetration (451).
- Brassicas generally contain more N than rye, scavenge N nearly as well and are less likely to tie up N because they decompose more rapidly.

Seed sources. See *Seed Suppliers* (p. 195). 

SORGHUM-SUDANGRASS HYBRIDS

Sorghum bicolor x S. bicolor var. sudanese

Also called: Sudex, Sudax

Type: summer annual grass

Roles: soil builder, weed and nematode suppressor, subsoil loosener

Mix with: buckwheat, sesbania, sunnhemp, forage soybeans or cowpeas

See charts, pp. 66 to 72, for ranking and management summary.

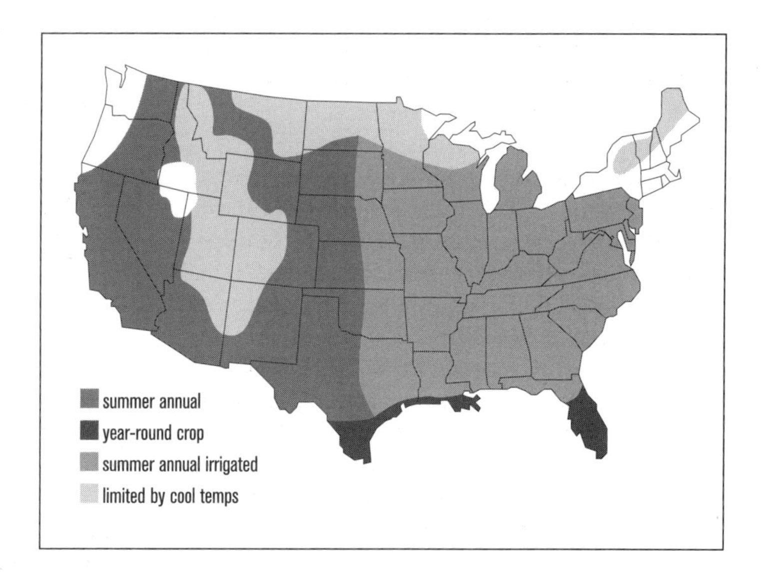

Sorghum-sudangrass hybrids are unrivaled for adding organic matter to worn-out soils. These tall, fast-growing, heat-loving summer annual grasses can smother weeds, suppress some nematode species and penetrate compacted subsoil if mowed once. Seed cost is modest. Followed by a legume cover crop, sorghum-sudangrass hybrids are a top choice for renovating overfarmed or compacted fields.

The hybrids are crosses between forage-type sorghums and sundangrass. Compared with corn, they have less leaf area, more secondary roots and a waxier leaf surface, traits that help them withstand drought (361). Like corn, they require good fertility—and usually supplemental nitrogen—for best growth. Compared with sudangrass, these hybrids are taller, coarser and more productive.

Forage-type sorghum plants are larger, leafier and mature later than *grain* sorghum plants. Compared with sorghum-sudangrass hybrids, they are shorter, less drought tolerant, and don't regrow as well. Still, forage sorghums as well as most forms of sudangrass can be used in the same cover-cropping roles as sorghum-sudangrass hybrids. All sorghum- and sudangrass-related species produce compounds that inhibit certain plants and nematodes. They are not frost tolerant, and should be planted after the soil warms in spring or in summer at least six weeks before first frost.

BENEFITS

Biomass producer. Sorghum-sudangrass grows 5 to 12 feet tall with long, slender leaves, stalks up to one-half inch in diameter and aggressive root systems. These features combine to produce ample biomass, usually about 4,000 to 5,000 lb. DM/A. Up to 18,000 lb. DM/A has been measured with multiple cuttings on fertile soils with adequate moisture.

Subsoil aerator. Mowing whenever stalks reach 3 to 4 feet tall increases root mass five to eight times compared with unmowed stalks, and forces the roots to penetrate deeper.

In addition, tops grow back green and vegetative until frost and tillering creates up to six new, thicker stalks per plant. A single mowing on New York muck soils caused roots to burrow 10 to 16 inches deep compared to 6 to 8 inches deep for unmowed plants. The roots of mowed plants fractured subsoil compaction with wormhole-like openings that improved surface drainage. However, four mowings at shorter heights caused plants to behave more like a grass and significantly decreased the mass, depth and diameter of roots (277, 450, 451).

Weed suppressor. When sown at higher rates than normally used for forage crops, sorghum-sudangrass hybrids make an effective smother crop. Their seedlings, shoots, leaves and roots secrete allelopathic compounds that suppress many weeds. The main root exudate, **sorgoleone**, is strongly active at extremely low concentrations, comparable to those of some synthetic herbicides (370). As early as five days after germination, roots begin secreting this allelochemical, which persists for weeks and has visible effects on lettuce seedlings even at 10 parts per million (440).

Sorghum-sudangrass hybrids suppress such annual weeds as velvetleaf, large crabgrass, barnyardgrass (126, 305), green foxtail, smooth pigweed (190), common ragweed, redroot pigweed and purslane (316). They also suppressed pine (214) and redbud tree seedlings in nursery tests (154). The residual weed-killing effects of these allelochemicals increased when sorghum-sudangrass hybrids were treated with the herbicides sethoxydim, glyphosate or paraquat, in descending order of magnitude (144).

Nematode and disease fighter. Planting sorghum-sudangrass hybrids instead of a host crop is a great way to disrupt the life cycles of many diseases, nematodes and other pests. For example, when sorghum-sudangrass or sorghum alone were no-tilled into endophyte-infected fescue pastures in Missouri that had received two herbicide applications, the disease was controlled nearly 100 percent. No-till reseeding with endophyte-free fescue completed this cost-effective renovation that significantly improved the rate of gain of yearling steers (16).

Renews farmed-out soils. The combination of abundant biomass production, subsoiling root systems, and weed and nematode suppression can produce dramatic results.

SORGHUM-SUDANGRASS (Sorghum bicolor X S. bicolor var. sudanese

On a low-producing muck field in New York where onion yields had fallen to less than a third of the local average, a single year of a dense planting of sorghum-sudangrass hybrid restored the soil to a condition close to that of newly cleared land (217).

Widely adapted. Sorghum-sudangrass hybrids can be grown throughout the U.S. wherever rainfall is adequate and soil temperature reaches 65° F to 70° F at least two months before frost. Once established, they can withstand drought by going nearly dormant. Sorghum-sudangrass hybrids tolerate pH as high as 9.0, and are often used in rotation with barley to reclaim alkaline soil (421). They tolerate pH as low as 5.0.

Quick forage. Sorghum-sudangrass is prized as summer forage. It can provide quick cover to prevent weeds or erosion where legume forages have been winterkilled or flooded out. Use care because these hybrids and other sorghums can produce prussic acid poisoning in livestock. Grazing poses the most risk to livestock when plants are young (up to 24 inches tall), drought stressed or killed by frost. Toxicity danger varies between cultivars.

MANAGEMENT

Establishment

Plant sorghum-sudangrass when soils are warm and moist, usually at least two weeks after the prime corn-planting date for your area. It will tolerate low-fertility, moderate acidity and high alkalinity, but prefers good fertility and near-neutral pH (361). Standard biomass production usually requires 75 to 100 lb. N/A .

With sufficient surface moisture, broadcast 40 to 50 lb./A, or drill 35 to 40 lb./A as deep as 2 inches to reach moist soil. These rates provide a quicker canopy to smother weeds than lower rates used for forage production, but they require mowing or grazing to prevent lodging. Herbicide treatment or a pass with a mechanical weeder may be necessary if germination is spotty or perennial weeds are a problem. New York on-farm tests show that a stale seedbed method—tilling, then retilling to kill the first flush of weeds just before planting—provides effective weed control.

These heat-loving plants are unrivaled for adding organic matter to soils.

Warm season mixtures. Plant sorghum-sudangrass in cover crop mixtures with buckwheat or with the legumes sesbania (*Sesbania exaltata*), sunnhemp (*Crotolaria juncea*), forage soybeans (*Glycine max*) or cowpeas (*Vigna unguiculata*). Broadcast these large-seeded cover crops with the sorghum-sudangrass, then incorporate about 1 inch deep. Fast-germinating buckwheat helps suppress early weeds. Sorghum-sudangrass supports the sprawling sesbania, forage soybeans and cowpeas. Sunnhemp has an upright habit, but could compete well for light if matched with a sorghum-sudangrass cultivar of a similar height.

Field Management

Plants grow very tall (up to 12 feet), produce tons of dry matter and become woody as they mature. This can result in an unmanageable amount of tough residue that interferes with early planting the following spring (277).

Mowing or grazing when stalks are 3 to 4 feet tall encourages tillering and deeper root growth, and keeps regrowth vegetative and less fibrous until frost. For mid-summer cuttings, leave at least 6 inches of stubble to ensure good regrowth and continued weed suppression. Delayed planting within seven weeks of frost makes mowing unnecessary and still allows for good growth before winterkilling (277, 361).

Disking while plants are still vegetative will speed decomposition. Make several passes with a heavy disk or combination tillage tool to handle the dense root masses (277). Sicklebar mowing or flail chopping before tillage will reduce the number of field operations required to incorporate the crop and speed decomposition. Sicklebar mowers cut more cleanly but leave the stalks whole. Using a front-mounted flail chopper avoids the problem of skips where tractor tires flatten the plants, putting them out of reach of a rear-mounted chopper.

Any operations that decrease the residue size shortens the period during which the decomposing residue will tie up soil nitrogen and hinder early planted crops in spring. Even when mowed, residue will become tough and slower to break down if left on the surface.

Flail chopping after frost or killing the cover crop with herbicide will create a suitable mulch for no-till planting, preserving soil life and soil structure in non-compacted fields.

Pest Management

Weeds. Use sorghum-sudangrass to help control nutsedge infestations, suggests Cornell Extension IPM vegetable specialist John Mishanec. Allow the nutsedge to grow until it's about 4 to 5 inches tall but before nutlets form, about mid-June in New York. Kill the nutsedge with herbicide, then plant the weed-smothering hybrid.

To extend weed suppressive effects into the second season, select a cultivar known for weed suppression and leave roots undisturbed when the stalks are mowed or grazed (440).

Beneficial habitat. Some related sorghum cultivars harbor beneficial insects such as seven-spotted lady beetles (*Coccinella septempunctata*) and lacewings (*Chrysopa carnea*) (421).

Nematodes. Sorghum-sudangrass hybrids and other sorghum-related crops and cultivars suppress some species of nematodes. Specific cultivars vary in their effectiveness on different races of nematodes. These high-biomass-producing crops have a general suppressive effect due to their organic matter contributions. But they also produce natural nematicidal compounds that chemically suppress some nematodes, many studies show.

Timing of cutting and tillage is very important to the effectiveness of nematode suppression. The cover crop needs to be tilled before frost while it is still green. Otherwise, the nematicidal effect is lost. For maximum suppression of soilborne diseases, cut or chopped sudangrass must be well incorporated immediately (308).

For suppressing root-knot nematodes in Idaho potato fields, rapeseed has proven slightly more effective and more dependable than sorghum-sudangrass hybrids (394).

In an Oregon potato trial, TRUDAN 8 sudangrass controlled Columbia root-knot nematodes (*Meloidogoyne Chitwoodi*), a serious pest of many vegetable crops. Control extended throughout the zone of residue incorporation. The cover crop's effect prevented upward migration of the nematodes into the zone for six weeks, working as well as the nematicide ethoprop. Both treatments allowed infection later in the season (285).

In the study, TRUDAN 8 sudangrass and the sorghum-sudangrass hybrid cultivars SORDAN 79 and SS-222 all reduced populations of root-knot nematodes. These cultivars are poor nematode hosts and their leaves—not roots—have a nematicidal effect. TRUDAN 8 should be used if the crop will be grazed due to its lower potential for prussic acid poisoning. The sorghum-sudangrass cultivars are useful if the cover crop is intended for anti-nematicidal effects only (285). In other Oregon and Washington trials, the cover crop suppression required supplemental chemical nematicide to produce profitable levels of U.S. No. 1 potatoes (285). These same sudangrass and sorghum-sudangrass hybrid cultivars failed to show any significant nematicidal effects in a later experiment in Wisconsin potato fields (249).

When faced with infestations of the nematodes *Meloidogoyne incognita* and *M. arenaria*, Oswego, N.Y., onion grower Dan Dunsmoor found that a well-incorporated sorghum-sudangrass cover crop was more effective than fumigation. Further, the nematicidal effect continued into the next season, while the conditions a year after fumigation seemed worse than before the application. He reports that the sorghum-sudangrass cover crop also controls onion maggot, thrips and Botrytis leaf blight (217).

These plants produce chemicals that inhibit certain weeds and nematodes.

Insect pests. Chinch bug (*Blissus leucopterus*), sorghum midge (*Contarinia sorghicola*), corn leaf aphid (*Rhopalosiphum maidis*), corn earworm (*Heliothis zea*), greenbugs (*Schizaphis graminum*) and sorghum webworm (*Celama sorghiella*) sometimes attack sorghum-sudangrass hybrids. Early planting helps control the first two pests, and may reduce damage from webworms. Some cultivars and hybrids are resistant to chinch bugs and some biotypes of greenbugs (361). In Georgia, some hybrids hosted corn leaf aphid, greenbug, southern green stinkbugs (*Nezara viridula*) and leaffooted bug (*Leptoglossus phyllopus*).

Crop Systems

There are several strategies for reducing nitrogen tie-up from residue:

- Interplant a legume with the sorghum-sudangrass hybrid.
- Plant a legume cover crop *after* the sorghum-sudangrass hybrid, in either late summer or the following spring.
- Apply nitrogen fertilizer or some other N source at incorporation and leave the land fallow for a few months when soil is not frozen to allow decomposition of the residue.

Summer Covers Relieve Compaction

A summer planting of sudangrass was the best single-season cover crop for relieving soil compaction in vegetable fields, a team of Cornell researchers found. Yellow mustard, HUBAM annual white sweetclover and perennial ryegrass also were effective to some extent in the multi-year study. "But sudangrass has proven the most promising so far," says project coordinator David Wolfe. "It has shown the fastest root growth."

"Sudangrass is best managed with one mowing during the season," Wolfe adds. Mowing promotes tillering and a deep, penetrating root system. Mowing also makes it easier to incorporate the large amount of biomass produced by this crop. With its high C:N ratio, it adds to soil organic matter.

Farmers and researchers have long known that alfalfa's deep root system is a great compaction-buster. But alfalfa does not establish easily on wet compacted fields, and most vegetable growers can't afford to remove land from production for two to three years to grow it, notes Wolfe. Many also lack the equipment to subsoil their fields, which is often only a temporary solution, at best. That's why Wolfe geared his study to identify cover crops that can produce results in a single season. In the case of heat-loving sudangrass, it also may be possible to squeeze a spring or fall cash crop into the rotation while still growing the cover during summer.

Heavy equipment, frequent tillage and lack of organic matter contribute to compaction problems for vegetable growers in the Northeast, where frequent rains often force growers into the fields when soils are wet. Compacted soils slow root development, hinder nutrient uptake, stunt plants, delay maturity and can worsen pest and disease problems (451). For example, the Cornell researchers found that slow-growing cabbages direct-seeded into compacted soils were vulnerable to flea beetle infestations (450).

Brassica cover crops such as yellow mustard were solid challengers to sudangrass as a compaction reliever, but it was sometimes difficult to establish these crops in the test. "We still have a lot to learn about how best to grow brassicas and fit them into rotations with vegetables," Wolfe says.

Wolfe and his team assessed the cover crops' effectiveness by measuring yields of subsequent crops and conducting a host of soil quality measurements, including infiltration rates, water-holding capacity, aggregate stability and organic matter levels.

For more information, contact David Wolfe, 607-255-7888; dww5@cornell.edu.

Updated in 2007 by Andy Clark

If you kill the cover crop early enough in fall, the residue will partially break down before cold temperatures slow biological action (361). Where possible, use sorghum-sudangrass ahead of later-planted crops to allow more time in spring for residue to decompose.

Planting sorghum-sudangrass every third year on New York potato and onion farms will rejuvenate soil, suppress weeds and may suppress soil pathogens and nematodes. Working a legume into the rotation will further build soil health and add nitrogen. Sorghum-sudangrass hybrids can provide needed soil structure benefits wherever intensive systems cause compaction and loss of soil organic matter reserves. See *Summer Covers Relieve Compaction*, above.

Grown as a summer cover crop that is cut once and then suppressed or killed, sorghum-sudangrass can reduce weeds in fall-planted alfalfa. Sorghum-sudangrass suppressed alfalfa root growth significantly in a Virginia greenhouse study (144), but no effect was observed on alfalfa germination when alfalfa was no-till planted into killed or living sorghum-sudangrass (145).

In Colorado, sorghum-sudangrass increased irrigated potato tuber quality and total marketable yield compared. It also increased nutrient uptake efficiency on the sandy, high pH soils. In this system, the sorghum-sudangrass is grown with limited irrigation, but with enough water so that the biomass could be harvested for hay or incorporated as green manure (112, 113).

In California, some table grape growers use sorghum-sudangrass to add organic matter and to reduce the reflection of light and heat from the soil, reducing sunburn to the grapes.

COMPARATIVE NOTES

Sorghum-sudangrass hybrids can produce more organic matter per acre, and at a lower seed cost, than any major cover crop grown in the U.S.

Incorporated sorghum-sudangrass residue reduces N availability to young crops more than oat residue but less than wheat residue (389).

Cultivars. When comparing sorghum-sudangrass cultivars, consider traits such as biomass yield potential, tillering and regrowth ability, disease resistance, insect resistance (especially if greenbugs are a problem) and tolerance to iron deficiency chlorosis.

If you plan to graze the cover crop, select sorghum-sudangrass hybrids and related crops with lower levels of dhurrin, the compound responsible for prussic acid poisoning. For maximum weed control, choose types high in sorgoleone, the root exudate that suppresses weeds. Sterile cultivars are best where escapes could be a problem, especially where crossing with johnsongrass (*Sorghum halpense*) is possible.

Seed sources. See *Seed Suppliers* (p. 195). 

WINTER WHEAT

Triticum aestivum

Type: winter annual cereal grain; can be spring-planted

Roles: prevent erosion, suppress weeds, scavenge excess nutrients, add organic matter

Mix with: annual legumes, ryegrass or other small grains

See charts, pp. 66 to 72, for ranking and management summary.

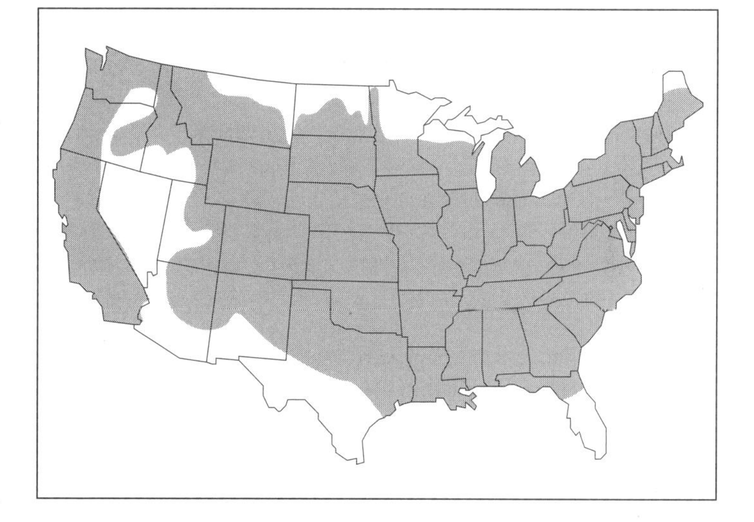

Although typically grown as a cash grain, winter wheat can provide most of the cover crop benefits of other cereal crops, as well as a grazing option prior to spring tiller elongation. It's less likely than barley or rye to become a weed and is easier to kill. Wheat also is slower to mature than some cereals, so there is no rush to kill it early in spring and risk compacting soil in wet conditions. It is increasingly grown instead of rye because it is cheaper and easier to manage in spring.

Whether grown as a cover crop or for grain, winter wheat adds rotation options for underseeding a legume (such as red clover or sweet-clover) for forage or nitrogen. It works well in no-till or reduced-tillage systems, and for weed control in potatoes grown with irrigation in semiarid regions.

BENEFITS

Erosion control. Winter wheat can serve as an overwintering cover crop for erosion control in most of the continental U.S.

Nutrient catch crop. Wheat enhances cycling of N, P and K. A heavy N feeder in spring, wheat takes up N relatively slowly in autumn. It adds up, however. A September-seeded stand absorbed 40 lb. N/A by December, a Maryland study showed (46). As an overwintering cover rather than a grain crop, wheat wouldn't need fall or spring fertilizer.

A 50-bushel wheat crop can take up 20 to 25 lb. P_2O_5 and 60 lb. K_2O per acre by boot stage. About 80 percent of the K is recycled if the stems and leaves aren't removed from the field at harvest. All the nutrients are recycled when wheat is managed as a cover crop, giving it a role in scavenging excess nitrogen.

"Cash and Cover" crop. Winter wheat can be grown as a cash crop or a cover crop, although you should manage each differently. It provides a cash-grain option while also opening a spot for a winter annual legume in a corn>soybean or similar rotation. For example:

- In the Cotton Belt, wheat and crimson clover would be a good mix.
- In Hardiness Zone 6 and parts of Zone 7, plant hairy vetch after wheat harvest, giving the legume plenty of time to establish in fall. Vetch growth in spring may provide most of the N necessary for heavy feeders such as corn, or all of the N for sorghum, in areas northward to southern Illinois, where early spring warm-up allows time for development.
- In much of Zone 7, cowpeas would be a good choice after wheat harvest in early July or before planting winter wheat in fall.
- In the Corn Belt and northern U.S., undersow red clover or frostseed sweetclover into a wheat nurse crop if you want the option of a year of hay before going back to corn. With or without underseeding a legume or legume-grass mix, winter wheat provides great grazing and nutritional value and can extend the grazing season.
- In Colorado vegetable systems, wheat reduced wind erosion and scavenged N from 5 feet deep in the profile (111, 114).
- In parts of Zone 6 and warmer, you also have a dependable double-crop option. See *Wheat Boosts Income and Soil Protection* (p. 113).

WINTER WHEAT
(Triticum aestivum)

Weed suppressor. As a fall-sown cereal, wheat competes well with most weeds once it is established (71). Its rapid spring growth also helps choke weeds, especially with an underseeded legume competing for light and surface nutrients.

Soil builder and organic matter source. Wheat is a plentiful source of straw and stubble. Wheat's fine root system also improves topsoil tilth. Although it generally produces less than rye or barley, the residue can be easier to manage and incorporate.

When selecting a locally adapted variety for use as a cover crop, you might not need premium seed. A Maryland study of 25 wheat cultivars showed no major differences in overall biomass production at maturity (92). Also in Maryland, wheat produced up to 12,500 lb. biomass/A following high rates of broiler litter (87).

In Colorado, wheat planted in August after early vegetables produced more than 4,000 lb. biomass/A, but if planted in October, yielded only one-tenth as much biomass and consequently scavenged less N (114).

Wheat Boosts Income *and* Soil Protection

Wheat is an ideal fall cover crop that you can later decide to harvest as a cash crop, cotton farmer Max Carter has found. "It's easier to manage than rye, still leaves plenty of residue to keep topsoil from washing away—and is an excellent double crop," says Carter.

The southeastern Georgia farmer no-till drills winter wheat at 2 bushels per acre right after cotton harvest, without any seedbed preparation. "It gives a good, thick stand," he says.

"We usually get wheat in by Thanksgiving, but as long as it's planted by Christmas, I know it'll do fine," he adds. After drilling wheat, Carter goes back and mows the cotton stalks to leave some field residue until the wheat establishes.

Disease or pests rarely have been a problem, he notes.

"It's a very easy system, with wheat always serving as a fall cover crop for us. It builds soil and encourages helpful soil microorganisms. It can be grazed, or we can burn some down in March for planting early corn or peanuts anytime from March to June," he says.

For a double crop before 2-bale-an-acre cotton, Carter irrigates the stand once in spring with a center pivot and harvests 45- to 60-bushel wheat by the end of May. "The chopper on the rear of the combine puts the straw right back on the soil as an even blanket and we're back planting cotton on June 1."

"It sure beats idling land and losing topsoil."

If weed control is important in your system, look for a regional cultivar that can produce early spring growth. To scavenge N, select a variety with good fall growth before winter dormancy.

MANAGEMENT

Establishment & Fieldwork

Wheat prefers well-drained soils of medium texture and moderate fertility. It tolerates poorly drained, heavier soils better than barley or oats, but flooding can easily drown a wheat stand. Rye may be a better choice for some poor soils.

Biomass production and N uptake are fairly slow in autumn. Tillering resumes in late winter/early spring and N uptake increases quickly during stem extension.

Adequate but not excessive N is important during wheat's early growth stages (prior to stem growth) to ensure adequate tillering and root growth prior to winter dormancy. In low-fertility or light-textured soils, consider a mixed seeding with a legume (80). See *Wheat Offers High Value Weed Control, Too* (p. 114).

A firm seedbed helps reduce winterkill of wheat. Minimize tillage in semiarid regions to avoid pulverizing topsoil (358) and depleting soil moisture.

Winter annual use. Seed from late summer to early fall in Zone 3 to 7—a few weeks earlier than a rye or wheat grain crop—and from fall to early winter in Zone 8 and warmer. If you are considering harvesting as a grain crop, you should wait until the Hessian fly-free date, however. If cover crop planting is delayed, consider sowing rye instead.

Drill 60 to 120 lb./A (1 to 2 bushels) into a firm seedbed at a $^1/_2$- to $1^1/_2$-inch depth or broadcast 60 to 160 lb./A (1 to 2.5 bushels) and disk lightly or cultipack to cover. Plant at a high rate if seeding late, when overseeding into soybeans at the leaf-yellowing stage, when planting into a dry seedbed or when you require a thick, weed-suppressing stand. Seed at a low to medium rate when soil moisture is plentiful (71).

After cotton harvest in Zone 8 and warmer, no-till drill 2 bushels of wheat per acre without any seedbed preparation. In the Southern Plains, 1 bushel is sufficient if drilling in a timely fashion (302).

With irrigation or in humid regions, you could harvest 45- to 60-bushel wheat, then double crop

Wheat Offers High-Value Weed Control

Pairing a winter wheat cover crop with a reduced herbicide program in the inland Pacific Northwest could provide excellent weed control in potatoes grown on light soils in irrigated, semiarid regions. A SARE-funded study showed that winter wheat provided effective competition against annual weeds that infest irrigated potato fields in Washington, Oregon and Idaho.

Banding herbicide over the row when planting potatoes improved the system's effectiveness, subsequent research shows, says project coordinator Dr. Charlotte Eberlein at the University of Idaho's Aberdeen Research and Extension Center. "In our initial study, we were effectively no-tilling potatoes into the Roundup-killed wheat," says Eberlein. "In this study we killed the cover crop and planted potatoes with a regular potato planter, which rips the wheat out of the potato row." A grower then can band a herbicide mixture over the row and depend on the wheat mulch to control between-row weeds.

"If you have sandy soil to start with and can kill winter wheat early enough to reduce water-management concerns for the potatoes, the system works well," says Eberlein.

"Winter rye would be a slightly better cover crop for suppressing weeds in a system like this," she notes. "Volunteer rye, however, is a serious problem in wheat grown in the West, and wheat is a common rotation crop for potato growers in the Pacific Northwest."

She recommends drilling winter wheat at 90 lb./A into a good seedbed, generally in mid-September in Idaho. "In our area, growers can deep rip in fall, disk and build the beds (hills), then drill wheat directly into the beds," she says. Some starter N (50 to 60 lb./A) can help the wheat establish. If indicated by soil testing, P or K also would be fall-applied for the following potato crop.

The wheat usually does well and shows good winter survival. Amount of spring rainfall and soil moisture and the wheat growth rate determine the optimal dates for killing wheat and planting potatoes.

Some years, you might plant into the wheat and broadcast Roundup about a week later. Other years, if a wet spring delays potato planting, you could kill wheat before it gets out of hand (before the boot stage), then wait for better potato-planting conditions.

Moisture management is important, especially during dry springs, she says. "We usually kill the wheat from early to mid May—a week or two after planting potatoes. That's soon enough to maintain adequate moisture in the hills for potatoes to sprout."

An irrigation option ensures adequate soil moisture—for the wheat stand in fall or the potato crop in spring, she adds. "You want a good, competitive wheat stand and a vigorous potato crop if you're depending on a banded herbicide mix and wheat mulch for weed control," says Eberlein. That combination gives competitive yields, she observes, based on research station trials.

with soybeans, cotton or another summer crop. See *Wheat Boosts Income and Soil Protection* (p. 113). You also could overseed winter wheat prior to cotton defoliation and harvesting.

Another possibility for Zone 7 and cooler: Plant full-season soybeans into wheat cover crop residue, and plant a wheat cover crop after bean harvest.

Mixed seeding or nurse crop. Winter wheat works well in mixtures with other small grains or with legumes such as hairy vetch. It is an excellent nurse crop for frostseeding red clover or sweetclover, if rainfall is sufficient. In the Corn Belt, the legume is usually sown in winter, before wheat's vegetative growth resumes. If frostseeding, use the full seeding rates for both species, according to recent work in Iowa (34). If you sow sweetclover in fall with winter wheat, it could outgrow the wheat. If you want a grain option, that could make harvest difficult.

Spring annual use. Although it's not a common practice, winter wheat can be planted in the spring as a weed-suppressing companion crop or early forage. You sacrifice fall nutrient scavenging, however. Reasons for spring planting include winter kill or spotty overwintering, or when you just didn't have time to fall-seed it. It won't have a chance to vernalize (be exposed to extended cold after germination), so it will not head out and usually dies on its own within a few months, without setting seed. This eliminates the possibility of it becoming a weed problem in subsequent crops. By sowing when field conditions permit in early spring, within a couple months you could have a 6- to 10-inch tall cover crop into which you can no-till your cash crop. You might not need a burndown herbicide, either.

Early spring planting of *spring* wheat, with or without a legume companion, is an option, especially if you have a longer rotation niche available.

Field Management

You needn't spring fertilize a winter wheat stand being grown as a cover crop rather than a grain crop. That would defeat the primary purpose (N-scavenging) of growing a small grain cover crop. As with any overwintering small grain crop, however, you will want to ensure the wheat stand doesn't adversely affect soil moisture or nutrient availability for the following crop.

Killing

Kill wheat with a roller crimper at soft-dough stage or later, with a grass herbicide, or by plowing, disking or mowing before seed matures. As with other small grain cover crops, it is safest to kill about 2-3 weeks before planting your cash crop, although this will depend on local conditions and your killing and tillage system.

Because of its slower spring growth, there is less need to rush to kill wheat in spring as is sometimes required for rye. That's one reason vegetable grower Will Stevens of Shoreham, Vt., prefers wheat to rye as a winter cover on his heavy, clay-loam soils. The wheat goes to seed slower and can provide more biomass than an earlier killing of rye would, he's found. With rye, he has to disk two to three weeks earlier in spring to incorporate the biomass, which can be a problem in wet conditions. "I only chisel plow wheat if it's really rank," he notes.

Wheat is less likely than barley or rye to become a weed, and is easier to kill.

Pest Management

Wheat is less likely than rye or barley to become a weed problem in a rotation, but is a little more susceptible than rye or oats to insects and disease. Managed as a cover crop, wheat rarely poses an insect or disease risk. Diseases can be more of a problem the earlier wheat is planted in fall, especially if you farm in a humid area.

Growing winter wheat could influence the buildup of pathogens and affect future small-grain cash crops, however. Use of resistant varieties and other IPM practices can avoid many pest problems in wheat grown for grain. If wheat diseases or pests are a major concern in your area, rye or barley might be a better choice as an overwintering cover crop that provides a grain option, despite their lower grain yield.

Other Options

Choosing wheat as a small-grain cover crop offers the flexibility in late spring or early summer to harvest a grain crop. Spring management such as removing grazing livestock prior to heading and topdressing with N is essential for the grain crop option.

Seed sources. See *Seed Suppliers* (p. 195). 

OVERVIEW OF LEGUME COVER CROPS

Commonly used legume cover crops include:

- Winter annuals, such as crimson clover, hairy vetch, field peas, subterranean clover and many others
- Perennials like red clover, white clover and some medics
- Biennials such as sweetclover
- Summer annuals (in colder climates, the winter annuals are often grown in the summer)

Legume cover crops are used to:

- Fix atmospheric nitrogen (N) for use by subsequent crops
- Reduce or prevent erosion
- Produce biomass and add organic matter to the soil
- Attract beneficial insects

Legumes vary widely in their ability to prevent erosion, suppress weeds and add organic matter to the soil. In general, legume cover crops do not scavenge N as well as grasses. If you need a cover crop to take up excess nutrients after manure or fertilizer applications, a grass, a brassica or a mixture is usually a better choice.

Winter-annual legumes, while established in the fall, usually produce most of their biomass and N in spring. Winter-annual legumes must be planted earlier than cereal crops in order to survive the winter in many regions. Depending on your climate, spring management of legumes will often involve balancing early planting of the cash crop with waiting to allow more biomass and N production by the legume.

Perennial or biennial legumes can fit many different niches, as described in greater detail in the individual sections for those cover crops. Sometimes grown for a short period between cash crops, these forage crops also can be used for more than one year and often are harvested for feed during this time. They can be established along with—or overseeded into—other crops such as wheat or oats, then be left to grow after cash crop harvest and used as a forage. Here they are functioning more as a rotation crop than a cover crop, but as such provide many benefits including erosion and weed control, organic matter and N production. They also can break weed, disease and insect cycles.

Summer-annual use of legume crops includes, in colder climates, the use of the winter-annual crops listed above, as well as warm-season legumes such as cowpeas. Grown as summer annuals, these crops produce N and provide ground cover for weed and erosion control, as well as other benefits of growing cover crops. Establishment and management varies widely depending on climate, cropping system and the legume itself. These topics will be covered in the individual sections for each legume.

Legumes are generally lower in carbon and higher in nitrogen than grasses. This lower C:N ratio results in faster breakdown of legume residues. Therefore, the N and other nutrients contained in legume residues are usually released faster than from grasses. Weed control by legume residues may not last as long as for an equivalent amount of grass residue. Legumes do not increase soil organic matter as much as grasses.

Mixtures of legume and grass cover crops combine the benefits of both, including biomass production, N scavenging and additions to the system, as well as weed and erosion control. Some cover crop mixtures are described in the individual cover crop sections.

COVER CROP MIXTURES EXPAND POSSIBILITIES

Mixtures of two or more cover crops are often more effective than planting a single species. Cover crop mixtures offer the best of both worlds, combining the benefits of grasses and legumes, or using the different growth characteristics of several species to fit your needs.

You can use cover crop mixtures to improve:

- Winter survival
- Ground cover
- Use of solar energy
- Biomass and N production
- Weed control
- Duration of active growing period
- Range of beneficial insects attracted
- Tolerance of adverse conditions
- Forage options
- Response to variable soil traits

Disadvantages of cover crop mixtures may include:

- Higher seed cost
- Too much residue
- More complicated management
- Difficult to seed

Crop mixtures can reduce risk in cropping systems because each crop in the mix may respond differently to soil, pest and weather conditions. In forage or grazing systems, for example, a mix of rye, wheat and barley is more nutritious, can be grazed over a longer period of time and is less likely to be devastated by a single disease.

Using drought-tolerant plants in a perennial mix builds in persistence for dry years. Using a number of cover crops with "hard seed" that takes many months to germinate also improves coverage over a broader range of conditions.

Mixing cultivars of a single species with varied maturity dates and growth habits maintains optimum benefits for a longer time. Orchardists in California mix subclovers to keep weeds at bay all season. One cultivar comes on early, then dies back as two later cultivars—one tall and one short—come on strong. Because they reseed themselves, the cooperative trio persists year after year.

Sometimes you don't know how much N may be left after cash crop harvest. Do you need a grass to scavenge leftover N, or a legume to provide fixed N? A grass/legume cover crop mixture adjusts to the amount of available soil N: If there is a lot of N, the grass dominates; if there is not much available soil N, the legume will tend to dominate a mixture. In either case, you get the combined benefit of N scavenging by the grass cover crop and N additions from the legume cover crop.

Mixing low-growing and taller crops, or fast-starting grasses and slow-developing legumes, usually provides better erosion control because more of the ground is covered. The vegetation intercepts more raindrops before they can dislodge soil particles. Sunlight is used more efficiently because light that passes through the tall crop is captured by the low-growing crop.

Adding grasses to a fall-seeded legume improves soil coverage over winter and increases the root mass to stabilize topsoil. A viny crop like vetch will climb a grass, so it can get more light and fix more N, or so it can be harvested more easily for seed. A faster-growing crop serves as a nurse crop for a slow-growing crop, while covering the ground quickly for erosion control. The possibilities are endless!

Mixtures can complicate management, however. For example:

- They may cost more to seed. Seeding rates for each component of the mix are usually lower than for sole-crop plantings, but the total seed cost may still be more.
- The best time to kill one crop may not be the best for another crop, so a compromise date may be used.
- If you use herbicides, your choices may be limited when you plant a mixture of legumes and nonlegumes.
- Sometimes you can end up with more residue than your equipment can handle.

The benefits of a mixture will usually outweigh these disadvantages, but you need to be prepared to manage the mixture carefully to prevent problems.

Each cover crop chapter gives examples of specific mixtures that have been tested and work well. Try some of the proven cover crop mixtures, and create your own tailor-made mixtures. Remember that adding another crop increases the diversity on your farm, and is likely to increase the many proven benefits of rotations over monocropping.

BERSEEM CLOVER

Trifolium alexandrinum

Also called: Egyptian clover

Type: summer annual or winter annual legume

Roles: suppress weeds, prevent erosion, green manure, chopped forage, grazing

Mix with: oats, ryegrass, small grains as nurse crops; as nurse crop for alfalfa

See charts, p. 66 to 72, for ranking and management summary.

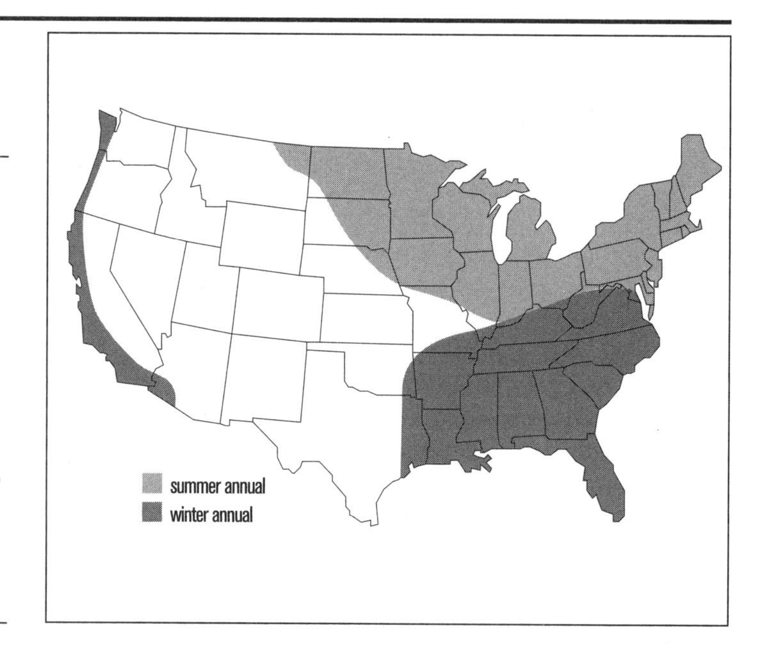

A fast-growing summer annual, berseem clover can produce up to 8 tons of forage under irrigation. It's a heavy N producer and the least winter hardy of all true annual clovers. This makes it an ideal winterkilled cover before corn or other nitrogen-demanding crops in Corn Belt rotations. Berseem clover draws down soil N early in its cycle. Once soil reserves are used up, it can fix 100 to 200 lb. N/A or more. It establishes well with an oat nurse crop, making it an excellent cover for small grain>corn>soybean rotations in the Midwest.

In Iowa, the cultivar BIGBEE compares favorably with alfalfa in its regrowth following small grain harvest, its feed value and its tolerance to drought and excess moisture (156). As a winter annual in California, irrigation usually is needed to allow berseem to achieve its full potential. Its peak growth period during the West Coast's rainy season and its highly efficient water use compare favorably to alfalfa as a high-producing forage and green manure.

BENEFITS

Green manure. Berseem clover is the fertility foundation of agriculture in the Nile Delta, and has nourished soils in the Mediterranean region for millennia. MULTICUT berseem clover averaged 280 lb. N/A in a six-year trial in California with six cuttings per year (162), and grew faster than BIGBEE in one Iowa report (155). Berseem is less prone to possible N leaching if grown to maturity without cutting, when it produces 100 to 125 lb. N/A. Top N fixation occurs when soils have less than 150 lb. N/A (162). A single cutting can yield

50 to 100 lb. topgrowth N/A. Berseem's dry matter N concentration is about 2.5 percent (162).

Biomass. Berseem clover produced the most biomass (6,550 lb./A) of five winter annual legumes in a two-year Louisiana test, and came in second to arrowleaf clover (*Trifolium vesiculosum*) in N, accumulating 190 lb. N/A to arrowleaf's 203 lb. N/A. Also tested were TIBBEE crimson clover, WOOGENELUP subterraneum clover and WOODFORD bigflower vetch. All but arrowleaf clover were able to set seed by May 13 and regrow in the fall, despite the herbicides used to suppress them in spring and to control weeds during summer (36).

In Alberta legume trials, berseem clover averaged 3,750 lb. dry matter/A over three years at a site where hairy vetch and field peas produced 5,300 and 4,160 lb./A, respectively. With irrigation, berseem clover topped 19 other legumes at the same site with a mean yield of 5,500 lb. DM/A.

Smother crop. Planted with oats or annual ryegrass, berseem clover suppresses weeds well during establishment and regrowth after oat harvest.

Companion crop. Planted with oat, the two crops can be harvested together as silage, haylage or hay, depending on the crop's development stage. Berseem/oat haylage has very high feed quality if cut at oats' boot stage (157). Dry seasons favor development of an oat grain crop, after which berseem clover can be cut one, two or three times in the Midwest.

Quick growing. At 60° F, berseem clover will be ready to cut about 60 days after planting.

Legume nurse crop. Because of its quick germination (seven days), quick growth and winterkilling tendency, berseem clover can be used as **a nurse crop for alfalfa.**

Seed crop. Berseem produces up to 1,000 lb. seed/A if it is left to mature. Only BIGBEE berseem clover has hard seed that allows natural self-reseeding, and it reseeds too late for timely planting of most summer crops (103).

Grazing and forage crop. At 18 to 28 percent protein, young berseem clover is comparable to or better than crimson clover or alfalfa as feed. No cases of bloat from grazing berseem clover have been reported (158, 278). Forage quality remains acceptable until the onset of seed production. BIGBEE berseem clover and TIBBEE crimson produce more fall and winter growth than do other winter annual clovers in the South. BIGBEE continues producing longer into the spring than other legumes, extending cuttings into late May or early June in Mississippi (225).

MULTICUT berseem clover averaged 280 lb. N/A in a six-year California trial.

MANAGEMENT

Establishment

Berseem prefers slightly alkaline loam and silty soils but grows in all soil types except sands. Soil phosphorus can limit berseem clover growth. Fertilize with 60 to 100 lb. P_2O_5/A if soil tests below 20 ppm (162). Boron also may limit growth, so test soil to maintain levels (278). Berseem tolerates saline conditions better than alfalfa and red clover (120). Use R-type inoculant suitable for berseem clover and crimson clovers.

Broadcast or drill berseem seed alone or with spring grains onto a firm, well-prepared seedbed or closely cropped sod so that it is 1/4-inch deep with a light soil covering. To improve seed-soil contact and to maintain seed-zone moisture, cultipack or roll soil before and after broadcast seeding (162). Dry, loose soil will suppress germination.

Recommended seeding rates are 8 to 12 lb./A drilled or 15 to 20 lb./A broadcast. Excessive rates will create an overly thick stand that prevents tillering and spreading of the root crowns. Montana trials set the optimum seeding rate at about 8 lb./A drilled in 12-inch rows, with a higher rate in narrower rows where herbicides are not used to control weeds (442).

BERSEEM CLOVER (Trifolium alexandrinum)

Midwest. Seed after April 15 to avoid crop loss due to a late frost. Berseem frostseeded at 15 lb./A yields well in the upper Midwest. In southern Michigan, frostseeded berseem clover produced 1.5 T dry matter/A and 85 lb. N/A (373, 376), but frost risk is significant.

Iowa tests over four years showed that inter-seeded berseem and oats averaged 76 percent more dry matter (ranging from 19 to 150 percent) than oats alone. Underseeding berseem clover did not significantly reduce oat yields in another Iowa study. Seed early- to mid-April in Iowa (159).

When seeding a mixture, harvest goals affect variety selection and seeding rates, Iowa researchers have found. If establishment of an optimum berseem clover stand **for green manure** is most important, oat or other small grain crop seeded at about 1 bushel per acre will protect the young clover and help to break the soil crust. If **early forage** before green manuring is the goal, seed a mixture of 4 bu. oats and 15 lb.berseem/A. If **biomass quantity** is foremost, use a short-stalked, long-season oat. If **oat grain** production is primary, keep oat seeding rate the same, but select a short-season, tall variety to reduce the likelihood of berseem clover interfering with grain harvest (156).

Berseem clover also can be a late-summer crop. Planted in mid-August in the Corn Belt, it should grow about 15 inches before frost, provide winter erosion protection and break down quickly in spring to deliver N from its topgrowth and roots.

You can overseed berseem clover into standing small-grain crops, a method that has worked well in a series of on-farm tests in Iowa (155). Plant the berseem as late as three weeks after the grain crop germinates or after the tillering stage of winter-seeded small grains. Use a heavy seeding rate to compensate for reduced seed-soil contact. Frostseeding in late winter into winter wheat has not worked in several attempts in Pennsylvania (361) and Iowa.

Southeast. Fall planting in mild regions provides effective weed control as well as N and organic matter for a spring crop. Seed Aug. 25 to Oct. 15 in Mississippi or up to Dec.1 in Florida. For a cool-season grass mixture, plant 12 lb. berseem clover seed with 10 lb. orchardgrass or 20 lb. annual or perennial ryegrass/A (225).

West. Berseem does best in California's Central Valley when planted by the first or second week of October. If planting is delayed until November, seedlings will start more slowly in the cool of winter (162).

Field Management

Mowing for green manure. Clip whenever plants are 12 to 15 inches tall and basal shoots begin to grow. This will be 30 to 60 days after planting, depending on weather, field and moisture conditions. Mow again every 25 to 30 days to encourage growth of up to 4 T/A. Keep stubble height at least 3 to 4 inches tall, because plants regrow from lower stem branches.

To maximize dry matter production, cut as soon as basal bud regrowth reaches 2 inches (162). At the latest, clip before early flowering stage or plants will not regrow. Berseem clover responds best when field traffic is minimized (156).

Mowed berseem clover left in the field as green manure can hinder regrowth of the legume from its lower stems. To lessen this problem, flail or sickle-bar mow then rake or fluff with a tedder at intervals until regrowth commences.

Remember that berseem clover has a tap root and shallow 6- to 8-inch feeder root system (156). In thin plantings or well-drained soils, it can be susceptible to drought, a trait that could trigger mowing, grazing or killing earlier than originally planned (186).

Abundant soil N will restrict N fixation by berseem clover, but moderate amounts up to 150 lb. N/A did not limit annual fixation in north central California. Researchers explain that berseem clover draws heavily on soil N during early growth. When soil N was depleted in this test, berseem began fixing N rapidly until it produced seed and died (447).

Berseem made its N contribution to soil in the final third of its cutting cycle—regardless of initial soil N availability—in all six years of the study. Nitrogen fixation was closely correlated to a drop in water-use efficiency in the trial. After producing from 400 to 640 lb. of dry matter per acre-inch of water in the first four cuttings, production dropped to 300 lb. DM/A-in. for the final two cuttings (447).

Small grain companion. Underseeded berseem clover provided about 1.2 T forage dry matter/A after oat harvest in Iowa. Removing the forage decreases the soil-saving ground cover and N contribution (159), trading soil and N benefits for attractive near-term income.

In the Midwest, greenchop an oat/berseem clover mixture when oat is at the pre-boot stage to avoid berseem clover going to seed early and, therefore, not producing maximum nitrogen. Oats have high crude protein at this stage. Monitor carefully during warm periods to avoid nitrogen toxicity.

A Montana study found that spring plantings of berseem clover will produce the most legume dry matter and N if clear seeded. If, however, you wish to maximize *total* dry matter and protein, seeding with oats is recommended. The oat nurse crop suppressed weeds well and increased total dry matter production by 50 to 100 percent regardless of whether plots were cut two, three or four times (434).

Killing

Berseem dies when exposed to temperatures below 20° F for several days, making winterkill a virtual certainty in Zone 7 and colder. This eliminates the need for herbicides or mechanical killing after a cold winter, and hastens delivery of nutrients to the soil.

The least winter hardy of the true annual clovers, berseem can be planted with oats in the Corn Belt and produce abundant N before winterkilling.

To kill berseem clover ahead of fall-planted crops, wait for it to die after blooming, use multiple diskings or apply herbicides. In mild areas, berseem clover grows vigorously through late spring. BIGBEE berseem clover remained vegetative until early May or later in an experiment at a northern Mississippi (Zone 7) site. Until it reaches full bloom, it will require either tillage or a combination of herbicides and mechanical controls to kill it.

In a northern Mississippi mechanical control study, BIGBEE berseem clover added the most dry matter after mid-April compared to hairy vetch, MT. BARKER subterranean clover and TIBBEE crimson clover. Berseem and hairy vetch remained vegetative until mid-May, but by early May, berseem clover and crimson had a considerable amount of stems laying down (105).

Rolling with 4-inch rollers killed less berseem clover than hairy vetch or crimson when the legumes had more than 10 inches of stem laying on the ground. Kill rate was more than 80 percent for the latter two crops, but only 53 percent for berseem clover. Without an application of atrazine two weeks prior to either flail mowing or rolling with coulters, the mechanical controls failed to kill more than 64 percent of the berseem clover until early May, when flailing achieved 93-percent control. Atrazine alone reduced the stand by 68 percent in early April, 72 percent in mid-April and 88 percent in early May (105).

Nodulation: Match Inoculant to Maximize N

With the help of nitrogen-fixing bacteria, legume cover crops can supply some or all of the N needed by succeeding crops. This nitrogen-producing team can't do the job right unless you carefully match the correct bacterial inoculant with your legume cover crop species.

Like other plants, legumes need nitrogen to grow. They can take it from the soil if enough is present in forms they can use. Legume roots also seek out specific strains of soil-dwelling bacteria that can "fix" nitrogen gas from the air for use by the plant. While many kinds of bacteria compete for space on legume roots, the root tissues will only begin this symbiotic N-fixing process when they encounter a specific species of rhizobium bacteria. Only particular strains of rhizobia provide optimum N production for each group of legumes.

When the root hairs find an acceptable bacterial match, they encircle the bacteria to create a nodule. These variously shaped lumps on the root surfaces range in size from a BB pellet to a kernel of corn. Their pinkish interiors are the visible sign that nitrogen fixation is at work.

Nitrogen gas (N_2) from air in the spaces between soil particles enters the nodule. The bacteria contribute an enzyme that helps convert the gas to ammonia (NH_3). The plant uses this form of N to make amino acids, the building blocks for proteins. In return, the host legume supplies the bacteria with carbohydrates to fuel the N-fixation process.

The rate of N fixation is determined largely by the genetic potential of the legume species and by the amount of plant-available N in the soil. Other environmental factors such as heat and moisture play a big role, as well. Fueling N fixation is an expensive proposition for the legume host, which may contribute up to 20 percent of its carbohydrate production to the root-dwelling bacteria. If the legume can take up free N from the soil, it won't put as much energy into producing nodules and feeding bacteria to fix nitrogen from the air.

Perennial legumes fix N during any time of active growth. In annual legumes, N fixation peaks at flowering. With seed formation, it ceases and the nodules slough from the roots. Rhizobia return to the soil environment to await their next encounter with legume roots. These bacteria remain viable in the soil for three to five years, but often at too low a level to provide optimum N-fixation when legumes return to the field.

If legume roots don't encounter their ideal bacterial match, they work with the best strains they can find. They just don't work as efficiently together and they produce less N. Inoculating seeds with the correct strain before planting is inexpensive insurance to make sure legumes perform up to their genetic potential. Clover inoculum, for example, costs just a few cents per pound of seed treated, or more for an enhanced sticker that buffers and feeds the seedling.

Pest Management

Avoid direct seeding small-seeded vegetables into fields where you have incorporated berseem clover within the past month, due to allelopathic compounds in the residue. Berseem clover, crimson clover and hairy vetch residue incorporated directly into the seed zone may suppress germination and seedling development of onion, carrot and tomato, based on lab tests (40).

Lygus bugs have been a serious problem in California seed production, and virus outbreaks can cause serious damage during wet springs where berseem grows as a winter annual. Where virus is a concern, use JOE BURTON, a resistant cul-

While they are alive, legumes release little or no nitrogen to the soil. The N in their roots, stalks, leaves and seeds becomes available when the plants die naturally or are killed by tillage, mowing or herbicide. This plant material becomes food for microbes, worms, insects and other decomposers.

Microorganisms mineralize, or convert, the complex "organic" forms of nitrogen in the plant material into inorganic ammonium and nitrate forms, once again making the N available to plants. How quickly the mineralization of N occurs is determined by a host of environmental and chemical factors. These will affect how much of that legume N is available to the next crop or has the potential to leach from the soil.

For more information about mineralization and how much you can reduce your N fertilizer rate for crops following legumes, see *How Much N?* (p. 22).

To get the most from your legume/bacteria combination:

- **Choose appropriate legume** species for your climate, soils and cropping system. Also, consider the amount of N it can deliver when you will need it.
- **Match inoculant** to the species of legume you are growing. See Chart 3B, *Planting* (p. 70) to determine the best inoculant to use.
- **Coat seed** with the inoculant just before planting. Use milk, weak sugar water or a commercial sticking agent to help the material stick to the seeds. Use only fresh inoculant (check the package's expiration date), and do not expose packages or inoculated seed to excessive heat or direct sunlight.

Mix the sticker with non-chlorinated water and add the inoculant to create a slurry, then thoroughly coat seeds. Seed should be dry enough to plant within half an hour.

Re-inoculate if you don't plant the seed within 48 hours. Mix small quantities in a five-gallon bucket or tub, either by hand or using a drill equipped with a paint-mixer attachment. For larger quantities, use special inoculant mixing hoppers or a cement mixer without baffles.

Gum arabic stickers with sugars and liming agents boost the chances for optimum nodulation over water-applied inoculant alone. Pre-inoculated ("rhizo-coated") seed weighs about one-third more than raw seed, so increase seeding rates accordingly.

Check nodulation as the plants approach bloom stage. Push a spade in the soil about 6 inches below the plant. Carefully lift the plant and soil, gently exposing roots and nodules. (Yanking roots from the soil usually strips off nodules). Wash gently in a bucket of water to see the extent of nodulation. Slice open nodules. A pink or reddish interior indicates active N-fixation. Remember, an overabundance of soil nitrogen from fertilizer, manure or compost can reduce nodulation.

For more information about nodulation, see two books by Marianne Sarrantonio: *Northeast Cover Crop Handbook* (361) and *Methodologies for Screening Soil-Improving Legumes* (360).

tivar. BIGBEE is susceptible to crown rot and other root diseases common to forage legume species (162).

Berseem, like other clovers, shows little resistance to root-knot nematode (*Meloidogyne* spp.). It seems to be particularly favored by rabbits (361).

Crop Systems

Flexible oats booster. In the Corn Belt, berseem clover seeded with oats helps diversify corn>soybean rotations, breaks pest cycles and provides some combination of grain and/or forage harvest, erosion control and N to the following corn crop. An added benefit is that it requires no tillage or herbicide to kill it in spring (159). Plant 4 bu. oats with 12 lb. berseem/A.

In a four-year Iowa study, planting berseem clover with oats increased net profit compared with oats alone. The clover was baled for forage and the underseeded oats were harvested for grain. Not calculated in the benefit were the 40 to 60 lb. N/A provided to the following corn crop or other soil-improvement benefits. The oats/ berseem mix produced 70 percent more biomass, increased subsequent corn yields by 10 percent and reduced weed competition compared with a year of oats alone (159).

Mow strips of berseem between vegetable rows and blow the clippings around the plants for mulch.

Pure berseem clover regrowth averaged 1.2 T dry matter/A, which can be used as forage or green manure. These options could help oats become an economically viable crop for Midwest crop/livestock farms in an era of decreasing government payments for corn and soybeans (159, 160).

Wheat companion. Berseem was one of six legume intercrops that improved productivity and profit of wheat and barley crops in low-N soils under irrigated conditions in northwestern Mexico. All of the legumes (including common and hairy vetch, crimson clover, New Zealand and Ladino white clover, and fava beans) provided multiple benefits without decreasing grain yield of 15 to 60 bu./A on the heavy clay soil.

Wheat and legumes were planted at normal monoculture rates with wheat in double rows about 8 inches apart atop 30-inch beds, and legumes in the furrows. In a second, related experiment, researchers found they could more than double total wheat productivity (grain and total dry matter) by interplanting 24-inch strips of berseem clover or hairy vetch with double rows of wheat 8 inches apart. Control plots showed wheat planted at a greater density did not increase yield (350).

Vegetable overseeding. Berseem can be overseeded into spring vegetables in northern climates where it thrives at moderate temperatures and moisture. Berseem is well suited to a "mow and blow" system where strips of green manure are chopped and transferred to adjacent crop strips as a green manure and mulch (361).

Boost the N plow-down potential of old pastures or winter-killed alfalfa by no-tilling or interseeding berseem clover. Or, broadcast seed then incorporate with light harrowing.

COMPARATIVE NOTES

Berseem clover is:

- Not as drought-tolerant as alfalfa. Some cultivars can tolerate more soil moisture—but not waterlogging—than alfalfa or sweet clover
- Similar in seed size to crimson clover
- Bee-friendly because its white or ivory blossoms have no tripping mechanism.
- Because of its short roots, berseem clover does not use phosphorus to the depth that mature, perennial alfalfa does.
- Winterkilled berseem allows for earlier spring planting than winter-hardy annuals. As a dead organic mulch, it poses no moisture depletion risk, but may slow soil warming and drying compared to erosion-prone bare fallow.

Cultivars. BIGBEE berseem clover was selected from other traditional cultivars for its cold-tolerance, which is similar to crimson clover. Some of the strong winter production tendency found in non-winter hardy berseem clover was sacrificed to obtain BIGBEE's winter hardiness (162). Mature BIGBEE plants hold their seeds well and produce adequate hard seed for reseeding. Other berseem clover cultivars have less hard seed and will not dependably reseed (278).

California tests show MULTICUT berseem clover produces 20 to 25 percent more dry matter than BIGBEE. It has greater N-fixing ability, blooms later, and has a longer growing period than other varieties, but is not as cold tolerant as BIGBEE (162). JOE BURTON, developed from MULTICUT, is more cold tolerant.

In California, BIGBEE begins to flower in mid-May, about two weeks ahead of MULTICUT. MULTICUT grows faster and produces more dry matter in California conditions, averaging about 1.6 T/A more in a six-year study. When the five or six cuttings per year were clipped and removed, MULTICUT was about 6 inches taller at each clipping than other varieties (447). In Montana tests, BIGBEE out-yielded MULTICUT in eight of 13 locations (381).

Seed sources. See *Seed Suppliers* (p. 195). 

COWPEAS

Vigna unguiculata

Also called: southern peas, black-eye peas, crowder peas

Type: summer annual legume

Roles: suppress weeds, N source, build soil, prevent erosion, forage

Mix with: sorghum-sudangrass hybrid or foxtail hay-type millet for mulch or plow-down before vegetables; interseeded with corn or sorghum

See charts, pp. 66 to 72, for ranking and management summary.

Cowpeas are the most productive heat-adapted legume used agronomically in the U.S. (275). They thrive in hot, moist zones where corn flourishes, but require more heat for optimum growth (263). Cowpea varieties have diverse growth habits. Some are short, upright bush types. Taller, viny types are more vigorous and better suited for use as cover crops. Cowpeas protect soil from erosion, smother weeds and produce 100 to 150 lb. N/A. Dense residue helps to improve soil texture but breaks down quickly in hot weather. Excellent drought resistance combined with good tolerance of heat, low fertility and a range of soils make cowpeas viable throughout the temperate U.S. where summers are warm or hot but frequently dry.

Cowpeas make an excellent N source ahead of fall-planted crops and attract many beneficial insects that prey on pests. Used in California in vegetable systems and sometimes in tree crops, cowpeas also can be used on poor land as part of a soil-building cover crop sequence.

BENEFITS

Weed-smothering biomass. Drilled or broadcast cowpea plantings quickly shade the soil to block out weeds. Typical biomass production is 3,000 to 4,000 lb./A (361). Cowpeas produced about 5,100 lb. dry matter/A in a two-year Nebraska screening of cover crops while soybeans averaged about 7,800 lb. DM/A in comparison plots (332).

Thick stands that grow well can outcompete bermudagrass where it does not produce seed and has been plowed down before cowpea planting (263). In New York, both cowpea and soybean provided some weed-suppressing benefits. Neither adequately controlled weeds, but mixing with buckwheat or sorghum-sudangrass improved performance as a weed management tool (43).

In California, cowpea mulch decreased weed pressure in fall-planted lettuce, while incorporated cowpea was less effective. An excellent desert cover crop, cowpea also reduced weed pressure in California pepper production (209).

Cowpeas thrive under hot, moist conditions, but also tolerate drought and low soil fertility.

The weed-suppressing activity of cowpea may be due, in part, to allelopathic compounds in the residue. The same compounds could adversely impact your main crop. Be sure to consult local information about impacts on cash crops.

Quick green manure. Cowpeas nodulate profusely, producing an average of about 130 lb. N/A in the East, and 200 lb. N/A in California. Properly inoculated in nitrogen deficient soils, cowpeas can produce more than 300 lb. N/A (120). Plowdown often comes 60 to 90 days after planting in California (275). Higher moisture and more soil N favor vegetative growth rather than seed production. Unlike many other grain legumes, cowpeas can leave a net gain of nitrogen in the field even if seed is harvested (361).

IPM insectary crop. Cowpeas have "extrafloral nectaries"—nectar-release sites on petioles and leaflets—that attract beneficial insects, including many types of wasps, honeybees, lady beetles, ants and soft-winged flower beetles (422). Plants have long, slender round pods often borne on bare petioles above the leaf canopy.

Intercropping cotton with cowpeas in India increased levels of predatory ladybugs and parasitism of bollworms by beneficial wasps. Intercropping with soybeans also increased parasitism of the bollworms compared with plots intercropped with onions or cotton without an intercrop. No effects on overall aphid, leafhopper or bollworm populations were observed (422).

Companion crop. Thanks to its moderate shade tolerance and attractiveness to beneficial insects, cowpeas find a place in summer cover crop mixtures in orchards and vineyards in the more temperate areas of California. Avoid use under a heavy tree canopy, however, as cowpeas are susceptible to mildew if heavily shaded (263). As in much of the tropical world where cowpeas are a popular food crop, they can be underseeded into corn for late-season weed suppression and post-harvest soil coverage (361).

Seed and feed options. Cowpea seed (yield range 350 to 2,700 lb./A) is valued as a nutritional supplement to cereals because of complementary protein types. Seed matures in 90 to 140 days. Cowpeas make hay or forage of highest feed value when pods are fully formed and the first have ripened (120). A regular sickle-bar mower works for the more upright-growing cultivars (120, 422). Crimping speeds drying of the rather fleshy stems to avoid over-drying of leaves before baling.

Low moisture need. Once they have enough soil moisture to become established, cowpeas are a rugged survivor of drought. Cowpeas' delayed leaf senescence allows them to survive and recover from midseason dry spells (21). Plants can send taproots down nearly 8 feet in eight weeks to reach moisture deep in the soil profile (107).

Cultivars for diverse niches. Cover crop cultivars include CHINESE RED, CALHOUN and RED RIPPER, all viny cultivars noted for superior resistance to rodent damage (317). IRON CLAY, a mixture of two formerly separate cultivars widely used in the Southeast, combines semi-bushy and viny plants and resistance to rootknot nematodes and wilt.

Most of the 50-plus commercial cowpea cultivars are horticultural. These include "crowder peas" (seeds are crowded into pods), grown throughout the temperate Southeast for fresh processing, and "blackeye peas," grown for dry seed in California. Watch for the release of new varieties for cover crop use.

Use leafy, prostrate cultivars for the best erosion prevention in a solid planting. Cultivars vary significantly in response to environmental conditions. Enormous genetic diversity in more than 7,000 cultivars (120) throughout West Africa, South America and Asia suggests that breeding for forage production would result in improved cultivars (21, 422) and cover crop performance.

Easy to establish. Cowpeas germinate quickly and young plants are robust, but they have more difficulty emerging from crusted soils than soybeans.

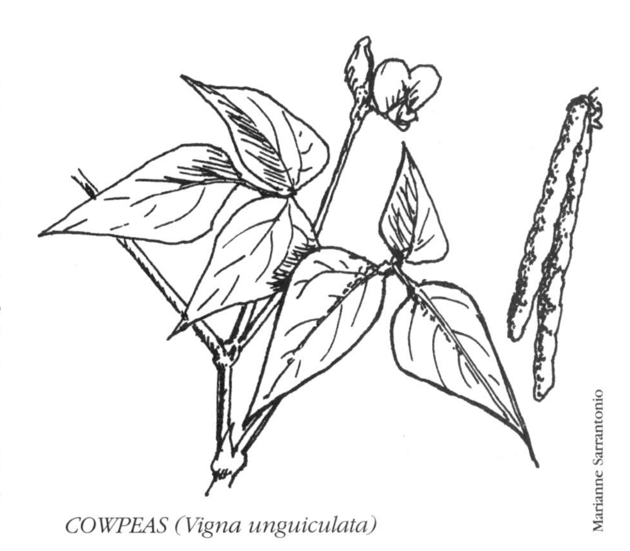

COWPEAS (Vigna unguiculata)

Marianne Sarrantonio

MANAGEMENT

Establishment

Don't plant cowpeas until soil temperature is a consistent 65° F and soil moisture is adequate for germination—the same conditions soybeans need. Seed will rot in cool, wet soils (107). Cowpeas for green manure can be sown later in summer (361), until about nine weeks before frost. Cowpeas grow in a range of well-drained soils from highly acid to neutral, but are less well adapted to alkaline soils. They will not survive in waterlogged soils or flooded conditions (120).

In a moist seedbed, drill cowpeas 1 to 2 inches deep at about 30 to 90 lb./A, using the higher rate in drier or cooler areas or for larger-seeded cultivars (361, 422). While 6- to 7-inch row spacings are best for rapid groundcover or a short growing season, viny types can be planted in 15- to 30-inch rows. Pay particular attention to pre-plant weed control if you go with rows, using pre-cultivation and/or herbicides.

If you broadcast seed, increase the rate to about 100 lb./A and till lightly to cover seed. A lower rate of 70 lb./A can work with good moisture and effective incorporation (361). Broadcast seeding usually isn't as effective as drilling, due to cowpeas' large seed size. You can plant cowpeas after harvesting small grain, usually with a single disking if weed pressure is low. No-till planting is also an option. Use special "cowpea" inoculant which also is used for sunn hemp (*Crotolaria juncea*), another warm-season annual legume. See *Up-and-Coming Cover Crops* (p. 191).

Field Management

Cowpea plants are sometimes mowed or rolled to suppress regrowth before being incorporated for green manure. It's best to incorporate cowpeas while the entire crop is still green (361) for quickest release of plant nutrients. Pods turn cream or brown upon maturity and become quite brittle. Stems become more woody and leaves eventually drop.

Crop duration and yield are markedly affected by night and day temperatures as well as day length. Dry matter production peaks at temperatures of 81° F day and 72° F night (120).

Killing

Mowing at any point stops vegetative development, but may not kill plants without shallow tillage. Mowing and rolling alone do not consistently kill cowpeas (95). Herbicides can also be used. If allowed to go to seed, cowpeas can volunteer in subsequent crops.

Pest Management

Farmers using cowpeas as cover crops do not report problems with insects that are pests in commercial cowpea production, such as Lygus bugs and 11-spotted cucumber beetle (95, 83). Insect damage to cowpea cover crops is most likely to occur at the seedling stage.

Cowpeas Provide Elegant Solution to Awkward Niche

PARTRIDGE, Kan.—Cowpeas fill a rotational rough spot between milo (grain sorghum) and wheat for Jim French, who farms about 640 acres near Partridge, Kan.

"I miss almost a full season after we take off the milo in late October or November until we plant wheat the following October," says French. "Some people use cash crops such as oats or soybeans. But with cowpeas, I get wind erosion control, add organic matter to improve soil tilth, save on fertilizer and suppress weeds for the wheat crop. Plus I have the options of haying or grazing."

He chisel plows the milo stubble in late April, disks in May and field cultivates just before planting about the first week of June. He drills 30 to 40 lb./A of CHINESE RED cowpeas 1 to 2 inches deep when soil temperature reaches 70° F. Growth is rapid, and by early August he kills the cowpeas by making hay, having his cattle graze them off or by incorporating them for maximum soil benefit.

French says cowpeas usually produce about 90 to 120 lb. N/A—relatively modest for a legume cover—but he feels his soil greatly benefits from the residue, which measured 8,000 lb./A in one of his better fields. He disks the sprawling, leafy legume once, then does a shallow chisel plowing to stop growth and save moisture. Breakdown of the somewhat tough stems depends on moisture.

When he leaves all the cowpea biomass in the field, he disks a second time to speed decomposition. He runs an S-tine field cultivator 1 to 2 inches deep just before planting wheat to set back fall weeds, targeting a 20 to 25 percent residue cover. The cowpeas improve rainfall infiltration and the overall ability of the soil to hold moisture.

French observes that the timing of rainfall after cowpea planting largely determines the weediness of the cover crop. "If I get a week to 10 days of dry weather after I plant into moisture, the cowpeas will out-compete the weeds. But if I get rain a few days after planting, they'll be weedy."

French manages his legumes to stay in compliance with USDA farm program provisions. The Freedom to Farm Act allows vegetables used as green manure, haying or grazing to be planted on program acres, but prohibits planting vegetables for seed harvest on those acres. The rules list cowpeas as a vegetable, even though different cultivars are used for culinary production. Use of grain legumes such as lentils, mung beans and dry peas (including Austrian winter peas) is not restricted by the act, opening flexible rotation options.

French cooperated with Rhonda Janke of Kansas State University to define soil health more precisely. He can tell that covers improve the "flow" of his soil, and he is studying root growth after covers. But he feels her work measuring enzymes and carbon dioxide levels will give farmers new ways to evaluate microbial activity and overall soil health.

Once cowpea plants form pods, they may attract stinkbugs, a serious economic pest in parts of the lower Southeast. However, no significant stinkbug presence was reported in three years of screening in North Carolina. If stinkbugs are a concern, remember these points:

- Flail mowing or incorporating cowpeas at pod set will prevent a stinkbug invasion. By that time, cowpeas can provide good weed suppression and about 90 percent of their nitrogen contribution. However, waiting too long before mowing or incorporation will flush stinkbugs into adjacent crops. Leaving remnant strips of cowpeas to attract stinkbugs may reduce movement into other crops, as long as the cowpeas keep producing enough new pods until the cash crop is no longer threatened.

- Plan crop rotations so the preceding, adjacent and succeeding crops are not vulnerable or are resistant to stinkbugs.
- If you plan to use an insecticide to control another pest, the application may also help manage stinkbugs.

No cowpea cultivar is resistant to root rot, but there is some resistance to stem rot. Persistent wet weather before development of the first true leaf and crowding of seedlings due to poor seed spacing may increase damping off. To reduce disease and nematode risks, rotate with four or five years of crops that aren't hosts. Also plant seed into warm soils and use certified seed of tolerant varieties (107). IRON and other nematode-tolerant cowpea cultivars reduced soybean cyst and root-knot nematode levels in greenhouse experiments (422). Despite some research (422) showing an increased nematode risk after cowpeas, California farmers report no such problem.

Crop Systems

Cowpeas' heat-loving nature makes them an ideal mid-summer replenisher of soil organic matter and mineralizable nitrogen. Cowpeas set pods over a period of several weeks. Viny varieties continue to increase dry matter yields during that time.

A mix of 15 lb. cowpeas and 30 lb. buckwheat/A makes it possible to incorporate the cover crop in just six weeks while still providing some nitrogen. Replacing 10 percent of the normal cowpea seeding rate with a fast-growing, drought-tolerant sorghum-sudangrass hybrid increases dry matter production and helps support the cowpea plants for mowing. Cowpeas also can be seeded with other tall annual crops such as pearl millet. Overseeding cowpeas into nearly mature spring broccoli in June in Zones 5 and 6 of the Northeast suppresses weeds while improving soil (361). Planting cowpeas in late June or early July in the upper Midwest after spring canning peas provides green manure or an emergency forage crop (422).

Cowpeas can fill a mid-summer fallow niche in inland North Carolina between spring and summer vegetable crops. A mix of IRON AND CLAY cowpeas (50 lb./A) and German millet (15 lb./A) planted in late June can be killed mechanically before no-till transplanted fall broccoli. In several years of screening trials at the same sites, cowpea dry matter (3,780 lb./A) out yielded soybeans (3,540 lb. DM/A), but plots of sesbania (*Sesbania exaltata*) had top yields at about 5,000 lb. DM/A (95).

Unlike many grain legumes, cowpeas can leave a net N gain even if seeds are harvested.

COMPARATIVE NOTES

Cowpeas are more drought tolerant than soybeans, but less tolerant of waterlogging (361) and frost (263). Sown in July, the cowpea canopy closed more rapidly and suppressed weeds better than lespedeza (*Lespedeza cuneata*), American jointvetch (*Aeschynomene americana*), sesbania and alyceclover (*Alysicarpus* spp.), the other warm-season legumes tested (422). Cowpeas perform better than clovers and alfalfa on poor or acid soils. Cowpea residue breaks down faster than white sweetclover (361) but not as fast as Austrian winter peas.

Warm-season alternatives to cowpeas include two crops that retain some cowpea benefits. Buckwheat provides good beneficial habitat and weed control without attracting stinkbugs. Velvetbeans (*Mucuna deeringiana*) provide nitrogen, soil protection and late-season forage in hot, long-season areas. They do not attract stinkbugs and are resistant to nematodes (107).

Cultivars. See *Cultivars for diverse niches* (p. 126).

Seed sources. See *Seed Suppliers* (p. 195). 

CRIMSON CLOVER

Trifolium incarnatum

Type: winter annual or summer annual legume

Roles: N source, soil builder, erosion prevention, reseeding inter-row ground cover, forage

Mix with: rye and other cereals, vetches, annual ryegrass, subclover, red clover, black medic

See charts, p. 66 to 72, for ranking and management summary.

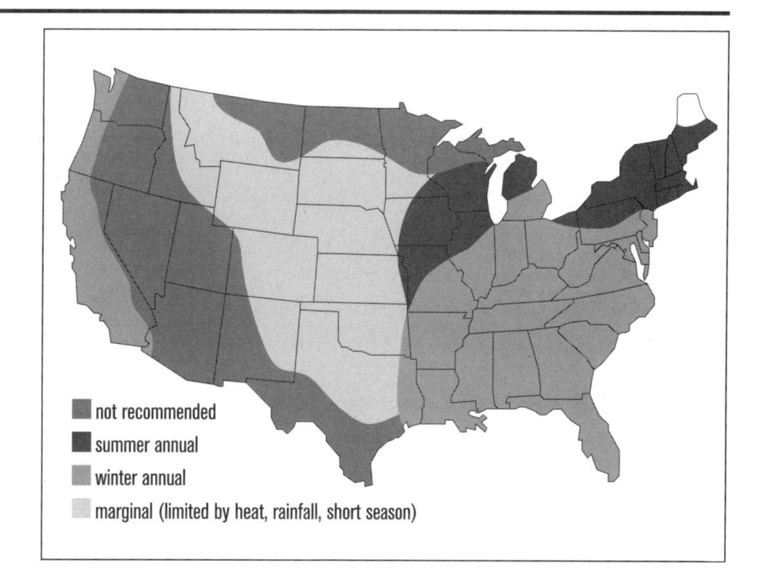

With its rapid, robust growth, crimson clover provides early spring nitrogen for full-season crops. Rapid fall growth, or summer growth in cool areas, also makes it a top choice for short-rotation niches as a weed-suppressing green manure. Popular as a staple forage and roadside cover crop throughout the Southeast, crimson clover is gaining increased recognition as a versatile summer-annual cover in colder regions. Its spectacular beauty when flowering keeps it visible even in a mix with other flowering legumes, a common use in California nut groves and orchards. In Michigan, it is used successfully between rows of blueberries.

BENEFITS

Nitrogen source. Whether you use it as a spring or fall N source or capitalize on its vigorous reseeding ability depends on your location. Growers in the "crimson clover zone"—east of the Mississippi, from southern Pennsylvania and southern Illinois south—choose winter annual crimson clover to provide a strong, early N boost. In Hardiness Zone 8—the warmer half of the Southeast—crimson clover will overwinter dependably with only infrequent winterkill. Its N contribution is 70 to 150 lb./A.

Reseeding cultivars provide natural fertility to corn and cotton. Crimson clover works especially well before grain sorghum, which is planted later than corn. It is being tested extensively in no-till and zone-till systems. One goal is to let the legume reseed yearly for no-cost, season-long erosion control, weed suppression and nitrogen banking for the next year.

Along the northern edge of the "crimson clover zone," winterkill and fungal diseases will be more of a problem. Hairy vetch is the less risky overwintering winter annual legume, here and in northern areas. Crimson clover often can survive winters throughout the lower reaches of Zone 6, especially from southeastern Pennsylvania northeast to coastal New England (195).

Crimson clover is gaining popularity as a winter-killed annual, like oats, in Zones 5 and colder. Planted in late summer, it provides good groundcover and weed control as it fixes nitrogen from the atmosphere and scavenges nitrogen from the soil. Its winterkilled residue is easy to manage in spring.

Biomass. As a winter annual, crimson clover can produce 3,500 to 5,500 lb. dry matter/A and fix 70 to 150 lb. N/A by mid-May in Zone 8 (the inland Deep South). In a Mississippi study, crimson clover had produced mature seed by April 21, as well as 5,500 lb. DM and 135 lb. N/A.The study concluded that crimson clover is one of several winter annual legumes that can provide adequate but not excessive amounts of N for southern grain sorghum production (22, 36, 105). Crimson clover has produced more than 7,000 lb. DM/A several times at a USDA-ARS site in Beltsville, Md., where it produced 180 lb. N and 7,800 lb. DM/A in 1996 (412).

In field trials of six annual legumes in Mississippi, crimson clover was found to produce the most dry matter (5,600 to 6,000 lb./A) compared to hairy vetch, bigflower vetch, berseem clover, arrowleaf clover (*Trifolium vesiculosum*) and winter peas. It produced 99 to 130 lb. N/A and is recommended for soil erosion control because of its high early-autumn dry matter production (426).

As a summer annual in lower Michigan, a midsummer planting of crimson clover seeded at 20 lb./A produced 1,500–2,000 lb. dry matter and 50-60 lb. N/A by late November (270).

Mixtures. Crimson clover grows well in mixtures with small grains, grasses and other clovers. An oats crop is a frequent companion, either as a nurse crop to establish a clear stand of crimson clover, or as a high-biomass, nutrient-scavenging partner. In California, crimson clover is planted with rose clover and medics in orchards and nut groves to minimize erosion and provide some N to tree crops (422).

Beneficial habitat and nectar source. Crimson clover has showy, deep red blossoms 1/2 to 1 inch long. They produce abundant nectar, and are visited frequently by various types of bees. The blooms may contain many minute pirate bugs, an important beneficial insect that preys on many small pests, especially thrips (422). In Michigan, crimson increased blueberry pollination when planted in row middles. Georgia research shows that crimson clover sustains populations of pea aphids and blue alfalfa aphids. These species are not pests of pecans, but provide alternative food for beneficial predators such as lady beetles, which later attack pecan aphids.

CRIMSON CLOVER (Trifolium incarnatum)

Nutrient cycler. Crimson clover adds to the soil organic N pool by scavenging mineralized N and by normal legume N fixation. The scavenging process, accomplished most effectively by grasses, helps reduce the potential for N leaching into groundwater during winter and spring (181, 265). Mixed with annual ryegrass in a simulated rainfall study, crimson clover reduced runoff from the herbicide lactofen by 94 percent and norflurazon and fluometuron by 100 percent (346). The grass/legume mixture combines fibrous surface roots with short tap roots.

MANAGEMENT

Establishment & Fieldwork

Crimson clover will grow well in about any type of well drained soil, especially sandy loam. It may fare poorly on heavy clay, waterlogged, extremely acid or alkaline soils. Once established, it thrives in cool, moist conditions. Dry soil often hinders fall plantings in the South.

Inoculate crimson clover if it hasn't been grown before. Research in Alabama showed that deficiencies of phosphorus or potassium—or strongly acidic soil with a pH of less than 5.0—can virtually shut down N fixation. Nodules were not even formed at pH 5.0 in the test. Phosphorus deficiency causes many small but inactive nodules to form (188).

Winter annual use. Seed six to eight weeks before the average date of first frost at 15 to 18 lb./A drilled, 22 to 30 lb./A broadcast. As with other winter legumes, the ideal date varies with elevation. In North Carolina, for example, the recommended seeding dates are three weeks later along the coast than in the mountains.

Don't plant too early or crimson clover will go to seed in the fall and not regrow in spring until the soil warms up enough to germinate seeds. Early to mid-August seeding is common in the northern part of crimson clover's winter-annual range. In southern Michigan (Zone 5b - 6a) crimson clover, no-tilled into wheat stubble in mid-July, not only grew well into fall, but thrived the following spring, performing nearly as well as hairy vetch (270).

In Hardiness Zone 5 and colder, crimson clover can provide a winterkilled mulch.

While October plantings are possible in the lower Mississippi Delta, an August 15 planting in a northern Mississippi test led to higher yields than later dates (228). In the lower Coastal Plain of the Gulf South, crimson clover can be planted until mid-November.

Nutrient release from crimson clover residue—and that of other winter annual legumes—is quicker if the cover crop is tilled lightly into the soil. Apart from erosion concerns, this fertility enhancing step adds cost and decreases the weed-suppression effect early in the subsequent crop's cycle.

Summer annual use. In general, plant as soon as all danger of frost is past. Spring sowing establishes crimson clover for a rotation with potatoes in Maine. In Michigan, researchers have successfully established crimson clover after short-season crops such as snap beans (229, 270).

In Northern corn fields, Michigan studies showed that crimson clover can be overseeded at final cultivation (layby) when corn is 16 to 24 inches tall. Crimson clover was overseeded at 15 lb./A in 20-inch bands between 30-inch rows using insecticide boxes and an air seeder. The clover established well and caused no corn yield loss (295). Crimson clover has proved to be more promising in this niche than black medic, red clover or annual ryegrass, averaging 1,500 lb. DM/A and more than 50 lb. N/A (270).

In Maine, spring-seeded crimson clover can yield 4,000 to 5,000 lb. DM/A by July, adding 80 lb. N/A for fall vegetables. Mid-July seedings have yielded 5,500 lb./A of weed-suppressing biomass by late October. Summer-annual use is planned with the expectation of winterkill. It sometimes survives the winter even in southern Michigan (270), however, so northern experimenters should maintain a spring-kill option if icy winds and heaving don't do the job.

In California, spring sowing often results in stunting, poor flowering and reduced seed yield, and usually requires irrigation (422).

Rotations. In the South, crops harvested in early fall or sown in late spring are ideal in sequence with crimson clover. Timely planting of crimson clover and its rapid spring growth can enable it to achieve its maximum N contribution, and perhaps reseed. While corn's early planting date and cotton's late harvest limit a traditional winter-annual role for crimson clover, strip planting and zone tillage create new niches. By leaving unkilled strips of crimson clover to mature between zone-tilled crop rows, the legume sets seed in May. The majority of its hard seed will germinate in fall.

Kill crimson clover before seed set and use longer season cultivars where regrowth from hard seed would cause a weed problem.

Researchers have successfully strip-tilled into standing crimson clover when 25 to 80 percent of the row width is desiccated with a herbicide or mechanically tilled for the planting area. Narrower strips of crimson clover increased weed pressure but reduced moisture competition, while wider strips favored reseeding of the cover (236).

In a crimson clover-before-corn system, growers can optimize grain yields by no-tilling into the crimson clover and leaving the residue on the surface, or optimize total forage yield by harvesting the crimson clover immediately before planting corn for grain or silage (204). In Mississippi, sweet potatoes and peanuts suffered no yield or quality penalty when they were no-tilled into killed crimson clover. The system reduced soil erosion and decreased weed competition (35).

In Ohio, crimson clover mixed with hairy vetch, rye and barley provided a fertility enhancing mulch for no-till processing tomato transplants. Use of a prototype undercutter implement with a rolling harrow provided a good kill. Because the wide blades cut just under the soil surface on raised beds, they do not break stalks, thus lengthening residue durability. The long-lasting residue gave excellent results, even under organic management without the herbicides, insecticides or fungicides used on parallel plots under different management regimes. Nancy Creamer at the University of North Carolina is continuing work on the undercutter and on cover crops in organic vegetable systems (96).

Mixed seeding. For cover crop mixtures, sow crimson clover at about two-thirds of its normal rate and the other crop at one third to one-half of its monoculture rate. Crimson clover development is similar to tall fescue. It even can be established with light incorporation in existing stands of aggressive grasses after they have been closely mowed or grazed.

Reseeding. Overwintered crimson clover needs sufficient moisture at least throughout April to produce seed (130). Cultivar selection is critical when early spring maturity is needed.

DIXIE and CHIEF are full-season standards. AU ROBIN and FLAME beat them by about two weeks; a new cultivar, AU SUNRISE, is reportedly 1-3 weeks earlier; the popular TIBBEE is about a week ahead of the standards. Price varies more by seasonal supply than by cultivar.

Killing. Its simple taproot makes crimson clover easy to kill mechanically. Mowing after early bud stage will kill crimson clover. Maximum N is available at late bloom or early seed set, even before the plant dies naturally. Killing earlier yields less N—up to 50 lb. N/A less at its late vegetative stage, which is about 30 days before early seed set (342).

A rolling stalk chopper flattens a mix of crimson clover, hairy vetch and rye ahead of no-till vegetable transplanting at Steve Groff's farm in southeastern Pennsylvania. The crimson is killed completely if it is in full bloom; and even early bloom is killed better than vegetative crimson.

Pest Management

Crimson clover is a secondary host to plant pests of the *Heliothus* species, which include corn earworm and cotton bollworm. Despite its known benefits, crimson clover has been eradicated from many miles of roadsides in Mississippi at the request of some Delta farmers who suspect it worsens problems from those pests (106).

In Mississippi, crimson clover produced mature seed and 135 lb. N/A by April 21.

Crimson clover doesn't significantly increase risk of Southern corn rootworm in no-till corn, while hairy vetch does (67). It is more resistant to diseases (422) and to some nematodes than other clovers (337). Crimson clover is said to tolerate viral diseases, but it succumbed to virus in July plantings in Mississippi (228) and to *Sclerotinia* in fall plantings in Maryland (108).

In lab tests, crimson clover, berseem clover and hairy vetch have been shown to inhibit germination and seedling development of onion, carrot and tomato (40). However, this interference hasn't been observed in North Carolina field crops where strips are mechanically tilled, or in other studies with crimson clover as part of a killed organic mulch. No-till vegetable transplanting has been done successfully on the same day as mechanically killing the cover crop mix on Steve Groff's Lancaster County, Pa., farm with no negative effects.

Wait two to three weeks after killing before planting seeds, to allow the biomass to begin to decompose and the soil biological life to stabilize. During this time, a flush of bacteria such as *Pythium* and *Rhizoctonia* attack rapidly decaying plants. These bacteria also can attack seedling crops. To plant more quickly, mow the clover and use row cleaners to clear the tops from the seed zone. The mow/wait/plant cycle also may be influenced by the need to wait for rain to increase seedbed moisture.

Mixed with hairy vetch, crimson clover attracts beneficial insects, provides nitrogen and suppresses weeds in Oklahoma's native and plantation pecan groves. Both legumes go to seed and then are harvested for forage. Arrowleaf clover provided more biomass and N, but didn't work as well for insect pest management and is very susceptible to root knot nematode.

Crimson clover harbors flower thrips and is a more likely host for tarnished plant bug than hairy vetch or subterranean clover (56). Intensive screenings show less abundant arthropod herbivores and predators on crimson clover than on hairy vetch (206).

Tillage practices and residue management variations (no-till, incorporate, removal) of cover cropped lupin, rye, hairy vetch or crimson clover had little consistent effect on nematodes in north Florida corn fields (264).

Other Options

Pasture and hay crop. Crimson clover is excellent for grazing and haying. It will regrow if grazed or mowed no lower than 3 or 4 inches before the early bud stage. Mixing with grass reduces its relatively low bloat risk even further. Timely mowing four to six weeks before bloom improves growth, reduces lodging and will cause more uniform flowering and seed ripening on highly fertile soils (120, 422).

Crimson clover can be grazed lightly in the fall, more intensively in the spring and still be left to accumulate N and/or set seed with little reduction in its soil N contribution, provided livestock are removed before flowering (80).

COMPARATIVE NOTES

Crimson clover is:

- less tolerant of mowing than are subclovers or medics (422)
- similar to hairy vetch and Austrian winter pea in the Southeast for total N production
- a better weed suppressor in fall than hairy vetch
- earlier to mature in spring than hairy vetch

Cultivars. See *Reseeding* (p. 154) for cultivar comparisons.

Seed Sources. See *Seed Suppliers* (p. 195). 

FIELD PEAS

Pisum sativum subsp. arvense

Also called: Austrian winter peas (black peas), Canadian field peas (spring peas)

Type: summer annual and winter annual legume

Roles: plow-down N source, weed suppressor, forage

Mix with: strong-stemmed wheat, rye, triticale or barley for vertical support

See charts, pp. 66 to 72, for ranking and management summary.

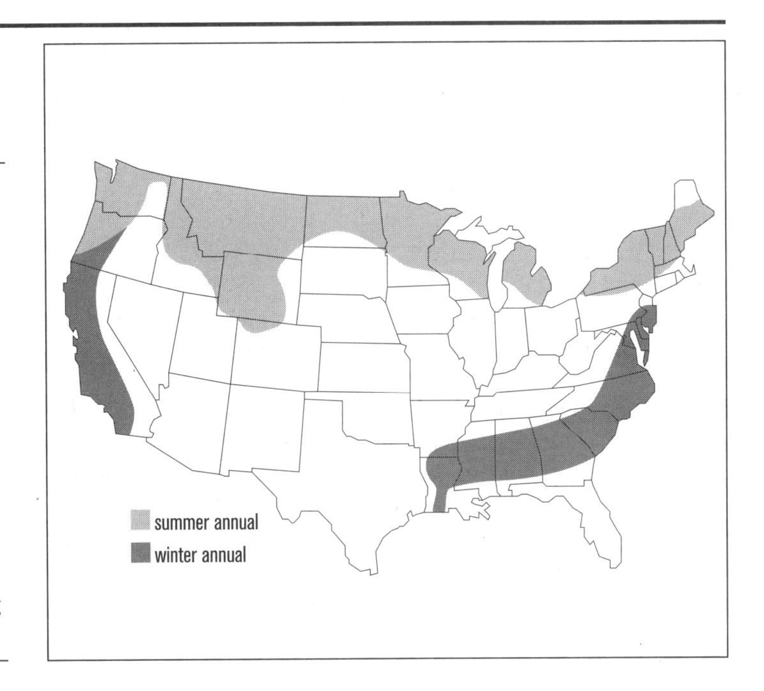

High N-fixers, field peas produce abundant vining forage and contribute to short-term soil conditioning. Succulent stems break down easily and are a quick source of available N (361). Field peas grow rapidly in the cool, moist weather they encounter as winter annuals in the South and in parts of Idaho, and as early-sown summer annuals in the Northeast, North Central and Northern Plains areas. Harvest options as high-quality forage and seed increase their value.

Winter-hardy types of field peas, especially **Austrian winter peas**, can withstand temperatures as low as 10° F with only minor injury, but they don't overwinter consistently in areas colder than moderate Hardiness Zone 6. They are sensitive to heat, particularly in combination with humidity. They tend to languish in mid-summer even in the cool Northeast (361), where average summers have fewer than 30 days exceeding 86° F. Temperatures greater than 90° F cause flowers to blast and reduce seed yield. On humus-rich black soils, field peas will produce abundant viny growth with few seed pods.

Use in the East and Southeast is limited by field peas' susceptibility to *Sclerotinia* crown rot, which can destroy whole fields during winter in the mid-Atlantic area. Risk of infection increases if pea crops are grown on the same land in close rotation.

Canadian field peas are a related strain of vining pea. These annual "spring peas" can outgrow *spring*-planted *winter* peas. They often are seeded with triticale or another small grain. Spring peas have larger seeds, so there are fewer seeds per pound and seeding rates are higher, about 100 to 160 lb./A. However, spring pea seed is a bit less expensive than Austrian winter pea seed. TRAPPER is the most common Canadian field pea cultivar.

This section focuses on the widely grown Austrian winter pea. "Field peas" refers to both the winter and spring types.

BENEFITS

Bountiful biomass. Under a long, cool, moist season during their vegetative stages, Austrian winter peas produce more than 5,000 lb. dry matter/A, even when planted in spring in colder climates. Idaho farmers regularly produce 6,000 to 8,000 lb. DM/A from fall-planted Austrian winter peas. Because the residue breaks down quickly, only peas in the high-production areas build up much long-term organic matter. Peas do not make a good organic mulch for weed control (361).

Nitrogen source. Austrian winter peas are top N producers, yielding from 90 to 150 lb. N/A, and at times up to 300 lb. N/A.

Plowed down as green manure, fall-planted legume crops of Austrian winter pea, alfalfa and hairy vetch each produced enough N for the production of high-quality muskmelons under plastic mulch and drip irrigation in a Kansas study. Melon yields produced with the legumes were similar to those receiving synthetic fertilizer at 63 and 90 lb. N/A. The winter peas in the experiment produced 96 lb. N/A the first year and 207 lb. N/A the second (387).

Austrian winter peas harvested as hay then applied as mulch mineralized N at more than double the rate of alfalfa hay. The N contribution was measured the summer after a fall plowdown of the residue. The estimated N recovery of Austrian winter pea material 10 months after incorporation was 77 percent—58 percent through spring wheat and 19 percent in the soil (254).

Austrian winter pea green manure provided the highest spring wheat yield the following year in a Montana trial comparing 10 types of medics, seven clovers, yellow biennial sweet clover and three grains. Crops that produced higher tonnage of green manure usually had a negative effect on the subsequent wheat crop due to moisture deficiency that continued over the winter between the crops (381). Field peas can leave 80 lb. N/A if terminated at mid-season in lieu of summer fallow in dryland areas, or leave more than 30 lb. N/A after pea harvest at season's end (74).

A winter pea green manure consistently resulted in higher malting barley protein content than that following other legumes or fallow in a Montana trial. Annual legumes harvested for seed left less soil N than did plots in fallow. Also tested were fava bean, lentil, chickpea, spring pea, winter pea hay and dry bean (262).

Rotational effects. Pulse crops (grain legumes such as field peas, fava beans and lentils) improved sustainability of dryland crop rotations by providing disease suppression, better tilth and other enhancements to soil quality in a Saskatchewan study. Even at rates of 180 lb. N/A, fertilizer alone was unable to bring yields of barley planted into barley residue to the maximum achieved from these pulse residues (163).

Water thrifty. In a comparison of water use alongside INDIANHEAD lentils and GEORGE black medic, Austrian winter pea was the most moisture-efficient crop in producing biomass. Each crop had used 4 inches of water when Austrian winter pea vines were 16 inches long, the lentils were 6 to 8 inches tall and the black medic central tillers were 4 inches tall (383).

Austrian winter peas grown in a controlled setting at 50° F recorded more than 75 percent of its N_2 fixed per unit of water used by the 63rd day of growth. White clover, crimson clover and hairy vetch reached the same level of water efficiency, but it took 105 days (334).

Quick growing. Rapid spring growth helps peas out compete weeds and make an N contribution in time for summer cash crops in some areas.

Forage booster. Field peas grown with barley, oat, triticale or wheat provide excellent livestock forage. Peas slightly improve forage yield, but significantly boost protein and relative feed value of small grain hay.

Seed crop. Seed production in Montana is about 2,000 lb./A and about 1,500 lb./A in the Pacific Northwest. Demand is growing for field peas as food and livestock feed (74).

Long-term bloomer. The purple and white blossoms of field peas are an early and extended source of nectar for honeybees.

Chill tolerant. Austrian winter pea plants may lose some of their topgrowth during freezes, but can continue growing after temperatures fall as low as 10° F. Their shallow roots and succulent stems limit their overwintering ability, however. Sustained cold below 18° F without snow cover usually kills Austrian winter pea (202). To maximize winter survival:

- Select the most winter-hardy cultivars available—GRANGER, MELROSE and COMMON WINTER.
- Seed early enough so that plants are 6 to 8 inches tall before soil freezes, because peas are shallow rooted and susceptible to heaving. Try to plant from mid-August to mid-September in Zone 5.
- Plant into grain stubble or a rough seedbed, or interseed into a winter grain. These environments protect young pea roots by suppressing soil heaving during freezing and thawing. Trapped snow insulates plants, as well.

Marianne Sarrantonio

FIELD PEAS (Pisum sativum subsp. *arvense)*

MANAGEMENT

Establishment & Fieldwork

Peas prefer well-limed, well-drained clay or heavy loam soils, near-neutral pH or above and moderate fertility. They also do well on loamy sands in North Carolina. Field peas usually are drilled 1 to 3 inches deep to ensure contact with moist soil and good anchoring for plants.

If you broadcast peas, incorporation will greatly improve stands, as seed left exposed on the surface generally does not germinate well. Long-vined plants that are shallow-seeded at low seeding rates tend to fall over (lodge), lay against the soil and rot. Combat this tendency by planting with a small grain nurse crop such as oats, wheat, barley, rye or triticale. Reduce the pea seeding rate by about one quarter—and grain by about one third—when planting a pea/grain mix.

Planted at 60 to 80 lb./A in Minnesota, Austrian winter peas make a good nurse crop for alfalfa.

Field pea seed has a short shelf life compared with other crops. Run a germination test if seed is more than two years old and adjust seeding rate accordingly. If you haven't grown peas in the seeded area for several years, inoculate immediately before seeding.

West. In mild winter areas of California and Idaho, fall-plant for maximum yield. In those areas, you can expect *spring-planted* winter peas to produce about half the biomass as those that are fall-planted. Seed by September 15 in Zone 5 of the Inter-Mountain region in protected valleys where you'd expect mild winter weather and good, long-term snow cover. October-planted Austrian winter pea in the Zone 9 Sacramento Valley of California thrive on cool, moist conditions and can contribute 150 lb. N/A by early April.

The general rule for other parts of the semi-arid West where snow cover is dependable is to plant peas in the fall after grain harvest. In these dry regions of Montana and Idaho, overseed peas at 90 to 100 lb./A by "frostseeding" any time soils have become too cold for pea germination. Be sure residue cover is not too dense to allow seed to work into the soil through freeze/thaw cycles as the soil warms (383).

In the low-rainfall Northern Plains, broadcast clear stands of peas in early spring at a similar rate for the "Flexible Green Manure" cropping system (below). Seeding at about 100 lb./A compensates somewhat for the lack of incorporation and provides strong early competition with weeds (383). Plant as soon as soil in the top inch reaches 40° F to make the most of spring moisture (74).

A mixture of Austrian winter peas and a small grain is suitable for dryland forage production because it traps snow and uses spring moisture to produce high yields earlier than spring-seeded annual forages (74). With sufficient moisture, spring peas typically produce higher forage yields than Austrian winter peas.

East. Planted as a companion crop in early spring in the Northeast, Austrian winter peas may provide appreciable N for summer crops by Memorial Day (361). In the mid-Atlantic, Austrian winter peas and hairy vetch planted October 1 and killed May 1 produced about the same total N and corn yields (108).

Southeast. Seed by October 1 in the inland Zone 8 areas of the South so that root crowns can become established to resist heaving. Peas produce more biomass in the cooler areas of the South than where temperatures rise quickly in spring (74, 361). Peas planted in late October in South Carolina's Zone 8 and terminated in mid- to late April produce 2,700 to 4,000 lb. dry matter/A (23).

Killing

Peas are easily killed any time with herbicides, or by disking or mowing after full bloom, the stage of maturity that provides the optimum N contribution. Disk lightly to preserve the tender residue for some short-term erosion control.

Winter pea residue breaks down and releases N quickly.

The downside to the quick breakdown of pea vines is their slimy condition in spring if they winterkill, especially in dense, pure stands. Planting with a winter grain provides some protection from winterkill and reduces matting of dead pea vegetation.

Pest Management

Winter peas break crop disease cycles, Ben Burkett of Petal, Miss., has found. *Septoria* leaf spot problems on his cash crops are reduced when he plants Austrian winter pea in fall after snap beans and ahead of collards and mustard greens the next summer. Between October 15 and November 15, Burkett broadcasts just 50 lb./A then incorporates the seed with a shallow pass of his field cultivator. They grow 3 to 6 inches tall before going dormant in late December in his Zone 8 location about 75 miles north of the Gulf of Mexico. Quick regrowth starts about the third week in January. He kills them in mid-April by disking, then shallow plows to incorporate the heavy residue (202).

Farmers and researchers note several IPM cautions, because Austrian winter peas:

- Host some races of nematodes
- Are susceptible to winter *Sclerotinia* crown rot, *Fusarium* root rot as well as seed rot and blights of the stem, leaf or pod
- Are variably susceptible to the *Ascochyta* blight (MELROSE cultivar has some resistance)
- Host the pathogen *Sclerotinia* minor. There was a higher incidence of leaf drop in California lettuce planted after Austrian winter peas in one year of a two-year test (232).

Austrian winter peas were heavily damaged by *Sclerotinia trifoliorum* Eriks in several years of a four-year study in Maryland, but the crop still produced from 2,600 to 5,000 lb. dry matter/A per year in four out of five years. One year DM production was only 730 lb./A. Mean N contribution despite the disease was 134 lb. N/A. Overall, Austrian winter peas were rated as being more suited for Maryland Coastal Plain use than in the Piedmont, due to harsher winters in the latter location (204).

To combat disease, rotate cover crops to avoid growing peas in the same field in successive years. To minimize disease risk, waiting several years is best. To minimize risk of losing cover crop benefits to *Sclerotinia* disease in any given season, mix with another cover crop such as cereal rye.

Crop Systems

Northern Plains. Austrian winter peas (and other grain legumes) are increasingly used instead of fallow in dryland cereal rotations. The legumes help prevent saline seeps by using excess soil moisture between cereal crops. They also add N to the system. The legume>cereal sequence starts with a spring- or fall-planted grain legume (instead of fallow), followed by a small grain.

Peas work well in this system because they are shallow-rooted and therefore do not extract deep soil moisture. The pea crop is managed according to soil moisture conditions. Depending on growing season precipitation, the peas can be grazed, terminated or grown to grain harvest. Growers terminate the crop when about 4 inches of plant-available water remains in the soil, as follows:

- **Below-normal rainfall**—terminate the grain legume early.
- **Adequate rainfall**—terminate the grain legume when about 4 inches of soil water remains. Residue is maintained for green manure, moisture retention and erosion prevention.
- **Above-average rainfall**—grow the crop to maturity for grain harvest.

In conventional fallow systems, fields are left unplanted to accumulate soil moisture for the cash crop. Weeds are controlled using tillage or herbicides.

Grain legumes provide a soil-protecting alternative to fallow that can be managed to ensure adequate moisture for the cereal crop. Legumes provide long-term benefits by producing N for the subsequent crop, disrupting disease, insect and weed cycles and building soil.

Austrian winter peas work in these rotations where there is at least 18 inches of rain per year. INDIANHEAD lentils (*Lens culinaris* Medik), a specialty lentil for cover crop use, are also widely used in this system.

Montana research shows that when soil moisture is replenished by winter precipitation, annual legumes can substitute for fallow without significantly reducing the yield of the next barley crop. Montana rainfall averages 12-16 inches, so peas are planted but can only be taken to grain harvest in above-normal rainfall years. The legume can generate income from harvest of its hay or grain or through fertilizer N savings from the legume's contribution to the small grain crop (136).

In Idaho, fall-seeded Austrian winter peas harvested for seed provided income, residual N from the pea straw and soil disease suppression in a study of efficient uses of the legume cover. A crop rotation of Austrian winter pea (for grain)>winter wheat>spring barley produced similar wheat yields as did using the peas as green manure or leaving the field fallow in the first year. While neither Austrian winter pea green manure nor fallow produced income, the green manure improved soil organic matter and added more N for wheat than did summer fallow. Fallow caused a net soil capacity loss by "mining" finite soil organic matter reserves (253).

In a northern Alberta comparison of conventional (tilled), chemical (herbicide) and green (field pea) fallow systems, spring-planted field peas provided 72 lb. N/A, significantly more than the other systems. The field pea system was also more profitable when all inputs were considered, providing higher yield for two subsequent cash crops, higher income and improvement of soil quality (12).

In the Northeast, spring-planted peas can be incorporated by Memorial Day.

Southeast. Fall-seeded Austrian winter peas outproduced hairy vetch by about 18 percent in both dry matter and N production in a three-year test in the Coastal Plain of North Carolina. When legumes were grown with rye, wheat or spring oats, Austrian winter pea mixtures also had the highest dry matter yields. Over the three years, Austrian winter peas ranked the highest (dry-matter and N) in the legume-only trials and as the legume component of the legume/grain mixtures. In descending order after the peas were hairy vetch, common vetch and crimson clover. The peas were sown at 54 lb./A in the pure seedings and 41 lb./A in mixtures (344).

In the year of greatest N fixation, soil N in the Austrian winter pea mixture treatments was 50 percent greater than the average of all other treatments. Researchers noted that the bottom leaves of pea vines were more decomposed than other legumes, giving the crop an earlier start in N contribution. Further, soil N in the upper 6 inches of soil under the Austrian winter peas held 30 to 50 percent of the total soil inorganic N in the winter pea treatments, compared with levels of less than

Peas Do Double Duty for Kansas Farmer

PARTRIDGE, Kan.—Jim French figures Austrian winter peas provide free grazing, free nitrogen, or both. The vining legume produces just as much N for the following grain sorghum crop even if he lets his registered Gelbvieh herd eat all they want of the winter annual's spring growth.

French farms on flat, well-drained sandy loam soil near Partridge, Kan. He manages about 640 acres each of cash crops (winter wheat and grain sorghum) and forages (alfalfa, sudangrass, winter peas and cowpeas, and an equal area in grass pasture). Peas follow wheat in the three-year crop rotation on his south-central Kansas farm. He chisel plows the wheat stubble twice about 7 inches deep, disks once to seal the surface, then controls weeds as necessary with a light field cultivator.

Between mid-September and mid-October he inoculates about 30 lb./A of the peas and drills them with an old John Deere double-run disk drill in 8-inch rows. Establishment is usually good, with his only anxiety coming during freeze-thaw cycles in spring. "Each time the peas break dormancy, start to grow, then get zapped with cold again they lose some of their root reserves and don't have quite the resistance to freezing they did. They'll sprout back even if there's vegetative freeze damage as long as their food reserves hold out," French reports.

Ironically, this spring freezing is less of a problem further north where fields stay frozen longer before a slower thaw. This works as long as snow cover protects the peas from the colder early and mid-winter temperatures. In most years, he sets up temporary fence and turns his cattle into the peas about April 1 at the stocking rate of two animal units per acre. During the best years of mild weather and adequate moisture, "the cattle have a hard time keeping up," says French. Depending on his need for forage or organic matter, he leaves the cattle in until he incorporates the pea stubble, or gives it time to regrow.

One reason he gets about the same 90 to 120 lb. N/A contribution with or without grazing is that the winter pea plants apparently continue N fixation and root growth while being grazed. Soil tests show that 25 to 30 lb. N/A are available in the nitrate form at incorporation in late spring, with the balance in an organic form that mineralizes over the summer. Grazing the peas helps to contain cheatgrass, which tends to tie up N if it's incorporated just ahead of his sorghum crop.

French is sold on winter peas ahead of his grain sorghum because it provides N while reducing weed pressure from cheatgrass and pigweed and decreasing lodging from charcoal root rot. The option to use the peas as forage—while still achieving adequate sorghum yield—lets him buy less processed feed, improves livestock health and accelerates conversion of the peas' organic material into available soil nutrients.

"Winter peas work best where you integrate crops and livestock," says French. "They give you so many benefits."

30 percent in the top soil layer for all other treatments. In situations where the early-summer N release from peas could be excessive, mixing Austrian winter peas with a grain can moderate the N contribution and slow down its release into the soil (344).

The carbon to nitrogen (C:N) ratio of plant matter is an indication of how rapidly vegetation will break down. Mixtures of small grains with Austrian winter peas and the vetches had C:N values from 13 to 34, but were generally under 25 to 30, the accepted threshold for avoiding net immobilization of N (344).

Austrian winter peas and crimson clover can provide adequate N for conventionally planted cotton in South Carolina. In a three-year trial, fertilizer rates of up to 150 lb. N/A made no improvement to cotton yield on the pea plots. The evaluation showed that soil nitrate under Austrian winter peas peaked about nine weeks after incorporation (22).

Austrian winter peas achieved 50 to 60 percent groundcover when they were overseeded at about 75 lb./A into soybeans at leaf yellowing in southeastern Pennsylvania, where they can survive some winters. The peas produced nearly 2 tons of dry matter and 130 lb. N/A by May 20 in this test (191). Overseeding peas into corn at last cultivation is not recommended due to poor shade tolerance.

Austrian winter peas, like other hollow-stemmed succulent covers such as vetch and fava beans, do not respond well to mowing or cutting after they begin to bloom. In their earlier stages, Austrian winter peas will regrow even when grazed several times. See *Peas Do Double Duty for Kansas Farmer* (p. 140).

After three years of moisture testing, Kansas farmer Jim French can explain why he sees more soil moisture after spring grazing than when the peas are left to grow undisturbed. "There's decreasing overall transpiration because there's less leaf area to move moisture out of the soil into the air. Yet the root mass is about the same." Ungrazed peas pump more water as they keep growing.

Other Options.

Harvest field peas for hay when most of the pods are well formed. Use a mower with lifting guards and a windrow attachment to handle the sprawling vines.

COMPARATIVE NOTES

Field peas won't tolerate field traffic due to succulent stems (191). When selecting types, remember that long-vined varieties are better for weed control than short-vined types.

In dryland systems, winter peas produce abundant biomass with limited moisture.

Cultivars. MELROSE, known for its winter-hardiness, is a cultivar of the Austrian winter pea type. Planted the first week of September in Idaho, MELROSE peas yielded 300 lb. N/A and 6 tons of dry matter the next June. Planted in mid-April, the cultivar yielded "just" 175 lb. N/A and 3.5 T dry matter/A (202).

GRANGER is an improved winter pea that has fewer leaves and more tendrils, which are stiffer than standard cultivars. It is more upright and its pods dry more quickly than other winter pea types. MAGNUS field peas have out-produced Austrian winter peas in California and bloom up to 60 days earlier.

Seed sources. See *Seed Suppliers* (p. 195). 

HAIRY VETCH

Vicia villosa

Type: winter annual or summer annual legume

Roles: N source, weed suppressor, topsoil conditioner, reduce erosion

Mix with: small grains, field peas, bell beans, crimson clover, buckwheat

See charts, p. 66 to 72, for ranking and management summary.

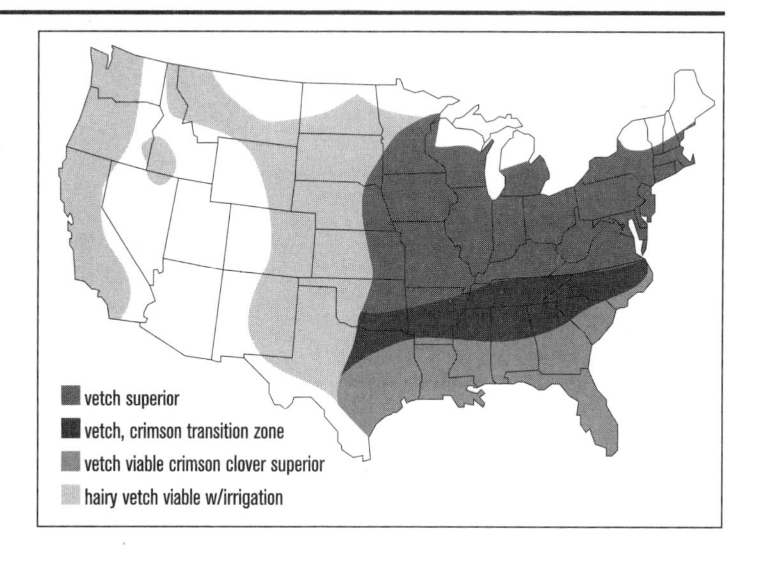

Few legumes match hairy vetch for spring residue production or nitrogen contribution. Widely adapted and winter hardy through Hardiness Zone 4 and into Zone 3 (with snow cover), hairy vetch is a top N provider in temperate and subtropical regions.

The cover grows slowly in fall, but root development continues over winter. Growth quickens in spring, when hairy vetch becomes a sprawling vine up to 12 feet long. Field height rarely exceeds 3 feet unless the vetch is supported by another crop. Its abundant, viney biomass can be a benefit and a challenge. The stand smothers spring weeds, however, and can help you replace all or most N fertilizer needs for late-planted crops.

BENEFITS

Nitrogen source. Hairy vetch delivers heavy contributions of mineralized N (readily available to the following cash crop). It can provide sufficient N for many vegetable crops, partially replace N fertilizer for corn or cotton and increase cash crop N efficiency for higher yield.

In some parts of California and the East in Zone 6, hairy vetch provides its maximum N by safe corn planting dates. In Zone 7 areas of the Southeast, the fit is not quite as good, but substantial N from vetch is often available before corn planting.

Corn planting date comparison trials with cover crops in Maryland show that planting *as late* as May 15 (the very end of the month-long local planting period) optimizes corn yield and profit from the system. Spring soil moisture was higher under the vetch or a vetch-rye mixture than under cereal rye or no cover crop. Killed vetch left on the surface conserved summer moisture for improved corn production (80, 82, 84, 85, 173, 243).

Even without crediting its soil-improving benefits, hairy vetch increases N response and produces enough N to pay its way in many systems. Hairy vetch without fertilizer was the preferred option for "risk-averse" no-till corn farmers in Georgia, according to calculations comparing costs, production and markets during the study. The economic risk comparison included crimson clover, wheat and winter fallow. Profit was higher, but less predictable, if 50 pounds of N were added to the vetch system (310).

Hairy vetch ahead of no-till corn was also the preferred option for risk averse farmers in a three-

---

Note: To roughly estimate hairy vetch N contribution in pounds per acre, cut and weigh fresh vetch top growth from a 4-foot by 4-foot area. Multiply pounds of fresh vetch by 12 to gauge available N, by 24 to find total N (377). For a more accurate estimate, see *How Much N?* (p. 22).

year Maryland study that also included fallow and winter wheat ahead of the corn. The vetch-corn system maintained its economic advantage when the cost of vetch was projected at maximum historic levels, fertilizer N price was decreased, and the herbicide cost to control future volunteer vetch was factored in (173). In a related study on the Maryland Coastal Plain, hairy vetch proved to be the most profitable fall-planted, spring desiccated legume ahead of no-till corn, compared with Austrian winter peas and crimson clover (243).

In Wisconsin's shorter growing season, hairy vetch planted after oat harvest provided a gross margin of $153/A in an oat/legume/corn rotation (1995 data). Profit was similar to using 160 lb. N/A in continuous corn, but with savings on fertilizer and corn rootworm insecticide (400).

Hairy vetch provides yield improvements beyond those attributable to N alone. These may be due to mulching effects, soil structure improvements leading to better moisture retention and crop root development, soil biological activity and/or enhanced insect populations just below and just above the soil surface.

Soil conditioner. Hairy vetch can improve root zone water recharge over winter by reducing runoff and allowing more water to penetrate the soil profile through macropores created by the crop residue (143). Adding grasses that take up a lot of water can reduce the amount of infiltration and reduce the risk of leaching in soils with excess nutrients. Hairy vetch, especially an oats/hairy vetch mix, decreased surface ponding and soil crusting in loam and sandy loam soils. Researchers attribute this to dual cover crop benefits: their ability to enhance the stability of soil aggregates (particles), and to decrease the likelihood that the aggregates will disintegrate in water (143).

Hairy vetch improves topsoil tilth, creating a loose and friable soil structure. Vetch doesn't build up long-term soil organic matter due to its tendency to break down completely. Vetch is a succulent crop, with a relatively "low" carbon to nitrogen ratio. Its C:N ratio ranges from 8:1 to 15:1, expressed as parts of C for each part of N. Rye C:N ratios range from 25:1 to 55:1, showing why it persists much longer under similar conditions than does vetch. Residue with a C:N ratio of 25:1 or more tends to immobilize N. For more information, see *How Much N?* (p. 22), and the rest of that section, *Building Soil Fertility and Tilth with Cover Crops* (p. 16).

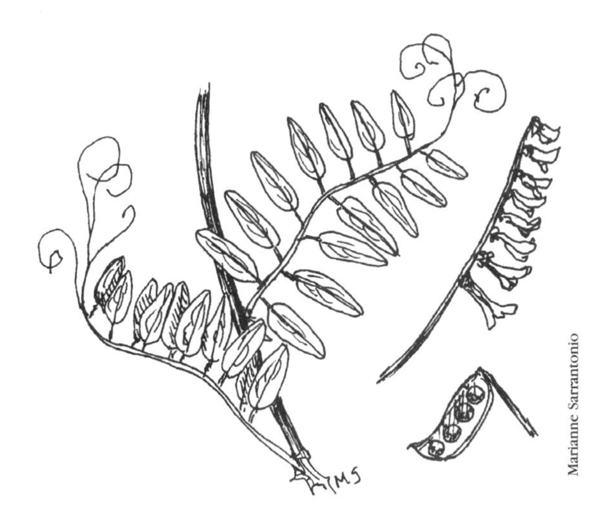

HAIRY VETCH (Vicia villosa)

Early weed suppression. The vigorous spring growth of fall-seeded hairy vetch out-competes weeds, filling in where germination may be a bit spotty. Residue from killed hairy vetch has a weak allelopathic effect, but it smothers early weeds mostly by shading the soil. Its effectiveness wanes as it decomposes, falling off significantly after about three or four weeks. For optimal weed control with a hairy vetch mulch, select crops that form a quick canopy to compensate for the thinning mulch or use high-residue cultivators made to handle it.

Mixing rye and crimson clover with hairy vetch (seeding rates of 30, 10, and 20 lb./A, respectively) extends weed control to five or six weeks, about the same as an all-rye mulch. Even better, the mix provides a legume N boost, protects soil in fall and winter better than legumes, yet avoids the potential crop-suppressing effect of a pure rye mulch on some vegetables.

Good with grains. For greater control of winter annual weeds and longer-lasting residue, mix hairy vetch with winter cereal grains such as rye, wheat or oats.

Growing grain in a mixture with a legume not only lowers the overall C:N ratio of the combined residue compared with that of the grain, it may actually lower the C:N ratio of the small grain residue as well. This internal change causes the grain residue to break down faster, while accumulating the same levels of N as it did in a monoculture (344).

Moisture-thrifty. Hairy vetch is more drought-tolerant than other vetches. It needs a bit of moisture to establish in fall and to resume vegetative growth in spring, but relatively little over winter when above-ground growth is minimal.

Phosphorus scavenger. Hairy vetch showed higher plant phosphorus (P) concentrations than crimson clover, red clover or a crimson/ryegrass mixture in a Texas trial. Soil under hairy vetch also had the lowest level of P remaining after growers applied high amounts of poultry litter prior to vegetable crops (121).

Fits many systems. Hairy vetch is ideal ahead of early-summer planted or transplanted crops, providing N and an organic mulch. Some Zone 5 Midwestern farmers with access to low-cost seed plant vetch after winter grain harvest in mid-summer to produce whatever N it can until it winterkills or survives to regrow in spring.

Widely adapted. Its high N production, vigorous growth, tolerance of diverse soil conditions, low fertility need and winter hardiness make hairy vetch the most widely used of winter annual legumes.

MANAGEMENT

Establishment & Fieldwork

Hairy vetch can be no-tilled, drilled into a prepared seedbed or broadcast. Dry conditions often reduce germination of hairy vetch. Drill seed at 15 to 20 lb./A, broadcast 25 to 30 lb./A. Select a higher rate if you are seeding in spring, late in the fall, or into a weedy or sloped field. Irrigation will help germination, particularly if broadcast seeded.

Plant vetch 30 to 45 days before killing frost for winter annual management; in early spring for summer growth; or in July if you want to kill or incorporate it in fall or for a winter-killed mulch.

Hairy vetch has a relatively high P and K requirement and, like all legumes, needs sufficient sulfur and prefers a pH between 6.0 and 7.0. However, it can survive through a broad pH range of 5.0 to 7.5 (120).

An Illinois farmer successfully no-tills hairy vetch in late August at 22 lb./A into closely mowed stands of fescue on former Conservation Reserve Program land (417). Using a herbicide to kill the fescue is cheaper than mowing, but it must be done about a month later when the grass is actively growing for the chemical to be effective. Vetch also can be no-tilled into soybean or corn stubble (50, 80).

In Minnesota, vetch can be interseeded into sunflower or corn at last cultivation. Sunflower should have at least 4 expanded leaves or yield will be reduced (221, 222).

Farmers in the Northeast's warmer areas plant vetch by mid-September to net 100 lb. available N/A by mid-May. Sown mid-August, an oats/hairy vetch mix can provide heavy residue (180).

Rye/hairy vetch mixtures mingle and moderate the effects of each crop. The result is a "hybrid" cover crop that takes up and holds excess soil nitrate, fixes N, stops erosion, smothers weeds in spring and on into summer if not incorporated, contributes a moderate amount of N over a longer period than vetch alone, and offsets the N limiting effects of rye (81, 83, 84, 86, 377).

Seed vetch/rye mixtures, at 15-25 lb. hairy vetch with 40-70 lb. rye/A (81, 361).

Overseeding (40 lb./A) at leaf-yellowing into soybeans can work if adequate rainfall and soil moisture are available prior to the onset of freezing weather. Overseeding into ripening corn (40 lb./A) or seeding at layby has not worked as consistently. Late overseeding into vegetables is possible, but remember that hairy vetch will not stand heavy traffic (361).

Killing

Your mode of killing hairy vetch and managing residue will depend on which of its benefits are most important to you. Incorporation of hairy vetch vegetation favors first-year N contribution, but takes significant energy and labor. Keeping vetch residue on the surface favors weed suppression, moisture retention, and insect habitat, but may reduce N contribution. However, even in no-till systems, hairy vetch consistently provides very large N input (replacing up to 100 lb. N/A).

In spring, hairy vetch continues to add N through its "seed set" stage after blooming. Biomass and N increase until maturity, giving either greater benefit or a dilemma, depending on your ability to deal with vines that become more sprawling and matted as they mature.

Mulch-retaining options include strip-tilling or strip chemical desiccation (leaving vetch untreated between the strips), mechanical killing (rotary mowing, flailing, cutting, sub-soil shearing with an undercutter, or chopping/flattening with a roller/crimper) or broadcast herbicide application.

No-till corn into killed vetch. The best time for no-till corn planting into hairy vetch varies with local rainfall patterns, soil type, desired N contribution, season length and vetch maturity.

In **southern Illinois**, hairy vetch no-tilled into fescue provided 40 to 180 lb. N/A per year over 15 years for one researcher/farmer. He used herbicide to kill the vetch about two weeks before the area's traditional mid-May corn planting date. The 14-day interval was critical to rid the field of prairie voles, present due to the field's thick fescue thatch.

He kills the vetch when it is in its pre-bloom or bloom stage, nearing its peak N-accumulation capacity. Further delay would risk loss of soil moisture in the dry period customary there in early June. When the no-tilled vetch was left to grow one season until seed set, it produced 6 tons of dry matter and contributed a potentially polluting 385 lb. N/A (417). This high dose of N must be managed carefully during the next year to prevent leaching or surface runoff of nitrates.

A series of trials in **Maryland** showed a different mix of conditions. Corn planting in late-April is common there, but early killing of vetch to plant corn then had the surprising effect of *decreasing* soil moisture and corn yield, as well as predictably lowering N contribution. The earlier-planted corn had less moisture-conserving residue. Late April or early May kill dates, with corn no-tilled 10 days later, consistently resulted in higher corn yields than earlier kill dates (82, 83, 84, 85). With hairy vetch and a vetch/rye mixture, summer soil water conservation by the cover crop residue had a greater impact than spring moisture depletion by the growing cover crop in determining corn yield (84, 85).

Results in the other trials, which also included a pure rye cover, demonstrated the management flexibility of a legume/grain mix. Early killed rye protects the soil as it conserves water and N, while vetch killed late can meet a large part of the N requirement for corn. The vetch/rye mixture can conserve N and soil moisture while fixing N for the subsequent crop. The vetch and vetch/rye mixture accumulated N at 130 to 180 lb./A. The mixture contained as much N or more than vetch alone (85, 86).

In an **Ohio** trial, corn no-tilled into hairy vetch at mid-bloom in May received better early season weed control from vetch mulch than corn seeded into vetch killed earlier. The late planting date decreased yield, however (189), requiring calculation to determine if lower costs for tillage, weed control, and N outweigh the yield loss.

Once vetch reaches about 50% bloom, it is easily killed by any mechanical treatment. To mow-kill for mulch, rye grown with hairy vetch improves cutting by holding the vetch off the ground to allow more complete severing of stems from roots. Rye also increases the density of residue covering the vetch stubble to prevent regrowth.

Much quicker and more energy-efficient than mowing is use of a modified Buffalo rolling stalk chopper, an implement designed to shatter standing corn stubble. The chopper's rolling blades break over, crimp and cut crop stems at ground level, and handle thick residue of hairy vetch at 8 to 10 mph (169).

Cover Crop Roller Design Holds Promise For No-Tillers

However, timing of control and planting in a single pass could limit adoption; hope lies in breeding cover crops that flower in time for traditional planting window.

THE POSSIBILITY of using rollers to reduce herbicide use isn't new, but advances are being made to improve the machines in ways that could make them practical for controlling no-till cover crops.

Cover crop rolling is gaining visibility and credibility in tests by eight university/farmer research teams across the country. The test rollers were designed and contributed by The Rodale Institute (TRI), a Pennsylvania-based organization focused on organic agricultural research and education. The control achieved with the roller is comparable to a roller combined with a glyphosate application, according to TRI.

The Rodale crop rollers were delivered to state and federal cooperative research teams in Virginia, Michigan, Mississippi, North Dakota, Pennsylvania, Georgia, California and Iowa in Spring, 2005. Funding for the program comes from grants and contributions from the Natural Resources Conservation Service and private donors. I&J Manufacturing in Gap, Pa., fabricated the models distributed to the research teams.

"The requirement is that each research leader partners with a farmer cooperator to adapt the rollers to local conditions and cover cropping systems," explains Jeff Moyer, TRI's farm manager. "Our goal is to gain more knowledge about the soil building and weed management effects of cover crops while reducing the need for herbicides," he says.

Farmer Built. Moyer designed and built the first front-mounted TRI roller prototype in 2002 in conjunction with Pennsylvania farmer John Brubaker, whose land abuts the TRI property. The original 10-foot, 6-inch roller width is equal to 4 rows on 30-inch spacing, with a 3-inch overlap on each end. The original design has already been modified to include a 15-foot, 6-inch model suitable for use with a 6-row planter on 30-inch rows. It can be adapted to fit a 4-row planter on 38-inch rows, and a 5-foot version for 2-row vegetable planters.

"We realize that 6-row equipment is small by today's standards, and work is under way on a system that mounts one section of the roller in front of the tractor with the remainder mounted on the planter ahead of the row units. This design will allow as wide a roller system as a farmer needs," Moyer says.

Chevron Pattern. The chevron pattern on the face of the roller came about after the designers realized that mounting the roller blades in a straight line would cause excessive bouncing, while just curving the blades in a screw pattern would act like an auger and create a pulling effect. "If you were driving up a hill that might be fine, but we don't need help pulling our tractors down the steep slopes we farm. The chevron pattern neutralizes any forces that might pull the tractor in either direction," Moyer explains. It overcomes both the bounce of straight-line blades and the auguring effect of corkscrew blades.

"In addition, with the twisted blade design, only a very small portion of the blade touches the ground at any one time as it turns, so the full pressure of the roller is applied 1 inch at a time. This roller design works better than anything we've ever used," he adds.

Prior to settling on the TRI prototype, Moyer and Brubaker studied stalk choppers with nine rolling drums arranged in two parallel rows. This design required 18 bearings and provided lots of places for green plant material to bunch up. The TRI ground-driven roller has a single cylinder and two offset bearings inset 3 inches on either side and fronted with a shield. The blades are welded onto the 16-inch-diameter drum, but replacement blades can be purchased from the manufacturer and bolted on as needed. The 10-foot, 6-inch roller weighs 1,200 pounds empty and 2,000 pounds if filled with water.

Front Mount Benefits. The biggest advantage of the front-mounted roller is that the operator can roll the cover crop and no-till the cash crop in a single field pass, Moyer explains. In TRI trials, simultaneous rolling and no-till planting eliminated seven of the eight field passes usually necessary with conventional organic corn production, including plowing, discing, packing, planting, two rotary hoe passes and two cultivations.

Rolling the field and no-tilling in one pass also eliminates the problem of creating a thick green cover crop mat that makes it difficult to see a row marker line on a second pass for planting.

Also, planting in a second pass in the opposite direction from which the cover crop was rolled makes getting uniform seeding depth and spacing more difficult because the planter tends to stand the plant material back up. "Think of it as combing the hair on your dog backwards," Moyer says.

"Another disadvantage of rear-mounted machines like stalk choppers is that the tractor tire is the first thing touching the cover crop. If the soil is even a little spongy, the cover crop will be pushed into the tire tracks and because the roller is running flat, it can't crimp the depressed plant material. A week later the plants missed by the roller will be back up and growing again."

Crop Versatility. The TRI roller concept has been tested in a wide range of winter annual cover crops, including cereal rye, hairy vetch, wheat, triticale, oats, buckwheat, clover, winter peas and other species. Timing is the key to success, Moyer emphasizes, and a lot of farmers don't have the patience to make it work right.

"The bottom line is that winter annuals want to die anyway, but if you time it wrong, they're hard to kill," he says. "If you try to roll a winter annual before it has flowered—before it has physiologically reproduced—the plant will try to stand up again and complete the job of reproduction, the most important stage of its life cycle. But, if you roll it after it has flowered, it will dry up and die."

At least a 50 percent, and preferably a 75 to 100 percent bloom, is recommended before rolling. Moyer hopes to see plant breeders recognize the need to develop cover crop varieties with blooming characteristics that coincide with preferred crop planting windows.

"We really like to use hairy vetch on our farm, for example, because it's a great source of nitrogen and is a very suitable crop to plant corn into. The roller crimps the stem of the hairy vetch every 7 inches, closing the plant's vascular system and ensuring its demise.

"The problem is we would like it to flower a couple weeks earlier to fit our growing season. It's hard for farmers to understand when it's planting time and we're telling them to wait a couple more weeks for their cover crop to flower," he says.

"We need to identify the characteristics we want in cover crops and encourage plant breeders to focus on some of those. It should be a relatively easy task to get an annual crop to mature a couple weeks earlier, compared to some of the breakthrough plant breeding we've seen recently," Moyer says.

continued on page 148

For More Information. Updates on roller research, more farmer stories and plans for the TRI no-till cover crop roller can be accessed at www.newfarm.org/depts/notill. To ask questions of The Rodale Institute, e-mail to info@rodaleinst.org.

See also "Where can I find information about the mechanical roller-crimper used in no-till production?" http://attra.ncat.org/calendar/question.php/2006/05/08/p2221.

To contact the manufacturer of commercially available cover crop rollers, visit www.croproller.com.

Editor's Note: PURPLE BOUNTY, a new, earlier variety of hairy vetch, was released in 2006 by the USDA-Agricultural Research Service, Beltsville, MD in collaboration with the Rodale Institute, Pennsylvania State University and Cornell University.

—Ron Ross. Adapted with permission from www.no-tillfarmer.com

No-till vegetable transplanting. Vetch that is suppressed or killed without disturbing the soil maintains moisture well for transplanted vegetables. No-till innovator Steve Groff of Lancaster County, Pa., uses the rolling stalk chopper to create a killed organic mulch. His favorite mix is 25 lb. hairy vetch, 30 lb. rye and 10 lb. crimson clover/A.

Winter hardy through the warmer parts of Zone 4, few legumes can rival hairy vetch's N contributions.

No-till, delayed kill. Farmers and researchers are increasingly using a roller/crimper to kill hairy vetch and other cover crops (11). Jeff Moyer and others at the Rodale Institute in Kutztown, Pa., roll hairy vetch and other cover crops in late May or early June (at about 50% flower). The modified roller is front-mounted, and corn is no-tilled on the same pass (303). See *Cover Crop Roller Design Holds Promise For No-Tillers*, p. 146.

Also useful in killing hairy vetch on raised beds for vegetables and cotton is the improved prototype of an undercutter that leaves severed residue virtually undisturbed on the surface (96). The undercutter tool includes a flat roller attachment, which, by itself, usually provides only partial suppression unless used after flowering.

Herbicides will kill vetch in three to 30 days, depending on the material used, rate, growth stage and weather conditions.

Vetch incorporation. As a rule, to gauge the optimum hairy vetch kill date, credit vetch with adding two to three pounds of N per acre per sunny day after full spring growth begins. Usually, N contribution will be maximized by early bloom (10-25 percent) stage.

Cutting hairy vetch close to the ground at full bloom stage usually will kill it. However, waiting this long means it will have maximum top growth, and the tangled mass of mature vetch can overwhelm many smaller mowers or disks. Flail mowing before tillage helps, but that is a time and horsepower intensive process. Sickle-bar mowers should only be used when the vetch is well supported by a cereal companion crop and the material is dry (422).

Management Cautions

About 10 to 20 percent of vetch seed is "hard" seed that lays ungerminated in the soil for one or more seasons. This can cause a weed problem, especially in winter grains. In wheat, a variety of herbicides are available, depending on crop growth stage. After a corn crop that can utilize the vetch-produced N, you could establish a hay or pasture crop for several years.

Don't plant hairy vetch with a winter grain if you want to harvest grain for feed or sale.

Production is difficult because vetch vines will pull down all but the strongest stalks. Grain contamination also is likely if the vetch goes to seed before grain harvest. Vetch seed is about the same size as wheat and barley kernels, making it hard and expensive to separate during seed cleaning (361). Grain price can be markedly reduced by only a few vetch seeds per bushel.

A severe freeze with temperatures less than 5° F may kill hairy vetch if there is no snow cover, reducing or eliminating the stand and most of its N value. If winterkill is possible in your area, planting vetch with a hardy grain such as rye ensures spring soil protection.

Pest Management

In legume comparison trials, hairy vetch usually hosts numerous small insects and soil organisms (206). Many are beneficial to crop production, (see below) but others are pests. Soybean cyst nematode (*Heterodera glycines*) and root-knot nematode (*Meliodogyne* spp.) sometimes increase under hairy vetch. If you suspect that a field has nematodes, carefully sample the soil after hairy vetch. If the pests reach an economic threshold, plant nematode-resistant crops and consider using another cover crop.

Other pests include cutworms (361) and southern corn rootworm (67), which can be problems in no-till corn, tarnished plant bug, noted in coastal Massachusetts (56), which readily disperses to other crops, and two-spotted spider mites in Oregon pear orchards (142). Leaving unmowed remnant strips can lessen movement of disruptive pests while still allowing you to kill most of the cover crop (56).

Prominent among beneficial predators associated with hairy vetch are lady beetles, seven-spotted ladybeetles (56) and bigeyed bugs (*Geocaris* spp.). Vetch harbors pea aphids (*Acyrthosiphon pisum*) and blue alfalfa aphids (*Acyrthosiphon kondoi*) that do not attack pecans but provide a food source for aphid-eating insects that can disperse into pecans (58). Similarly, hairy vetch blossoms harbor flower thrips (*Frankliniella* spp.), which in turn attract important thrip predators such as insidious flower bugs (*Orius insidiosus*) and minute pirate bugs (*Orius tristicolor*).

Two insects may reduce hairy vetch seed yield in heavy infestations: the vetch weevil or vetch bruchid. Rotate crops to alleviate buildup of these pests (361).

CROP SYSTEMS

In no-till systems, killed hairy vetch creates a short-term but effective spring/summer mulch, especially for transplants. The mulch retains moisture, allowing plants to use mineralized nutrients better than unmulched fields. The management challenge is that the mulch also lowers soil temperature, which may delay early season growth (361). One option is to capitalize on high quality, low-cost tomatoes that capture the late-season market premiums. See *Vetch Beats Plastic* (p. 150).

How you kill hairy vetch influences its ability to suppress weeds. Durability and effectiveness as a light-blocking mulch are greatest where the stalks are left whole. Hairy vetch severed at the roots or sickle-bar mowed lasts longer and blocks more light than flailed vetch, preventing more weed seeds from germinating (96, 411).

Mix hairy vetch with cereal grains to reduce the risk of N leaching.

Southern farmers can use an overwintering hairy vetch crop in continuous no-till cotton. Vetch mixed with rye has provided similar or even increased yields compared with systems that include conventional tillage, winter fallow weed cover and up to 60 pounds of N fertilizer per acre. Typically, the cover crops are no-till drilled after shredding cotton stalks in late October. Covers are spray killed in mid-April ahead of cotton planting in May. With the relatively late fall planting, hairy vetch delivers only part

---

Note: An unmowed rye/hairy vetch mix sustained a population of aphid-eating predators that was six times that of the unmowed volunteer weeds and 87 times that of mown grass and weeds (57).

Vetch Beats Plastic

BELTSVILLE, Md.—Killed cover crop mulches can deliver multiple benefits for no-till vegetable crops (1, 2, 3, 4). The system can provide its own N, quell erosion and leaching, and displace herbicides. It's also more profitable than conventional commercial production using black plastic mulch. A budget analysis showed it also should be the first choice of "risk averse" farmers, who prefer certain although more modest profit over higher average profit that is less certain (224).

The key to the economic certainty of a successful hairy vetch planting is its low cost compared with the black plastic purchase, installation and removal.

From refining his own research and on-farm tests in the mid-Atlantic region for several years, Aref Abdul-Baki, formerly of the USDA's Beltsville (Md.) Agricultural Research Center, outlines his approach:

- Prepare beds—just as you would for planting tomatoes—at your prime time to seed hairy vetch.
- Drill hairy vetch at 40 lb./A, and expect about 4 inches of top growth before dormancy, which stretches from mid-December to mid-March in Maryland.
- After two months' spring growth, flail mow or use other mechanical means to suppress the hairy vetch. Be ready to remow or use herbicides to clean up trouble spots where hairy vetch regrows or weeds appear.
- Transplant seedlings using a minimum tillage planter able to cut through the mulch and firm soil around the plants.

The hairy vetch mulch suppresses early season weeds. It improves tomato health by preventing soil splashing onto the plants, and keeps tomatoes from soil contact, improving quality. Hairy vetch-mulched plants may need more water. Their growth is more vigorous and may yield up to 20 percent more than those on plastic. Completing harvest by mid-September allows the field to be immediately reseeded to hairy vetch. Waiting for vetch to bloom in spring before killing it and the tight fall turnaround may make this system less useful in areas with a shorter growing season than this Zone 7, mid-Atlantic site.

Abdul-Baki rotates season-long cash crops of tomatoes, peppers and cantaloupe through the same plot between fall hairy vetch seedings. He shallow plows the third year after cantaloupe harvest and seeds hairy vetch for flat-field crops of sweet corn or snap beans the following summer.

He suggests seeding rye (40 lb./A) with the vetch for greater biomass and longer-lasting mulch. Adding 10-12 lb./A of crimson clover will aid in weed suppression and N value. Rolling the covers before planting provides longer-lasting residue than does mowing them. Some weeds, particularly perennial or winter annual weeds, can still escape this mixture, and may require additional management (4).

of its potential N in this system. It adds cost, but supplies erosion control and long-term soil improvement (35).

Cotton yields following incorporated hairy vetch were perennial winners for 35 years at a northwestern Louisiana USDA site. Soil organic matter improvement and erosion control were additional benefits (276).

Other Options

Spring sowing is possible, but less desirable than fall establishment because it yields significantly less biomass than overwintering stands. Hot weather causes plants to languish.

Hairy vetch makes only fair grazing—livestock do not relish it.

Harvesting seed. Plant hairy vetch with grains if you intend to harvest the vetch for seed. Use a moderate seeding rate of 10-20 lb./A to keep the stand from getting too rank. Vetch seed pods will grow above the twining vetch vines and use the grain as a trellis, allowing you to run the cutter bar higher to reduce plugging of the combine. Direct combine at mid-bloom to minimize shattering, or swath up to a week later. Seed is viable for at least five years if properly stored (361).

If you want to save dollars by growing your own seed, be aware that the mature pods shatter easily, increasing the risk of volunteer weeds. To keep vetch with its nurse crop, harvest vetch with a winter cereal and keep seed co-mingled for planting. Check the mix carefully for weed seeds.

COMPARATIVE NOTES

Hairy vetch is better adapted to sandy soils than crimson clover (344), but is less heat-tolerant than LANA woollypod vetch. See *Woollypod Vetch* (p. 185).

Cultivars. MADISON—developed in Nebraska—tolerates cold better than other varieties. Hairy vetches produced in Oregon and California tend to be heat tolerant. This has resulted in two apparent types, both usually sold as "common" or "variety not stated" (VNS). One has noticeably hairy, bluish-green foliage with bluish flowers and is more cold-tolerant. The other type has smoother, deep-green foliage and pink to violet flowers.

A closely related species—LANA woollypod vetch (*Vicia dasycarpa*)—was developed in Oregon and is less cold tolerant than *Vicia villosa*. Trials in southeastern Pennsylvania with many accessions of hairy vetch showed big flower vetch (*Vicia grandiflora*, cv WOODFORD) was the only vetch species hardier than hairy vetch. EARLY COVER hairy vetch is about 10 days earlier than regular common seed. PURPLE BOUNTY, released in 2006, is a few days earlier and provides more biomass and better ground cover than EARLY COVER.

Seed sources. See *Seed Suppliers* (p. 195). 

MEDICS

Medicago spp.

Also called: black medic, burr (or bur) medic, burclover

Type: Winter annual or summer annual legume

Roles: N source, soil quality builder, weed suppressor, erosion fighter

Mix with: Other medics; clovers and grasses; small grains

See charts, p. 66 to 72, for ranking and management summary.

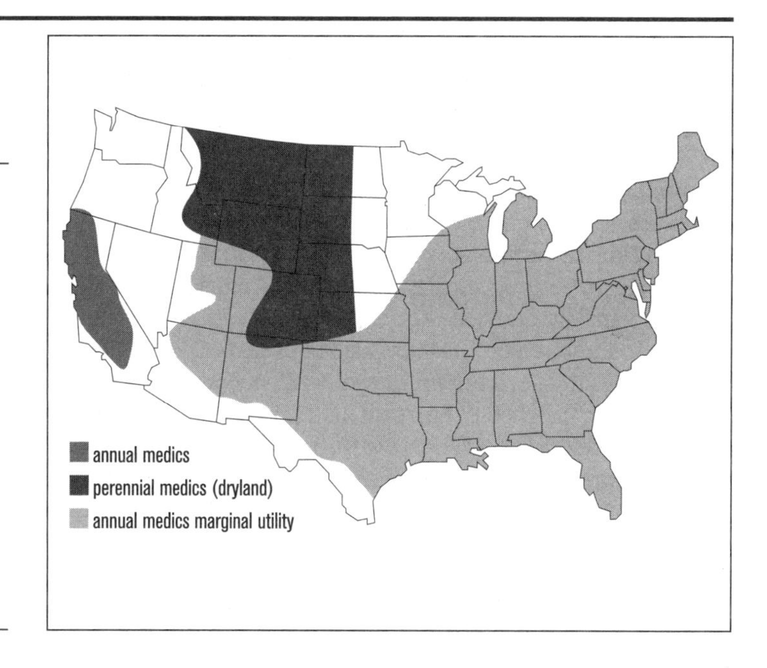

Once established, few other legumes outperform medics in soil-saving, soil-building and—in some systems—forage, when summer rainfall is less than 15 inches. They serve well in seasonally dry areas from mild California to the harsh Northern Plains. With more rainfall, however, they can produce almost as much biomass and N as clovers. Perennial medics are self-reseeding with abundant "hard seed" that can take several years to germinate. This makes medics ideal for long rotations of forages and cash crops in the Northern Plains and in cover crop mixtures in the drier areas of California.

Annual medics include 35 known species that vary widely in plant habit, maturity date and cold tolerance. Most upright varieties resemble alfalfa in their seeding year with a single stalk and short taproot. Medics can produce more than 100 lb. N/A in the Midwest under favorable conditions, but have the potential for 200 lb. N/A where the plants grow over winter. They germinate and grow quickly when soil moisture is adequate, forming a thick ground cover that holds soil in place. The more prostrate species of annual medic provide better ground cover.

Significant annual types include: **burr** medic (*M. polymorpha*), which grows up to 14 inches tall, is semi-erect or prostrate, hairless, and offers great seed production and N-fixing ability; barrel medic (*M. truncatula*), about 16 inches tall, with many mid-season cultivars; and snail medic (*M. scutellata*), which is a good biomass and N producer.

Southern spotted burr medic is a native *M. polymorpha* cultivar with more winterhardiness than most of the current burr medics, which are imported from Australia. See *Southern Burr Medic Offers Reseeding Persistence* (p. 154). Naturalized burr medic seed is traded locally in California.

Annual medics broadcast in spring over wheat stubble in Michigan reduced weed number and growth of spring annual weeds prior to no-till corn planting the following spring. Spring-planted annual medics produced dry matter yields similar to or greater than alfalfa by July (373, 376).

Black medic (*M. lupulina*) is usually called a perennial. It can improve soil, reduce diseases, save moisture and boost grain protein when grown in rotations with grains in the Northern Plains. GEORGE is the most widely used cultivar in dryland areas of the Northern Plains. Black medic produces abundant seed. Up to 96 percent of it is hard seed, much of it so hard seeded that it won't

Jess Counts on GEORGE for N and Feed

STANFORD, Mon.—Jess Alger can count on 13 inches of rainfall or less on his central Montana farm, occasional hail damage, too few solar units to raise safflower or millet, some bone-chilling winters without snow cover—and George. That's GEORGE black medic.

On-farm tests showed he got 87 lb. N/A and 3 percent organic matter on his Judith clay loam soils. He initially seeded the medic on 10-inch row spacings with barley at 10 lb./A, his standard rate and seeding method. He grazed the medic early in the second year, and then let it go to seed. In Year 3, he sprayed it with glyphosate in order to establish a sorghum-sudangrass hybrid as emergency forage on May 15. He had several inches of growth when frost hit about June 10 and killed the tender grass.

The medic came on strong. He let it mature to its full 12 inches to harvest it for seed. "It was already laying over, but the pickup guards on my combine helped to gather in about half the seed." The other half pumped up the seed bank for years ahead.

He did a comparison with side-by-side fields of spring wheat. One followed a spring wheat crop, the other he planted into a six-year-old stand of GEORGE medic. The medic/wheat interplant yielded 29 bushels per acre—six bushels less than the other field. But the interplanted grain tested at 15 percent protein, a full percentage point higher. Those are high yields for Alger's area, partly due to timely summer rain. "The yield drop with medic was mostly a weed problem with Persian darnel," Alger explains, "but I now have that mostly under control."

Jess continues to fine-tune his system to maximize income and weed management. He became certified organic in 1999. He maintains the medic seed bank with no-till plantings of GEORGE with a nurse crop of Austrian winter peas. He is experimenting successfully with rye instead of summer fallow.

If weed pressure is high, medic fields are grazed closely to prevent weeds from going to seed, then plowed. Otherwise, he no-tills winter wheat into standing medic so he can leave most of the medic in place, bury less seed and allow GEORGE to rest more securely in his field.

Updated in 2007 by Andy Clark

germinate for two years. Second-year growth may be modest, but coverage improves in years three and four after the initial seeding if competition is not excessive (422) and grazing management is timely.

BENEFITS

Good N on low moisture. In dryland areas, most legumes offer a choice between N production and excessive water use. Medics earn a place in dryland crop rotations because they provide N while conserving moisture comparable to bare-ground fallow (230, 380).

Fallow is the intentional resting of soil for a season so it will build up moisture and gain fertility by biological breakdown of organic matter. Black medic increased spring wheat yield by about 92 percent compared with spring wheat following fallow, and also appreciably raised the grain protein level (379). GEORGE grows in a prostrate to ascending fashion and overwinters well with snow cover in the Northern Plains.

April soil N value after black medic in one Montana test was 117 lb./A, about 2.5 times the fallow N level and the best of six cultivars tested, all of which used less water than the fallow treatment (378). In North Dakota, however, unrestricted medic growth depressed yield of a following wheat crop (73).

Great N from more water. Under normal dryland conditions, medics usually produce about 1 T dry matter/A, depending on available soil mois-

Southern Spotted Burr Medic Offers Reseeding Persistence

While annual medics, in general, are hard seeded, they usually cannot tolerate winters north of the Gulf South. Southern spotted burr medic (*Medicago arabica*) shows promise as a winter legume that can reseed for several years from a single seed crop in Hardiness Zone 7 of the Southeast.

Once as widely grown as hairy vetch in the mid-South region of the U.S., burr medic persists in non-cropland areas because it is well adapted to the region (326, 327). A local accession collected in northern Mississippi exhibits better cold hardiness and insect resistance than commercially available (Australian) annual medics.

In a replicated cold-hardiness trial spanning several states, spotted burr medic flowered in mid-March, about two weeks after SERENA, CIRCLE VALLEY or SANTIAGO burclover, but two weeks before TIBBEE crimson clover. The burr medic flowered over a longer period than crimson, matured seed slightly sooner than TIBBEE but generally did not produce as much biomass.

The big advantage of spotted burr medic over crimson clover was its ability to reseed for several years from a single seed crop. In studies in several states, the native medic successfully reseeded for at least two years when growth was terminated two weeks after TIBBEE bloomed. Only balansa clover (see *Up-and-Coming Cover Crops*, p. 191) reseeded as well as spotted burclover (105). The burr medic cultivar CIRCLE VALLEY successfully reseeded in a Louisiana no-till cotton field for more than 10 years without special management to maintain it (103).

Research in the Southeast showed that if Southern spotted burr medic begins blooming March 23, it would form viable seed by May 2, and reach maximum seed formation by May 12. By allowing the cover crop to grow until 40 to 50 days after first bloom and managing the cropping system without tillage that would bury burclover seeds too deeply, Southern spotted burclover should successfully reseed for several years.

Native medic seed is being increased in cooperation with the USDA-Natural Resources Conservation Service's Jamie Whitten Plant Materials Center, Coffeeville, Miss., for possible accelerated release to seed growers as a "source-identified" cover crop.

Insect pests such as clover leaf weevil (*Hypera punctata* Fabricius) and the alfalfa weevil (*Hypera postica* Gyllenhal) preferentially attack medics over other winter legume cover crops in the Southeast, and could jeopardize seed production. These insects are easily controlled with pyrethroid insecticides when weevils are in their second instar growth stage. While not usually needed for single-season cover crop benefits, insecticides may be warranted in the seeding year to ensure a reseeding crop for years to come.

ture and fertility. When moisture is abundant, medics can reach their full potential of 3 T/A of 3.5 to 4 percent plant-tissue nitrogen, contributing more than 200 lb. N/A (201, 422).

Fight weeds. Quick spring regrowth suppresses early weeds. Fall weeds are controlled by medic regrowth after harvest, whether the medic stand is overseeded or interplanted with the grain, or the grain is seeded into an established medic stand. In California orchards and vineyards where winters are rainy instead of frigid, medics mixed with other grasses and legumes provide a continuous cover that crowds out weeds. In those situations, medics help reduce weed seed production for the long-term.

Boost organic matter. Good stands of medics in well drained soil can contribute sufficient residue to build soil organic matter levels. One Indiana

Hard-seeded medics are ideal for reseeding systems in orchards and vineyards.

test reported a yield of more than 9,000 lb. dry matter/A from a spring-sown barrel medic (164).

Reduce soil erosion. Medics can survive in summer drought-prone areas where few other cultivated forage legumes would, thanks to their hard-seeded tendency and drought tolerance. Low, dense vegetation breaks raindrop impact while roots may penetrate 5 feet deep to hold soil in place.

Tolerate regular mowing. Medics can be grazed or mowed at intervals with no ill effects. They should be mowed regularly to a height of 3 to 5 inches during the growing season for best seed set and weed suppression. To increase the soil seed bank, rest medic from blooming to seed maturation, then resume clipping or grazing (285, 422, 435).

Provide good grazing. Green plants, dry plants and burs of burr medic provide good forage, but solid stands can cause bloat in cattle (422). The burs are concentrated nutrition for winter forage, but lower the value of fleece when they become embedded in wool. Annual medics overseeded into row crops or vegetables can be grazed in fall after cash crop harvest (376).

Reseeding. Black medic has a high percentage of hard seed. Up to 90 percent has an outer shell that resists the softening by water and soil chemicals that triggers germination (286). Scarified seed will achieve 95 percent germination, and 10-year old raw seed may still be 50 percent viable (422). Burr medic seed in the intact bur remains viable for a longer time than hulled seed (120).

Their status as a resilient, reseeding forage makes medics the basis for the "ley system" developed in dry areas of Australia. Medics or subterranean clover pastured for several years on Australian dry-lands help to store moisture and build up soil productivity for a year of small grain production before being returned to pasture. This use requires livestock for maximum economic benefit. GEORGE black medic is prostrate, allowing other grasses and forbs to become the overstory for grazing. It is well-suited to cold winter areas of Hardiness Zone 4, where it can stay green much of the winter (6).

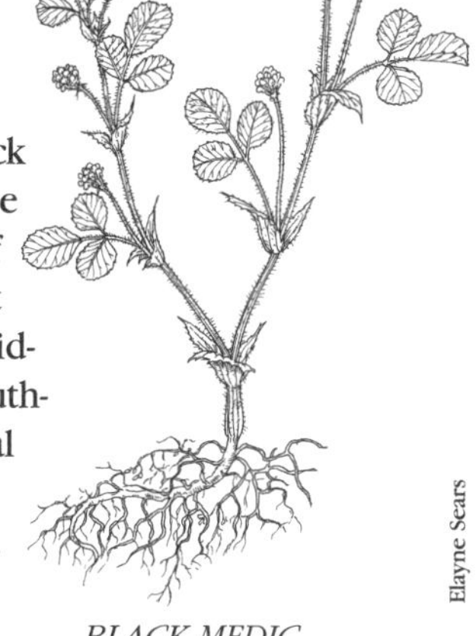

BLACK MEDIC
(Medicago lupulina)

Quick starting. Black medic can germinate within three days of planting (286). About 45 days after mid-April planting in southern Illinois, two annual medics were 20 inches tall and blooming. In the upper Midwest, snail and burr medics achieve peak biomass about 60 days after planting. An early August seeding of the annuals in southern Illinois germinated well, stopped growing during a hot spell, then restarted. Growth was similar to the spring-planted plots by September 29 when frost hit. The plants stayed green until the temperature dipped to the upper teens (201).

Widely acclimated. Species and cultivars vary by up to seven weeks in their estimated length of time to flowering. Be sure to select a species to fit your weather and crop rotation.

MANAGEMENT

Establishment

Annual medics offer great potential as a substitute for fallow in dry northern regions of the U.S. with longer day length. Annual medics need to fix as much N as winter peas or lentils and have a competitive establishment cost per acre to be as valuable as these better-known legume green manures (383).

Medics are widely adapted to soils that are reasonably fertile, but not distinctly acid or alkaline. Excessive field moisture early in the season can significantly reduce medic stands (373). Acid-tolerant rhizobial strains may help some cool-season medics, especially barrel medic, to grow on sites that otherwise would be inhospitable (422).

To reduce economic risk in fields where you've never grown medic, sow a mixture of medics with variable seed size and maturation dates. In dry areas of California, medic monocultures are planted at a rate of 2 to 6 lb./A, while the rate with grasses or clovers is 6 to 12 lb./A (422).

Establishment options vary depending on climate and crop system:

• **Early spring—clear seed.** Drill 1/4 to 1/2 inch deep (using a double-disk or hoe-type drill) into a firm seed bed as you would for alfalfa. Rolling is recommended before or after seeding to improve seed-soil contact and moisture in the seed zone. Seeding rate is 8 to 10 lb./A for black medic, 12 to 20 lb./A for larger-seeded (snail, gamma and burr) annual medics. In the arid Northern Plains, fall germination and winter survival are dependable, although spring planting also has worked.

Medics earn a place in dryland rotations because they provide N while conserving moisture.

• **Spring grain nurse crop.** Barley, oats, spring wheat and flax can serve as nurse crops for medic, greatly reducing weed pressure in the seeding year. The drawback is that nurse crops will reduce first-year seed production if you are trying to establish a black medic seed bank. To increase the soil seed reserve for a long-term black medic stand (germinating from hard seed), allow the medic to blossom, mature and reseed during its second year.

• **Corn overseed.** SANTIAGO burr medic and SAVA snail medic were successfully established in no-till corn three to six weeks after corn planting during a two-year trial in Michigan. Corn yield was reduced if medics were seeded up to 14 days after corn planting. Waiting 28 days did not affect corn yield, but medic biomass production was reduced by 50% (219).

Where medic and corn work together, such as California, maximize medic survival during the corn canopy period by seeding early (when corn is eight to 16 inches tall) and heavy (15 to 20 lb./A) to build up medic root reserves (47, 422).

• **After wheat harvest.** MOGUL barrel medic seeded after wheat harvest produced 119 lb. N/A in southern Michigan, more than double the N production of red clover seeded at the same time (373). In Montana, mid-season establishment of snail medic after wheat works only in years with adequate precipitation, when it smothers weeds, builds up N, then winterkills for a soil-holding organic mulch (72).

• **Autumn seeding.** Where winters are rainy in California, medics are planted in October as winter annuals (436). Plant about the same time as crimson clover in the Southeast, Zones 7 and 8.

Killing

Medics are easy to control by light tillage or herbicides. They reseed up to three times per summer, dying back naturally each time. Medics in the vegetative stage do not tolerate field traffic.

Field Management

Black medic>small grain rotations developed in Montana count on successful self-reseeding of medic stands for grazing by sheep or cattle. A month of summer grazing improves the economics of rotation by supplying forage for about one animal unit per acre. In this system, established self-reseeding black medic plowed down as green manure in alternate years improved spring wheat yield by about 50 percent compared to fallow (380).

Black medic is a dual-use legume in this adapted "ley" system. Livestock graze the legume in the "medic years" when the cover crop accumulates biomass and contributes N to the soil. Cash crops can be no-tilled into killed medic, or the legume can be incorporated.

A well-established black medic stand can reduce costs compared with annual crops by coming back for many years. However, without the livestock grazing benefit to supply additional utilization, water-efficient legumes such as lentils and Austrian winter peas will probably be more effective N sources. Further, the long-lived seed bank that black medic establishes may be undesirable for some cash crop rotations (383).

Use of medics for grain production in the upper Midwest has given inconsistent results. Berseem clover may be a better choice in many situations. In a series of trials in Ohio, Michigan, Wisconsin and Minnesota, medic sometimes reduced corn yield and did not provide enough weed control or N to justify its use under current cash grain prices, even when premiums for pesticide-free corn were evaluated (141, 219, 373, 374, 376, 456, 457). One Michigan farmer's situation is fairly typical. He established annual medic at 10 lb./A when his ridge-tilled corn was about knee-high. The legume germinated, but didn't grow well or provide weed suppression until after corn dry-down in mid-September. The medic put on about 10 inches of growth before winterkilling, enough for effective winter erosion protection (201).

Black medic and two annual medics produced 50 to 150 lb. N/A when interplanted with standard and semi-dwarf barley in a Minnesota trial. Annual MOGUL produced the most biomass by fall, but also reduced barley yields. GEORGE was the least competitive and fixed 55 to 120 lb. N/A. The taller barley was more competitive, indicating that taller small grain cultivars should be used to favor grain production over medic stand development (289).

Midwestern farmers can overseed annual medic or a medic/grass mixture into wheat in very early spring for excellent early summer grazing. With timely moisture, you can get a hay cutting within nine to 10 weeks after germination, and some species will keep working to produce a second cutting. Regrowth comes from lateral stems, so don't clip or graze lower than 4 or 5 inches if you want regrowth. To avoid bloat, manage as you would alfalfa (201).

Annual medics can achieve their full potential when planted after a short-season spring crop such as processing peas or lettuce. Wisconsin tests at six locations showed medic produced an average of 2.2 T/A when sown in the late June or early July (399). Early planting in this window with a late frost could give both forage and N-bearing residue, protecting soil and adding spring fertility. Take steps to reduce weed pressure in solid seedings, especially in early July.

In another Michigan comparison, winter canola (*Brassica napus*) yields were similar after a green manure comparison of two medics, berseem clover and NITRO annual alfalfa. All the covers were clear (sole-crop) seeded in early May after pre-plant incorporated herbicide treatment, and were plowed down 90 days later. Harvesting the medics at 60 days as forage did not significantly lessen their green manure value (373).

With abundant moisture, medics can produce more than 200 lb. N/A.

In the mid-Atlantic at the USDA Beltsville, Md. site, medics have been difficult to establish by over-seeding at vegetable planting or at final cultivation of sweet corn.

Pest Management

Under water logged conditions for which they are ill-suited, annual medics are susceptible to diseases like *Rhizoctonia, Phytophthora* and *Fusarium*.

Burr medic harbors abundant lygus bugs in spring. It also appears to be particularly prone to outbreaks of the two-spotted spider mite, a pest found in many West Coast orchards (422).

Pods and viable seeds develop without pollinators because most annual medics have no floral nectaries (120).

COMPARATIVE NOTES

Snail medic produced about the same biomass and N as red clover when both legumes were spring sown with an oats nurse crop into a disked seedbed in Wisconsin. Yields averaged over one wet year and one dry year were about 1 T dry matter and 60 lb. N/A (141).

Medics can establish and survive better than subterranean clover in times of low rainfall, and are more competitive with grasses. A short period of moisture will allow medic to germinate and send down its fast-growing taproot, while subclover needs more consistent moisture for its shallower, slower growing roots (422). Medics are more susceptible than subclover to seed production loss from closely mowing densely planted erect stalks. Burr and barrel medics are not as effective as subclover at absorbing phosphorus (422).

Medics may survive where true clovers (*Trifolium* spp.) fail due to droughty conditions (422) if there is at least 12 in. of rain per year (292).

Medics are easy to kill with light tillage or most herbicides.

Medics grow well in mixtures with grasses and clovers, but don't perform well with red clover (422, 263). Once established, black medic handles frost better than crimson or red clover.

GEORGE grows more slowly than yellow blossom sweetclover in spring of the second year, but it starts flowering earlier. It uses less water in the 2- to 4-foot depth than sweetclover, soybeans or hairy vetch seeded at the same time.

Annual Medic Cultivars. Species and cultivars of annual medic vary significantly in their dry matter production, crude protein concentration and total N. Check with local or regional forage specialists for cultivar recommendations

Burr medic (also called burclover) cultivars are the best known of the annual medics. They branch profusely at the base, and send out prostrate stems that grow more erect in dense stands (422). They grow quickly in response to fall California rains and fix from 55 to 90 lb. N/A, nearly as much as true clovers (294, 422). Most stands are volunteer and can be encouraged by proper grazing, cultivation or fertilization.

Selected cultivars include SERENA (an early bloomer), and CIRCLE VALLEY, both of which have fair tolerance to Egyptian alfalfa weevil (435). SANTIAGO blooms later than SERENA. Early burr medics flower in about 62 days in California, ranging up to 96 days for mid-season cultivars (422).

Naturalized and imported burr medic proved the best type of burclover for self-reseeding cover crops in several years of trials run from northern California into Mexico in the 1990s. While some of the naturalized strains have been self-reseeding for 30 years in some orchards, Extension specialists say the commercial cultivars may be preferable because they are widely available and better documented.

Established burr medic tolerates shade as a common volunteer in the understories of California walnut orchards, which are heavily shaded from April through November. However, in Michigan trials over several years, SANTIAGO (a burr medic with no spines on its burs) failed to establish satisfactorily when it was overseeded into corn and soybeans at layby. Researchers suspect the crop canopy shaded the medic too soon after planting, and that earlier overseeding may have allowed the medic to establish.

There are at least 10 cultivars of **barrel medic**. Dates of first flowering for barrel medics range from 80 to 105 days after germination, and seed count per pound ranges from 110,000 for HANAFORD to 260,000 for SEPHI (422). A leading new cultivar, SEPHI, flowers about a week earlier than JEMALONG, commonly used in California (251, 422).SEPHI, a mid-season cultivar, has a more erect habit for better winter production, is adapted to high- and low-rainfall areas, yields more seed and biomass than others, has good tolerance to Egyptian alfalfa weevil and high tolerance to spotted alfalfa aphid and blue green aphid. It is susceptible to pea aphid.

Snail medic (*M. scutellata*) is a prolific seed producer. Quick germination and maturity can lead to three crops (two reseedings) in a single season from a spring planting in the Midwest (373). MOGUL barrel medic grew the most biomass in a barley intercrop, compared with SANTIAGO burr medic and GEORGE black medic in a four-site Minnesota trial. It frequently reduced barley yields, particularly those of a semi-dwarf barley variety, but increased weed suppression and N and biomass production (289).

In a Michigan test of forage legumes for emergency forage use, MOGUL **barrel** medic produced 1.5 T dry matter/A compared to about 1 T/A for SAVA **snail** medic and SANTIAGO **burr medic** (*M. polymorpha*). Nitrogen production was 66 lb./A for MOGUL, 46 for SAVA and 22 for SANTIAGO. The seeding rate for SAVA medic is 29 lb./A, more than twice the 13 lb./A recommended for clear seedings of MOGUL and SANTIAGO (373, 376).

In a California pasture comparison of three annual medics, JEMALONG **barrel** had the highest level of seed reserves in the soil after six years, but didn't continue into the seventh year after the initial seeding. GAMMA medic (*M. rugosa*) had the highest first-year seed production but re-established poorly, apparently due to a low hard seed content. All the medics re-established better under permanent pasture than under any rotational system involving tillage (94, 422).

Seed sources. See *Seed Suppliers* (p. 195). 

RED CLOVER

Trifolium pratense

Also called: medium red clover (multi-cut, early blooming, June clover); mammoth clover (singlecut, late blooming, Michigan red)

Type: short-lived perennial, biennial or winter annual legume

Roles: N source, soil builder, weed suppressor, insectary crop, forage

Mix with: small grains, sweetclover, corn, soybeans, vegetables, grass forages

See charts, p. 66 to 72, for rankings and management summary.

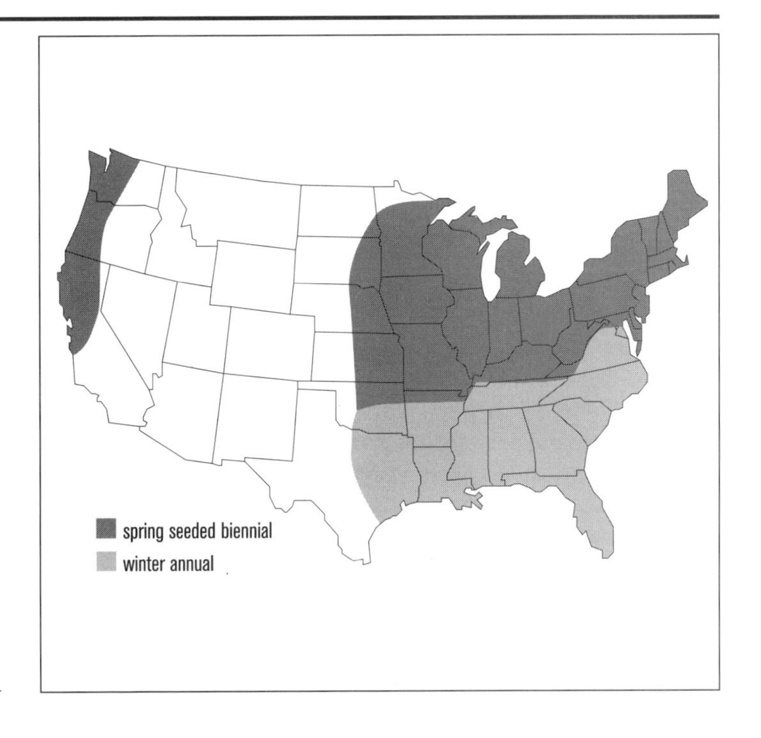

Red clover is a dependable, low-cost, readily available workhorse that is winter hardy in much of the U.S. (Hardiness Zone 4 and warmer). Easily overseeded or frostseeded into standing crops, it creates loamy topsoil, adds a moderate amount of N, helps to suppress weeds and breaks up heavy soil. Its most common uses include forage, grazing, seed harvest, plowdown N and, in warmer areas, hay. It's a great legume to frostseed or interseed with small grains where you can harvest grain as well as provide weed suppression and manage N.

BENEFITS

Crop fertility. As a cover crop, red clover is used primarily as a legume green manure killed ahead of corn or vegetable crops planted in early summer. Full-season, over-wintered red clover can produce 2 to 4 T dry matter/A and fix 70 to 150 lb. N/A. In Ohio, over-wintered mammoth and medium red clover contained about 75 lb. N/A by May 15, increasing to 130 lb. N by June 22 (366).

Two years of testing in Wisconsin showed that conventionally planted corn following red clover yielded the same as corn supplied with 160 lb. N/A, with less risk of post-harvest N leaching. Corn and the soil testing showed that 50 percent of the cover crop N was released in the first month after incorporation, corresponding well with corn's fertility demand. Post-harvest soil N levels in the clover plots were the same or less than the fertilized plots, and about the same as unfertilized plots (401).

Widely adapted. While many other legumes can grow quicker, produce more biomass and fix more nitrogen, few are adapted to as many soil types and temperate climatic niches as red clover. As a rule, red clover grows well wherever corn grows well. It does best in cool conditions.

Red clover can yield 2 to 3 tons of dry matter and 70 to 150 lb. N/A

In southern Canada and the northern U.S., and in the higher elevations of the Southeast and West, red clover grows as a biennial or short-lived perennial. At lower elevations in the Southeast, it grows as a winter annual, and at lower elevations in the West and Canada, it grows under irrigation as a biennial (120). It grows in any loam or clay soil, responding best to well-drained, fertile soils, but also tolerates less well-drained conditions.

Many economic uses. Red clover has been a popular, multi-use crop since European immigrant farmers brought it to North America in the 1500s. It remains an important crop thanks to its greater adaptability, lower seeding cost and easier establishment than alfalfa. It can produce up to 8,000 lb. biomass/A.

A red clover/small grain mix has been a traditional pairing that continues to be profitable. A rotation of corn and oats companion-seeded with red clover proved as profitable as continuous corn receiving 160 lb. N/A in a four-year Wisconsin study (400). For more information, see the Wisconsin Integrated Cropping Systems Trial (449) and the final report of this project, partially funded by SARE (328).

Red clover was the most profitable of five legumes under both seeding methods in the trial—sequentially planted after oats harvest or companion planted with oats in early spring. The companion seedings yielded nearly twice as much estimated fertilizer replacement value as the sequential seedings. The work showed that red clover holds great potential to reduce fertilizer N use for corn grown in rotation (401).

In Michigan, red clover frost-seeded into winter wheat suppressed common ragweed growth through wheat harvest and into the summer. The red clover did not provide complete ragweed control, but there was no adverse effect on wheat yield (297).

Red clover sown as a companion with spring oats outperformed the other legumes, which suffered from insect damage, mechanical damage during oat harvest and slow subsequent regrowth. The short season proved inadequate for sequentially seeded legumes with the exception of hairy vetch, which was nearly as profitable as the red clover (400).

The role of red clover's N contribution in the rotation grew more significant as N prices increased in the late 1990s (and 2007!), even though clover seed price also increased from the original 1989 calculations (398).

Soil conditioner. Red clover is an excellent soil conditioner, with an extensive root system that permeates the topsoil. Its taproot may penetrate several feet.

Attracts beneficial insects. Red clover earned a co-starring role with LOUISIANA S-1 white clover in pecan orchard recommendations from Oklahoma State University. Red clover attracts more beneficials than white clover, which features higher N fixation and greater flood tolerance than red clover (261).

Two Types

Two distinct types of red clover have evolved from the same species. Be sure you plant a multi-cut cultivar if you plan to make more than one green manure cutting, or to maintain the stand to prepare for a late-summer vegetable planting.

Medium red clover. Medium red (some call it multi-cut) grows back quickly, and can be cut once late in the seeding year and twice the following year. For optimum N benefit and flexible cropping options from the planting (allowing it to overwinter as a soil-protecting mulch), you can use it for hay, grazing or seed throughout the second season. Seed may be up to 25 percent more expensive than single-cut. See Chart 3B: *Planting* (p. 70).

Mammoth red clover produces significant biomass and as much N as medium red in a single first cutting, but does not produce as much total biomass and N as medium red's multiple cuttings over time. Use this "single-cut" red clover where a field will be all-clover just during the seeding year. Slow-growing mammoth doesn't bloom the establishment year and regrows quite slowly after cutting, but can provide good biomass by the end of even one growing season.

A single cutting of mammoth will give slightly more biomass—at a slightly lower cost—than a single cutting of medium red. Where multiple cuttings or groundcover are needed in the second season, medium red clover's higher seed cost is easily justified (197).

Some types of mammoth do better overseeded into wheat than into oats. ALTASWEDE (Canadian) mammoth is not as shade tolerant as MICHIGAN mammoth, but works well when seeded with oats. MICHIGAN mammoth shows the best vigor when frostseeded into wheat, but is not as productive as medium red (229).

MANAGEMENT

Establishment & Fieldwork

In spring in cool climates, red clover germinates in about seven days—quicker than many legumes—but seedlings develop slowly, similar to winter annual legumes. Traditionally it is drilled at 10 to 12 lb./A with spring-sown grains, using auxiliary or "grass seed" drill boxes. Wisconsin researchers who have worked for several years to optimize returns from red clover/oats interseedings say planting oats at 3 to 4 bu./A gives good stands of clover without sacrificing grain yield (398).

Red clover's tolerance of shade and its ability to germinate down to 41° F give it a remarkable range of establishment niches.

It can be **overseeded** at 10 to 12 lb./A into:

- **Dormant winter grains** before ground thaws. This "frostseeding" method relies on movement of the freeze-thaw cycle to work seed into sufficient seed-soil contact for germination. If the soil is level and firm, you can broadcast seed over snow cover on level terrain. You can seed the clover with urea if fertilizer application is uniform (229). Use just enough N fertilizer to support proven small-grain yields, because excess N application will hinder clover establishment. To reduce small grain competition with clover in early spring, graze or clip the small grain in early spring just before the stems begin to grow (120). Hoof impact from grazing also helps ensure seed-to-soil contact.
- **Summer annuals** such as oats, barley, spelt or spring wheat before grain emergence.
- **Corn at layby.** Wait until corn is 10 to 12 inches tall, and at least 6 weeks (check labels!) after application of pre-emergent herbicides such as atrazine. Clover sown earlier in favorable cooler conditions with more light may compete too much for water. Later, the clover will grow more slowly and not add substantial biomass until after corn harvest lets light enter (197). Dairy producers often broadcast red clover after corn silage harvest.
- **After wheat harvest.** Red clover logged a fertilizer replacement value of 36 lb. N/A in a two-year Michigan trial that used N isotopes to track nitrogen fixation. Red clover and three other legumes were no-till drilled into wheat stubble in August, then chemically killed by mid-May just ahead of no-till corn. Clover even in this short niche shows good potential to suppress weeds and reduce N fertilizer application (140).
- **Soybeans at leaf-yellowing.** Sowing the clover seed with annual or perennial ryegrass as a nurse crop keeps the soil from drying out until the clover becomes established (197).

Whenever possible, lightly incorporate clover seed with a harrow. Wait at least six weeks (check labels!) to establish a red clover stand in soil treated with pre-emergent herbicides such as atrazine.

Marianne Sarrantonio

RED CLOVER (Trifolium pratense)

Killing

For peak N contribution, kill red clover at about mid-bloom in spring of its second season. If you can't wait that long, kill it earlier to plant field corn or early vegetables. If you want to harvest the first cutting for hay, compost or mulch, kill the regrowth in late summer as green manure for fall vegetables (197). If avoiding escapes or clover regrowth is most important, terminate as soon as soil conditions allow.

Actively growing red clover can be difficult to kill mechanically, but light fall chisel plowing followed by a second such treatment has worked well in sandy loam Michigan soils.

To kill clover mechanically in spring, you can till, chop or mow it any time after blooming starts. You can also shallow plow, or use a moldboard plow. Chop (using a rolling stalk chopper), flail or sicklebar mow about seven to 10 days ahead of no-till planting, or use herbicides. Roundup Ready® soybeans can be drilled into living red clover and sprayed later.

A summer mowing can make it easier to kill red clover with herbicides in fall. Michigan recommendations call for mowing (from mid-August in northern Michigan to early September in southern Michigan), then allowing regrowth for four weeks before spraying. The daytime high air temperature should be above 60° F (so that the plants are actively growing). When soil temperature drops below 50° F, biological decomposition slows to the point that mineralization of N from the clover roots and top-growth nearly stops (229).

Field Evaluation

In Michigan, about half of the total N fixed by a legume will mineralize during the following growing season and be available to that season's crop (229). However, Wisconsin research shows release may be faster. There, red clover and hairy vetch released 70 to 75 percent of their N in the first season (401).

Rotations

Rotation niches for red clover are usually between two non-leguminous crops. Spring seeding with oats or frostseeding into wheat or barley are common options (34). The intersowing allows economic use of the land while the clover is developing. This grain/red clover combination often follows corn, but also can follow rice, sugar beets, tobacco or potatoes in two-year rotations. For three-year rotations including two full years of red clover, the clover can be incorporated or surface-applied (clipped and left on the field) for green manure, cut for mulch or harvested for hay (120).

Red clover in a corn>soybean>wheat/red clover rotation in a reduced-input system out-performed continuous corn in a four-year Wisconsin study. The legume cover crop system used no commercial fertilizer, no insecticides and herbicides on only two occasions—once to-spot spray Canada thistles and once as a rescue treatment for soybeans. Rotary hoeing and cultivating provided weed control.

Gross margins were $169 for the corn>soybeans>wheat/red clover and $126 for continuous corn using standard agricultural fertilizers, insecticides and herbicides. Top profit in the study went to a corn>soybean rotation with a gross margin of $186, using standard inputs (272, 398, 167).

Ohio farmer Rich Bennett frostseeds redclover (10-12 lb./A) into wheat in February. He gets a decent stand of clover that keeps weeds down in summer after wheat harvest. The clover overwinters and continues to grow in spring. He waits as long as possible, and then kills the clover with a disc and roll (two passes) in late April and plants corn. He doesn't add any fertilizer N and the corn averages 165 bu/A on his Ottokee fine sandy soil.

If summer annual grasses are a problem, red clover is not your best option because it allows the grasses to set seed, even under a mowing regime.

Pest Management

If poor establishment or winterkill leads to weed growth that can't be suppressed with clipping or grazing, evaluate whether the anticipated cover crop benefits warrant weed control. Take care to completely kill the cover crop when planting dry beans or soybeans after clover. Unless you are using herbicide-tolerant crops, you have limited herbicide options to control clover escapes that survive in the bean crop (229).

Root rots and foliar diseases typically kill common medium red clover in its second year, making it function more like a biennial than a perennial. Disease-resistant cultivars that persist three to four years cost 20 to 40 cents more per pound and are unnecessary for most green manure applications. When fertilizer N cost is high, however, remember that second-year production for some improved varieties is up to 50 percent greater than for common varieties.

Bud blight can be transmitted to soybeans by volunteer clover plants.

Other Management Options

Mow or allow grazing of red clover four to six weeks before frost in its establishment year to prepare it for overwintering. Remove clippings for green manure or forage to prevent plant disease. Red clover reaches its prime feeding value at five to 15 days after first bloom.

Under ideal conditions, medium red clover can be cut four times, mammoth only once. Maximum cutting of medium one year will come at the expense of second-year yield and stand longevity. Red clover and red clover/grass mixtures make good silage if wilted slightly before ensiling or if other preservative techniques are used (120).

If an emergency forage cut is needed, harvest red clover in early summer, then broadcast and lightly incorporate millet seed with a tine harrow or disk. Millet is a heat-loving grass used as a cover and forage in warm-soil areas of Zone 6 and warmer.

Few legumes are as widely adapted as red clover, which can be used as green manure, forage or seed crop.

COMPARATIVE NOTES

Medium red clover has similar upper-limit pH tolerance as other clovers at about 7.2. It is generally listed as tolerating a minimum pH of 6.0—not quite as low as mammoth, white or alsike (*Trifolium hybridum*) clovers at 5.5—but it is said to do well in Florida at the lower pH. Red clover and sweetclover both perform best on well-drained soils, but will tolerate poorly drained soils. Alsike thrives in wet soils.

Red clover has less tendency to leach phosphorus (P) in fall than some non-legume covers. It released only one-third to one-fifth the P of annual ryegrass and oilseed radish, which is a winter-annual brassica cover crop that scavenges large amounts of N. Figuring the radish release rates—even balanced somewhat by the erosion suppression of the covers—researchers determined that P runoff potential from a quick-leaching cover crop can be as great as for unincorporated manure (274).

For *early* fall plowdown, alsike clover may be a cheaper N source than mammoth, assuming similar N yields.

Red clover and alfalfa showed multi-year benefits to succeeding corn crops, justifying a credit of 90 lb. N/A the first year for red clover (197) and 50 lb. N/A the second year. The third legume in the trial, birdsfoot trefoil (*Lotus corniculatus*), was the only one of the three that had enough third-year N contribution to warrant a credit of 25 lb. N/A (148).

Cultivars. KENLAND, KENSTAR, ARLINGTON, and MARATHON are improved varieties of medium red clover with specific resistance to anthracnose and mosaic virus strains. They can persist three or even four years with ideal winter snow cover (90). CHEROKEE has performed well in Iowa (384), is suited to the Coastal Plain and lower South, and has superior resistance to rootknot nematode.

Seed sources. See *Seed Suppliers* (p. 195). 

SUBTERRANEAN CLOVERS

Trifolium subterraneum, T. yanninicum, T. brachycalcycinum

Also called: Subclover

Type: reseeding cool season annual legume

Roles: weed and erosion suppressor, N source, living or dying mulch, continuous orchard floor cover, forage

Mix with: other clovers and subclovers

See charts, p. 66 to 72, for ranking and management summary.

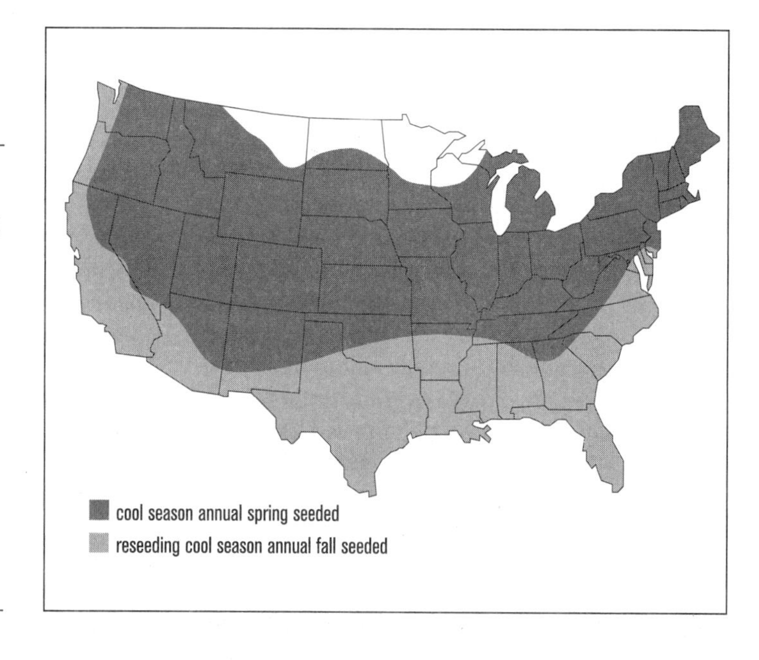

Subterranean clovers offer a range of low-growing, self-reseeding legumes with high N contribution, excellent weed suppression and strong persistence in orchards and pastures. Fall-planted subclovers thrive in Mediterranean conditions of mild, moist winters and dry summers on soils of low to moderate fertility, and from moderately acidic to slightly alkaline pH.

Subclover mixtures are used on thousands of acres of California almond orchards. It holds promise in the coastal mid-Atlantic and Southeast (Hardiness Zone 7 and warmer) as a killed or living mulch for summer or fall crops.

Most cultivars require at least 12 inches of growing-season rainfall per year. A summer dry period limits vegetative growth, but increases hard seed tendency that leads to self-reseeding for fall reestablishment (131).

Subclovers generally grow close to the ground, piling up their biomass in a compact layer. A Mississippi test showed that subclover stolons were about 6, 10 and 17 inches long when the canopy was 5, 7 and 9 inches tall, respectively (105).

Diversity of Types, Cultivars

Select among the many subclover cultivars that fit your climate and your cover crop goals. Identify your need for biomass (for mulch or green manure), time of natural dying to fit your spring-planting schedule and prominence of seed set for a persistent stand.

Subclovers comprise three *Trifolium* species:

- *T. subterraneum*. The most common cultivars that thrive in acid to neutral soils (pH=5.5-7.5) and a Mediterranean climate
- *T. yanninicum*. Cultivars best adapted to water-logged soils
- *T. brachycalcycinum*. Cultivars adapted to alkaline soils and milder winters

Primary differences between these species are their moisture requirements, seed production and days to maturity (21). Other variables include:

- Overall dry matter yield
- Dry matter yield at low moisture or low fertility
- Season of best growth (fall, winter or spring)
- Hard-seeding tendency
- Grazing tolerance

Subclover cultivars often are described by their days to maturity. Seed production is dependent on maturity and weather. The wetter it is during seed set, the lower the percentage of hard seed – important for reseeding systems (131).

• Short season subclovers tend to set seed quickly. They need only 8 to 10 inches of growing-season rainfall and set seed about 85 days after planting. Early subclovers tend to be less winter hardy (103).

• Intermediate types thrive with 14 to 20 inches of rain and set mature seed in about 100 days.

• Long-season cultivars perform best with 18 to 26 inches of rainfall, setting seed in about 130 days.

BENEFITS

Weed suppressor. Subclover can produce 3,000 to 8,500 lb. dry matter/A in a thick mat of stems, petioles (structures connecting leaves to stems) and leaves. Denser and less viny than hairy vetch, it also persists longer as a weed-controlling mulch.

Subclover mixtures help West Coast orchardists achieve season-long weed management. In Coastal California, fast-growing TRIKKALA, a midseason cultivar with a moderate moisture requirement, jumps out first to suppress weeds and produces about twice as much winter growth during January and February as the other subclovers. It dies back naturally as KOALA, (tall) and KARRIDALE (short) come on strong in March and April. The three cultivars complement each other spatially and temporally for high solar efficiency, similar to the interplanting of peas, purple vetch, bell beans and oats in California vegetable fields where a high-residue, high-N cover is desired.

In legume test plots along the Maryland shore, subclover mulch controlled weeds better than conventional herbicide treatments. The only weed to penetrate the subclover was a fall infestation of yellow nutsedge. The cover crop regrew in fall from hard seed in the second and third years of the experiment (31).

Subclovers thrive in Mediterranean climates of mild, moist winters and dry summers.

Green manure. In east Texas trials, subclover delivered 100 to 200 lb. N/A after spring plowdown. Grain sorghum planted into incorporated subclover or berseem clover with no additional N yielded about the same as sorghum planted into disked and fertilized soil without a cover crop in three out of four years. The fertilized fields had received 54 lb. N/A (243).

Versatile mulch. Subclover provides two opportunities for use as a mulch in vegetable systems. In spring, you can no-till early planted crops after subclover has been mechanically or chemically killed, or plant later, after subclover has set seed and dried down naturally (31). In fall, you can manage new growth from self reseeding to provide a green living mulch for cold-weather crops such as broccoli and cauliflower.

Conventionally tilled corn without a cover crop in a New Jersey test leached up to 150 lb. N/A

SUBTERRANEAN CLOVER (Trifolium subterraneum)

over winter while living subclover prevented N loss (128). Mowing was effective in controlling a living mulch of subclover in a two-year California trial with late-spring, direct-seeded sweet corn and lettuce. This held true where subclover stands were dense and weed pressure was low. Planting into the subclover mulch was difficult, but was done without no-till equipment (239).

Soil loosener. In an Australian study in compaction-prone sandy loam soil, lettuce yield doubled following a crop of subclover. Without the clover, lettuce yields were reduced 60 percent on the compacted soil. Soil improvement was credited to macropores left by decomposing clover roots and earthworms feeding on dead mulch (395).

Great grazing. Subclovers are highly palatable and relished by all livestock (120). Seeded with perennial ryegrass, tall fescue or orchardgrass, subclovers add feed value as they improve productivity of the grasses by fixing nitrogen. In California, subclover is used in pasture mixtures on non-irrigated hills. Perennial ryegrass is preferred for pasture through early summer, especially for sheep (309).

Insect pest protection. In the Netherlands, subclover and white clover in cabbage suppressed pest insect egg laying and larval populations enough to improve cabbage quality and profit compared with monocropped control plots. Eliminating pesticide costs offset the reduced weight of the cabbages in the undersown plots. Primary pests were *Mamestra brassicae, Brevicoryne brassicae* and *Delia brassicae*. Undersowing leeks with subclover in the Netherlands greatly reduced thrips that cannot be controlled by labeled insecticides, and slightly reduced leek rust, a disease that is difficult to control. While leek quality improved, the quantity of leeks produced was reduced considerably (415).

When tarnished plant bug (*Lygus lineolaris*) is a potential pest, subclover may be the legume cover crop of choice, based on a Georgia comparison among subclovers, hybrid vetches and crimson clover. MT. BARKER had particularly low levels, and nine other subclover cultivars had lower levels than the crimson (56).

Home for beneficial insects. In tests of eight cover crops or mixtures intercropped with cantalope in Georgia, MT. BARKER subclover had the highest population of big-eyed bugs (*Geocorus punctipes*), a pest predator. Subclover had significantly higher numbers of egg masses of the predator than rye, crimson clover and a polyculture of six other cover crops, but not significantly higher than for VANTAGE vetch or weedy fallow. While the covers made a significant difference in the predator level, they did not make a significant difference in control of the target pest, fall armyworm (*Spodoptera frugiperda*) (56).

Erosion fighter. Subclover's soil-hugging, dense, matted canopy is excellent for holding soil.

Disease-free. No major diseases restrict subclover acreage in the U.S. (21).

MANAGEMENT

Establishment

Subclovers grow best when they are planted in late summer or early autumn and grow until early winter. They go dormant over winter and resume growth in early spring. In late spring, plants flower and seeds mature in a bur at or below the soil surface (hence the name subterranean clover) as the plant dries up and dies. A dense mulch of dead clover leaves and long petioles covers the seeds, which germinate in late summer to establish the next winter's stand (127). Their persistence over many seasons justifies the investment in seed and careful establishment.

In California, sow in September or early October to get plants well established before cool weather (309). Planting continues through November in the most protected areas.

In marginally mild areas, establish with grasses for winter protection. Subclover stimulates the grasses by improving soil fertility. You can overseed pasture or range land without tillage, but you can improve germination by having livestock trample in the seed. Subclover often is aerially applied to burned or cleared land. Initial growth will be a little slower than that of crimson, but a little faster than white clover (120).

Broadcast at 20-30 lb./A in a firm, weed-free seedbed. Cover seed with a light, trailing harrow or with other light surface tillage to a depth of less than one-half inch. Add lime if soil is highly acid—below pH 5.5 (309). Soils low in pH may require supplemental molybdenum for proper growth, and phosphorus and sulfur may also be limiting nutrients. Only the *T. yanninicum* cultivars will tolerate standing water or seepage areas (21, 309).

Subclover often is planted with rose clover and crimson clover in California orchard mixes. Crimson and subclover usually dominate, but hard-seeded rose clover persists when dry weather knocks out the other two (447).

In the East, central Mississippi plantings are recommended Sept. 1 to Oct. 15, although earlier plantings produce the earliest foliage in spring (120). In coastal Maryland where MT. BARKER plants were tallest and most lush, winterkill (caused when the temperature dropped to 15° F or below) has been most severe. Planting in this area of Zone 7 should be delayed until the first two weeks of October. Plant at about 22 lb./A for cover crop use in the mid-Atlantic (31) and Southeast (103). This is about double the usual recommended rate for pastures in the warmer soils of the Southeast.

Small plants of ground-hugging subclover benefit more from heat radiating from the soil than larger plants, but are more vulnerable during times of freezing and thawing. Where frost heaving is expected, earlier planting and well-established plants usually survive better than smaller ones (103).

Killing

Subclover dies naturally in early summer after blooming and seed set. It is relatively difficult to kill without deep tillage *before* mid-bloom stage. After stems get long and seed sets, you can kill plants with a grain drill or a knife roller (95).

In northern Mississippi, subclover was the least controlled of four legumes in a mechanical kill test. The cover crops were rolled with coulters spaced 4 inches apart when the plants had at least 10 inches of prostrate growth. While hairy vetch and crimson clover were 80 to 100 percent controlled, berseem control was 53 percent and subclover was controlled only 26 to 61 percent (105).

Subclover mixes help keep weeds in check all season long in West Coast orchards.

Researchers in Ohio had no trouble killing post-bloom subclover with a custom-built undercutter. The specialized tool is made to slice 1 to 2 inches below the surface of raised beds. The undercutter consisted of two blades that are mounted on upright standards on either side of the bed and slant backward at 45 degrees toward the center of the bed. A mounted rolling harrow was attached to lay the cover crop flat on the surface after being cut (96). The tool severs stalks from roots while above-ground residue is undamaged, greatly slowing residue decomposition (95).

Subclover tolerance to herbicides varies with cultivar and growth stage. Generally, subclover is easier to kill after it has set some seed (104, 165).

Reseeding Management

The "over-summering" fate of reseeding subclover plantings is as critical to their success as is the over-wintering of winter-annual legumes. The thick mat of vegetation formed by dead residue can keep subclover seeds dormant if it is not disturbed by grazing, tillage, burning or seed harvest. Where cover crop subclover is to be grazed before another year's growth is turned under, intensive grazing management works best to reduce residue but to avoid excess seed bur consumption (309). Grazing or mowing in late spring or early summer helps control weeds that grow through the mulch (292).

You can improve volunteer regrowth of subclover in warm-season grass mixes by limiting N fertilization during summer, and by grazing the grass shorter until cold temperatures limit grass growth. This helps even though subclover seedlings may emerge earlier (21). Subclover flowers are inconspicuous and will go unnoticed without careful, eye-to-the-ground inspection (103).

After plants mature, livestock will eagerly eat seed heads (120). In dry years when you want to maintain the stand, limit grazing over summer to avoid over-consumption of seed heads and depletion of the seed bank. Close mowing or grazing can be done any time.

Management Challenges

Possible crop seedling suppression. The allelopathic compounds that help subclover suppress weeds also can hurt germination of some crops. To avoid problems with these crops, delay planting or remove subclover residue. No-till planters equipped with tine-wheel row cleaners can reduce the recommended 21-day waiting period that allows allelopathic compounds to drop to levels that won't harm crops (101). Kill subclover at least a year before planting peach trees to avoid a negative effect on seedling vigor. It's best to wait until August of the trees' second summer to plant subclover in row middles, an Arkansas study found (61).

The degree to which a cover crop mulch hinders vegetable seedlings is crop specific. Plant-toxic compounds from subclover mulch suppressed lettuce, broccoli and tomato seedlings for eight weeks, but not as severely or as long as did compounds from ryegrass (*Lolium rigidum* cv. WIMMERA) mulch. An alfalfa mulch showed no such allelopathic effect in an Australian study (395).

Guard against moisture competition from subclover at planting. Without irrigation to ensure crop seeds will have enough soil moisture to germinate in a dry year, be sure that the subclover is killed seven to 14 days prior to planting to allow rainfall to replenish soil moisture naturally (31).

Soil-borne crop seedling disease. In north Mississippi tests, residue and leachate from legume cover crops (including subclover) caused greater harm to grain sorghum seedlings, compared to nonlegumes. *Rhizoctonia solani*, a soil-borne fungus, infected more than half the sorghum seedlings for more than a month, but disappeared seven to 13 days after legume residues were removed (101).

N-leaching. The early and profuse nodulation of subclovers that helps grass pastures also has a downside—excess N in the form of nitrate can contaminate water supplies. Topgrowth of subclover, black medic and white clover leached 12 to 26 lb. N/A over winter, a rate far higher than red clover and berseem clover, which leached only 2 to 4 lb. N/A in a Swedish test (227).

Pest Management

Subclovers showed little resistance to root-knot nematodes in Florida tests on 134 subclover lines in three years of testing the most promising varieties (233).

Lygus species, important pests of field, row and orchard crops in California and parts of the Southeast, were notably scarce on subclover plants in a south Georgia comparison. Other legumes harboring more of the pests were, in descending order, CAHABA and VANTAGE vetch, hairy vetch, turnip and monoculture crimson clover (56).

Most cultivars imported to the U.S. are low in estrogen, which is present in sufficient levels in some Australian cultivars to reduce fertility in ewes, but not in goats or cattle. Confirm estrogen status of a cultivar if you plan to graze sheep on it (309).

Crop Systems

Interseeded with wheat. NANGEELA subclover provided 59 lb. N/A when it was grown as an interseeded legume in soft red winter wheat in eastern Texas. That extra N helped boost the wheat yield 283 percent from the previous year's yield when four subclover cultivars were first established and actually decreased yield, compared with a control plot. NANGEELA, MT. BARKER, WOOLGENELLUP and NUNGARIN cultivars boosted wheat yield by 24, 18, 18, and 11 bu./A, respectively, in the second year of the study. Over all three years, the four cultivars added 59, 51, 38 and 24 lb. N/A, respectively (44).

Plant breeder Gerald Ray Smith of Texas A&M University worked with several subclovers in eastern Texas. While the subclovers grew well the first year, he concluded that those cultivars need a prolonged dry period at maturity to live up to their reseeding performance in Australia and California. Surface moisture at seed set reduces seed hardening and increases seed decay. Midsummer rains cause premature germination that robs the subclover seed bank, especially in pastures where grasses tend to create moist soil. Most summer-germinating plants die when dry weather returns.

In Mississippi, subclover hard seed development has been quite variable from year to year. In dry years, close to 100 percent hard seed is developed. Dormancy of the seed breaks down more rapidly on bare soil with wider temperature swings than it does on mulched soils (133, 134). To facilitate reseeding or to seed into pastures, the grasses must be mowed back or grazed quite short for the subclover to establish (103).

Mix for persistence. California almond growers need a firm, flat orchard floor from which to pick up almonds. Many growers use a mix of moisture-tolerant TRIKKALA, alkaline-tolerant KOALA, and KARRIDALE, which likes neutral to acid soils. These blended subclovers give an even cover across moist swails and alkaline pockets.

Rice N-source. In Louisiana trials, subclover regrew well in fall when allowed to set seed before spring flooding of rice fields. Compared with planting new seed, this method yields larger seedling populations, and growth usually begins earlier in the fall. The flood period seems to enhance dormancy of both subclover and crimson clover, and germination is robust when the fields are drained (103). Formerly, some Louisiana rice farmers seeded the crop into dry soil then let it develop for 30 days before flooding. Early varieties such as DALKIETH and NORTHAM may make seed prior to the recommended rice planting date. In recent decades, "water planting" has been used to control red rice, a weedy relative of domestic rice. Water seeding into cover crop residues has not been successful (36).

Fertility, weed control for corn. In the humid mid-Atlantic region, grain and silage corn no-tilled into NANGEELA subclover did well in a six-year New Jersey trial. With no additional N, the subclover plots eventually out-yielded comparison plots of rye mulch and bare-soil that were conventionally tilled or minimum-tilled with fertilizer at up to 250 lb. N/A. The subclover contributed up to 370 lb. N/A (128), an N supply requiring careful management after the subclover dies to prevent leaching.

Control of fall panicum was poor in the first year, but much better the next two years. Control of the field's other significant weed, ivyleaf morning glory, was excellent in all years. Even though no herbicide was used in the subclover plots, weed biomass was lowest there (128).

Central New Jersey had mild winters during these experiments. Early spring thaws triggered subclover regrowth followed by plunging temperatures that dropped below 15° F. This weakened the plants and thinned the stands. The surviving plants, which formed dense stands at times, were mowed or strip-killed using herbicides or tillage. Mowing often induced strong regrowth, so strips at least 12 inches wide proved to be the best to prevent moisture competition between the subclover and the cabbage and zucchini transplants.

Sustainable sweet corn. On Maryland's Eastern Shore (one USDA hardiness zone warmer than New Jersey), University of Maryland weed specialist Ed Beste reported good reseeding in four consecutive years and no problems with stand loss from premature spring regrowth. Overwintering MT. BARKER plants sent out stolons across the soil surface to quickly re-establish a good stand ahead of sweet corn plantings (31).

Beste believes the sandy loam soil with a sand underlayer at his site is better for subclover than the heavier clay soils at the USDA Beltsville station some 80 miles north, where hairy vetch usually out-performs subclover as a killed organic mulch in transplanted vegetable systems. Winterkill reduced the subclover stand on top of bedded rows one year of the comparison, yet surviving plants between the beds produced nearly as much biomass per square foot as did hairy vetch (2).

Beste has worked with subclover at his Salisbury, Md., site for several years, seeding vegetables in spring, early summer and mid-summer into the killed or naturally dead cover crop mulch. For three years, subclover at Beste's sweet corn system comparison site yielded about 5,400 lb. DM/A. Without added N, the subclover plots yielded as much sweet corn as conventional plots receiving 160 lb. N/A. Weed suppression also was better than in the conventional plots. He sprayed glyphosate on yellow nutsedge in fall to prevent tuber formation by the grassy weed, the only weed that penetrated the subclover mulch (31).

Beste sprays paraquat twice to control subclover ahead of no-till, direct seeded zucchini in the first week of June. His MT. BARKER will set seed and die back naturally at the end of June—still in time to seed pumpkins, fall cucumbers, snap beans or fall zucchini planted without herbicides (31). Such a no-chemical/dying mulch/perpetually reseeding legume system is the goal of cultivar and system trials in California.

Seed production in subclovers normally is triggered by increasing day length in spring after the plant experiences decreasing fall day length. This explains why spring-planted subclover in Montana tests produced profuse vegetative growth, especially when fall rains began, but failed to set any seed (383). Stress from drought and heat also can trigger seed set.

COMPARATIVE NOTES

White and arrowleaf clovers have proved to be better self-reseeding clovers than subclover in the humid South because their seed is held in the air, giving them a better chance to harden. Top reseeding contenders are balansa clover (see *Up-and-Coming Cover Crops*, p. 191) and southern spotted burr medic (see *Southern Spotted Burr Medic Offers Reseeding Persistence*, p. 154).

While mid-season subclovers generally produced more dry matter and N than medics for dryland cereal-legume rotations in Montana (381), they did not set seed when grown as summer annuals in the region. Summer growth continued as long as moisture held up in trials there. Vegetative growth increased until frost, as cool, moist fall weather mimicks the Mediterranean winter conditions where subclover thrives (383).

CLARE is a cultivar of the subclover subspecies *brachycalycinum*. Compared with the more common subspecies subterranean (SEATON PARK and DALIAK), CLARE has vigorous seedlings, robust growth when mowed monthly and is said to tolerate neutral to alkaline soils. However, it appears to be less persistent than other types (61).

Subclover, rye and crimson clover provided grass weed control that was 46 to 61 percent better than a no-cover/no-till system at two North Carolina locations. Subclover topped the other covers in suppressing weeds in plots where no herbicides were used. None of the cover crop treatments eliminated the need for pre-emergent herbicides for economic levels of weed control (454).

Subclover creates a tighter mat of topgrowth than vetch (31) or crimson clover (103).

Cultivars. See *Comparative Notes*, above, and *Diversity of Types, Cultivars* (p. 165).

Seed sources. See *Seed Suppliers* (p. 195). 

SWEETCLOVERS

Yellow sweetclover *(Melilotus officinalis)* and white sweetclover *(M. alba)*

Also called: HUBAM (actually a cultivar of annual white sweetclover)

Type: biennial, summer annual or winter annual legume

Roles: soil builder, fertility source, subsoil aerator, weed suppressor, erosion preventer

Mix with: small grains, red clover

See charts, pp. 66 to 72, for ranking and management summary.

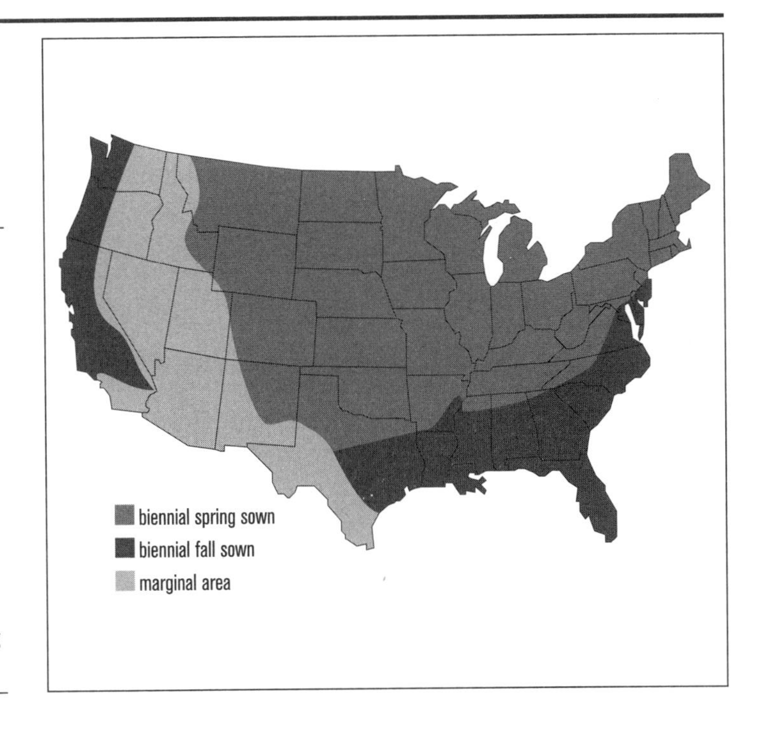

Within a single season on even marginally fertile soils, this tall-growing biennial produces abundant biomass and moderate amounts of nitrogen as it thrusts a taproot and branches deep into subsoil layers. Given fertile soils and a second season, it lives up to its full potential for nitrogen and organic matter production. Early in the second year it provides new top growth to protect the soil surface as its roots anchor the soil profile. It is the most drought-tolerant of forage legumes, is quite winter-hardy and can extract from the soil then release phosphorus, potassium and other micronutrients that are otherwise unavailable to crops.

Sweetclover thrives in temperate regions wherever summers are mild. *Annual* sweetclovers (HUBAM is the most well known) work best in the Deep South, from Texas to Georgia. There, they establish more quickly than the biennial types and produce more biomass in the seeding year in southern regions.

In this chapter, "sweetclover" refers to *biennial* types unless otherwise noted.

Sweetclover was the king of green manures and grazing legumes in the South and later throughout the Midwest in the first half of this century. Sweetclover is used as a cover crop most commonly now in the Plains region, with little use in California.

Types

Biennial *yellow* sweetclover can produce up to 24 inches of vegetative growth and 2.5 tons dry matter/A in its establishment year. During the second year, plants may reach 8 feet tall. Root mass and penetration (to 5 feet) are greatest at the end of dormancy in early spring, then gradually dissipate through the season (443).

A distinguishing sweetclover feature is bracts of tiny blooms through much of its second year. *White* biennial sweetclovers are taller, more coarsely stemmed, less drought tolerant, and produce less biomass in both the seeding and second years. White types bloom 10 to 14 days later than yellow, but bloom for a longer season. They reportedly establish more readily in New York (450). Tall, stemmy cultivars are better for soil improvement (120, 361, 422).

Marianne Sarrantonio

YELLOW SWEETCLOVER (Melilotus officinalis)

Both yellow and white sweetclover have cultivars bred for low levels of coumarin. This compound exists in bound form in the plant and poses no problem during grazing. However, coumarin can cause internal injury to cattle when they eat spoiled sweetclover hay or silage.

Annual sweetclover (*Melilotus alba var. annua*) is not frost tolerant, but can produce up to 9,000 lb. dry matter/A over a summer after being oversown into a grain crop or direct seeded with a spring grain nurse crop. The best-known annual sweetclover cultivar is HUBAM, a name often used for all annual white sweetclover. While its taproot is shorter and more slender than that of its biennial cousins, it still loosens subsoil compaction.

BENEFITS

Nutrient scavenger. Sweetclover appears to have a greater ability to extract potassium, phosphorus and other soil nutrients from insoluble minerals than most other cover crops. Root branches take in minerals from seldom-disturbed soil horizons, nutrients that become available as the tops and roots decompose (361).

Research in Saskatchewan during a 34-year period showed that phosphorus (P) availability increased in *subsoil* layers relative to *surface* layers, peaking at an 8-foot depth. Winter wheat and safflower, with deeper root systems than spring wheat, could tap the deep P buildup from the legume roots and fallow leaching, whereas spring wheat could not. The vesicular-arbuscular mycorrhizal (VAM) fungi associated with legume roots contribute to the increased P availability associated with sweetclover (69, 70).

N source. A traditional green manure crop in the upper Midwest before nitrogen fertilizer became widely available, sweetclover usually produces about 100 lb. N/A, but can produce up to 200 lb. N/A with good fertility and rainfall. In Ohio, it contained about 125 lb. N/A by May 15, increasing to 155 lb. by June 22. Illinois researchers reported more than 290 lb. N/A.

Abundant biomass. If planted in spring and then given two full seasons, biennial sweetclovers can produce 7,500 to 9,000 lb. dry matter/A (3,000 to 3,500 lb./A in the seeding year, and 4,500 to 5,500 lb./A the second). Second-year yields may go as high as 8,500 lb./A.

Hot-weather producer. Sweetclover has the greatest warm-weather biomass production of any legume, exceeding even alfalfa.

Soil structure builder. Kansas farmer Bill Granzow says sweetclover gives his soils higher organic matter, looser structure and better tilth. See *Sweetclover: Good Grazing, Great Green Manure* (p. 174). HUBAM annual sweetclover also improved soil quality and increased yield potential in 1996 New York trials (451).

Compaction fighter. Yellow sweetclover has a determinate taproot root up to 1 foot long with extensive branches that may penetrate 5 feet to aerate subsoils and lessen the negative effects of compaction on crops. White types have a strong tap root that is not determinate.

Drought survivor. Once established, sweetclover is the most drought tolerant of all cover crops that produce as much biomass. It is especially resilient in its second year, when it could do well in a dry spring during which it would be difficult to establish annual cover crops. The yellow type is less sensitive to drought and easier to establish in dry soils than the white type.

Attracts beneficial insects. Blossoms attract honeybees, tachinid flies and large predatory wasps, but not small wasps.

Widely acclimated. Self-reseeding sweetclover can be seen growing on nearly barren slopes, road rights-of-way, mining spoils and soils that have low fertility, moderate salinity or a pH above 6.0 (183). It also can tolerate a wide range of environments from sea level to 4,000 feet in altitude, including heavy soil, heat, insects, plant diseases (120) and as little as 6 inches of rain per year.

Livestock grazing or hay. If you need emergency forage, sweetclover has a first-year feed value similar to alfalfa, with greater volume of lesser quality in the second year.

MANAGEMENT

Establishment & Field Management

Sweetclover does well in the same soils as alfalfa. Loam soils with near-neutral pH are best. Like alfalfa, it will not thrive on poorly drained soils. For high yields, sweetclover needs P and K in the medium to high range. Deficient sulfur may limit its growth (153). Use an alfalfa/sweetclover inoculant.

In temperate areas of the Corn Belt, drill yellow sweetclover in pure stands at 8 to 15 lb./A or broadcast 15 to 20 lb./A, using the higher rate in dry or loose soils or if not incorporating.

In drier areas such as eastern North Dakota, trials of seeding rates from 2 to 20 lb./A showed that just 4 lb./A, broadcast or drilled, created an adequate sole-crop stand for maximum yield. Recommended rates in North Dakota are 4 to 6 lb./A drilled with small grains at small-grain planting, 5 to 8 lb./A broadcast and harrowed (sometimes in overseeding sunflowers), and 6 to 10 lb./A. broadcast without incorporating tillage (183).

An excessively dense stand will create spindly stalks that don't branch or root to the degree that plants do in normal seedings. Further, the plants will tend to lodge and lay over, increasing the risk of diseases. So for maximum effect of subsoil penetration or snow trapping, go with a lighter seeding rate.

Sweetclover produces 50 percent or more hard seed that can lie in soil for 20 years without germinating. Commercial seed is scarified to break this non-porous seedcoat and allow moisture to trigger germination. If you use unscarified seed, check hardseed count on the tag and do not count on more than 25 percent germination from the hardseed portion. The need for scarification to produce an adequate stand may be over-rated, however. The process had no effect on germination in six years of field testing in North Dakota—even when planting 70 percent hard seed still in seed pods.

Winter-hardy and drought tolerant, this biennial can produce up to 200 lb. N/A with good fertility and rainfall.

Seed at a depth of 1/4 to 1/2 inch in medium to heavy textured soils, and 1/2 to 1.0 inch on sandy soils. Seeding too deeply is a common cause of poor establishment.

Seed *annual* white sweetclover at 15 to 30 pounds per acre. Expect 70 to 90 lb. N/A from 4,000 to 5,000 lb. dry matter/A on well-drained, clay loam soils with neutral to alkaline pH.

A press-wheel drill with a grass seed attachment and a seed agitator is suitable for planting sweetclover into a firm seedbed. If the seedbed is too loose to allow the drill to regulate seeding depth, run the seed spouts from the grass and legume boxes to drop seed behind the double-disk opener and in front of the press wheels. Light, shallow harrowing can safely firm the seedbed and incorporate seed (183).

In the Canadian Northern Plains, dribble the seed through drill box hoses directly in front of the presswheels for quick and easy establishment (32).

If your press-wheel drill has no legume box or grass-seed attachment, you can mix the legume and small grain seed, but mix seed often due to settling. Reduce competition between the crops by seeding a part of the companion crop first, then seed a mix of the clover seed and the balance of the grain seed at right angles (183).

Sweetclover: Good Grazing, Great Green Manure

Bill Granzow taps biennial yellow sweetclover to enhance soil tilth, control erosion and prevent subsoil from becoming compacted. He uses common varieties, either from the elevator or one his father originally bought from a neighbor.

Granzow, of Herington, Kan., produces no-till grain and runs cattle in an area midway between Wichita and Manhattan in the east-central part of the state. Granzow overseeds sweetclover into winter wheat in December or January at 12 to 15 lb./A using a rotary broadcaster mounted on his pickup. Sometimes he asks the local grain cooperative to mix the seed with his urea fertilizer for the wheat. There's no extra charge for seed application. Alternately, Granzow plants sweetclover at the same rate with March-seeded oats.

Yellow sweetclover has overgrown Granzow's wheat only when the wheat stand is thin and abnormally heavy rains delay harvest. The minimal problem is even rarer in oats, he says.

He uses yellow sweetclover with the companion wheat crop in four possible ways, depending on what the field needs or what other value he wants to maximize. For each, he lets the clover grow untouched after wheat harvest for the duration of the seeding year. He used to disk the sweetclover at least twice to kill it. Now 100% no-till, he sprays with Roundup and "a little bit of 2,4-D." Second-year options include:

- **Grazing/green manure.** Turn in livestock when the clover reaches 4 inches tall, let them graze for several weeks, spray to kill, then plant grain sorghum within a couple of days. He feeds an anti-bloat medication to keep cattle healthy on the lush legume forage.
- **Quick green manure.** Spray after it has grown 3 to 4 inches, then plant sorghum. This method contributes about 60 pounds of N to the soil. He knocks back persistent re-growing sweetclover crowns in the sorghum by adding 2, 4-D or Banvel to the postemerge herbicide mix.
- **Green manure/fallow.** Kill at mid- to full bloom, leave fallow over summer, then plant wheat again in fall. This method provides about 120 lb. N/A, according to estimates from Kansas State University.
- **Seed crop.** He windrows the plants when about 50 percent of the seedpods have turned black, then runs the stalks through his combine. To remove all of the hulls, he runs the seed through the combine a second or third time.

Despite the heavy growth in the second year, yellow sweetclover matures and dies back naturally. If the residue is heavy, he sets the drill a bit deeper for planting.

He rates fall sweetclover hay from the seeding year as "acceptable forage." He's aware that moldy sweetclover hay contains coumarin, a compound that can kill cattle, but he's never encountered the problem. Second-year yellow sweetclover makes silage at initial to mid-bloom stage with 16 percent protein on a dry matter basis.

"Mixed with grass hay or other silage, it makes an excellent feed," he says, adding value to its cover crop benefits and giving him farming flexiblity.

Updated in 2007 by Andy Clark

Spring seeding provides yellow sweetclover ample time to develop an extensive root system and store high levels of nutrients and carbohydrates necessary for over-wintering and robust spring growth. It grows slowly the first 60 days (153). Where weeds would be controlled by mowing, no-till spring seeding in small grain stubble works well.

Broadcast seeding for pure sweetclover stands works in higher rainfall areas in early spring where soil moisture is adequate for seven to 10 days after planting. No-till seeding works well in small grain stubble.

Frostseeding into winter grains allows a harvest of at least one crop during the life cycle of the sweetclover and helps control weeds while the sweetclover establishes. Apply sweetclover seed before rapid stem elongation of the grain. Cut grain rate about one-third when planting the crops together.

Sweetclover spring seeded with oats exhibited poor regrowth after oat harvest in two years of a Wisconsin study. To establish a sweetclover cover crop in this way, the researchers found sweetclover did not fare well in years when the combine head had to be run low to pick up lodged oats. When oats remained upright (sacrificing some straw for a higher cut), sweetclover grew adequately (402).

You can plow down spring-planted yellow sweetclover in late fall of the planting year to cash in early on up to half its N contribution and a bit less than half its biomass.

Plant biennial sweetclover through late summer where winters are mild, north through Zone 6. Plant at least six weeks before frost so roots can develop enough to avoid winter heaving. In the Northern Plains into Canada, it should be planted by late August.

First-year management. Seeding year harvest or clipping is usually discouraged, because the energy for first-year regrowth comes directly from photosynthesis (provided by the few remaining leaves), not root reserves (361, 402).

Top growth peaks in late summer as the plant's main taproot continues to grow and thicken. Second-year growth comes from crown buds that form about an inch below the soil surface. Avoid mowing or grazing of sweetclover in the six- to seven-week period prior to frost when it is building final winter reserves. Root production practically doubles between Oct. 1 and freeze-up.

Sweetclover establishes well when sown with winter grains in fall, but it can outgrow the grain in a wet season and complicate harvest.

Second-year management. After it breaks winter dormancy, sweetclover adds explosive and vigorous growth. Stems can reach 8 feet before flowering, but if left to mature, the stems become woody and difficult to manage. Plants may grow extremely tall in a "sweetclover year" with high rainfall and moderate temperatures.

Nearly all growth the second year is topgrowth, and it seems to come at the expense of root mass. From March to August in Ohio, records show topgrowth increasing tenfold while root production *decreased* by 75 percent (443). All crown buds initiate growth in spring. If you want regrowth after cutting, leave plenty of stem buds on 6 to 12 inches of stubble. You increase the risk of killing the sweetclover plant by mowing heavier stands, at shorter heights, and/or at later growth stages, especially after bloom (183).

Before it breaks dormancy, sweetclover can withstand flooding for about 10 days without significant stand loss. Once it starts growing, however, flooding will kill the plants (183).

Killing

For best results ahead of a summer crop or fallow, kill sweetclover in the second year after seeding when stalks are 6 to 10 inches tall (183, 361). It can be killed by mowing, cultivating or disking once it reaches late bloom stage (32). Killing sweetclover before bud stage has several benefits: 80 percent of the potential N is present; N release is quick because the plant is still quite vegetative with a high N percentage in young stalks and roots; and moisture loss is halted without reducing N contribution. Sweetclover may regrow from healthy crowns if incorporated before the end of dormancy. For optimum full-season organic matter contribution, mow prior to blossom stage whenever sweetclover reaches 12 to 24 inches high before final incorporation or termination (361). Mowing or grazing at bloom can kill the plants.

In dryland areas, the optimum termination date for a green manure varies with moisture conditions. In a spring wheat>fallow rotation in Saskatchewan, sweetclover incorporated in mid-June of a dry year provided 80 percent more N the following spring than it did when incorporated in early July or mid-July—even though it yielded up to a third less biomass at the June date. Mineralization from sweetclover usually peaks about a year after it is killed. The potential rate of N release decreases as plants mature and is affected by soil moisture content (147).

During its second season, yellow sweetclover can grow 8 feet tall while roots penetrate 5 feet deep.

In this study, the differences in N release were consistent in years of normal precipitation, but were less pronounced. Little N mineralization occurred in the incorporation year. Nitrogen addition peaked in the following year, and has been shown to continue over seven years following yellow sweetclover (147).

In northern spring wheat areas of North Dakota, yellow sweetclover is usually terminated in early June just at the onset of bloom, when it reaches 2 to 3 feet tall. This point is a compromise between cover crop gain (in dry matter and N) and water consumption. A quick kill from tillage or haying is more expensive and labor-intensive than chemical desiccation, but it stops moisture-robbing transpiration more quickly (153).

Grazing is another way to manage second-year sweetclover before incorporation. Start early in the season with a high stocking rate of cattle to stay ahead of rapid growth. Bloat potential is slightly less than with alfalfa (153).

Pest Management

Sweetclover is a rather poor competitor in its establishment year, making it difficult to establish pure sweetclover in a field with significant weed pressure. Once established, it provides effective weed control during the first fall and spring of fallow, whether or not it is harvested for hay, incorporated or left on the soil surface (33).

Sweetclover residue is said to be allelopathic against kochia, Russia thistle, dandelion, perennial sowthistle, stinkweed and green foxtail. Repeated mowing of yellow sweetclover that is then left to mature is reported to have eradicated Canada thistle. Letting sweetclover bloom and go to seed dries out soil throughout the profile, depleting the root reserves of weeds.

Sweetclover weevil (*Sitonia cylindricollis*) is a major pest in some areas, destroying stands by defoliating newly emerged seedlings. Long rotations can reduce damage, an important factor for organic farmers who depend on sweetclover fertility and soil improvements. In the worst years of an apparent 12 to 15-year weevil cycle in his area, "every sweetclover plant across the countryside is destroyed," according to organic farmer David Podoll, Fullerton, N.D. "Then the weevil population crashes, followed by a few years where they're not a problem, then they begin to rebuild."

Cultural practices have not helped change the cycle, but planting early with a non-competitive nurse crop (flax or small grains) gives sweetclover plants the best chance to survive weevil foraging, Podoll says. Further research is needed to develop management techniques to control the weevil.

In a three-year Michigan trial of crop rotations to decrease economic losses to nematodes, a yellow sweetclover (YSC)>YSC>potato sequence out-yielded other combinations of rye, corn, sorghum-sudangrass and alfalfa. Two years of clover or alfalfa followed by potatoes led to a yield response equivalent to application of a nematicide for control of premature potato vine death (78). Legume-supplied N coupled with an overall nutrient balance and enhanced cation exchange capacity from the cover crop are thought to be involved in suppressing nematode damage (271).

Crop Systems

In the moderately dry regions of the central and northern Great Plains, "green fallow" systems with water-efficient legumes can be substituted for bare-ground or stubble mulch fallow. In fallow years, no cash crop is planted with the intent of recovering soil moisture, breaking disease or weed cycles and maximizing yields of following cash crops. The retained residue of "brown" fallow lessens the erosion and evaporation of tillage-intensive "black fallow," but "green fallow" offers even more benefits in terms of soil biological life, biodiversity, beneficial insect habitat, possible harvestable crops and alternate forages.

Rapeseed (*Brassica campestris*) is a summer annual cash crop in the dryland West that can serve as a nurse crop for sweetclover. A Saskatchewan study of seeding rates showed optimum clover yield came when sweetclover was sown at 9 lb./A and rapeseed was sown at 4.5 lb./A. The mixture allows an adequate stand of sweetclover that provides soil protection after the low-residue rapeseed (255).

Sole-cropped oilseed species (rapeseed, sunflower, crambe and safflower) require herbicides for weed control. Many of these materials are compatible with legumes, offering a post-emergent weed-control option if the covers do not adequately suppress weeds. The covers greatly reduce the erosion potential after oilseed crops, which leave little residue over winter (153).

Interplanting works with tall crops. A Wisconsin researcher reported success drilling sweetclover between the rows when corn was 6 to 12 inches tall. Overseeding sweetclover into sweet corn works even better due to greater light penetration.

Soil water availability at cover crop planting and depletion during growth are always a concern in semi-arid regions. The potential benefits must be balanced against possible negative effects on the cash crop.

Sweetclover overseeded into sunflowers at last cultivation succeed about half the time, North Dakota trials show. Dry conditions or poor seed-to-soil contact were the main reasons for not getting a stand. A heavier seeding rate or earlier planting will tend to increase stand. Band-seeding sweetclover over the row with an insecticide box at sunflower planting proved more successful in the trial. The method also permits between-row cultivation (153).

Even though legume green manures in another North Dakota study used about 2.8 inches (rainfall equivalent) more water than fallow, they led to a 1-inch equivalent increase over fallow in soil water content in the top 3 inches of soil the following spring (14).

Sweetclover is the best producing warm-season forage legume, even topping alfalfa.

One green fallow option is planting yellow sweetclover with spring barley or spring peas. This is challenging, however, because barley can be overly competitive while herbicide compatibility is a concern with the peas.

Further north into the Canadian Great Plains, sweetclover depleted soil moisture by September of year 1, but by May of year 2, soil moisture was greater due to snow trapping, increased infiltration and reduced evaporation (32).

Fred Kirschenmann of Windsor, N.D., controls spring weed flushes on his fallow after sunflowers with an initial shallow chisel plowing then a rod weeder pass or two before planting sweetclover with a nurse crop of buckwheat or oats (or millet, if there is less soil moisture). He harvests buckwheat, hoping for a 900 lb./A yield, then lets the clover grow and overwinter. In early summer, when he begins to see yellow blossoms, he disks the cover, lets it dry, then runs a wide-blade sweep plow just below the surface to cut apart the crowns. The biomass contribution of the sweetclover fallow builds up organic matter, he says, in contrast to the black-fallow route of burning up organic matter to release N. Preventing humus depletion holds back the dreaded kochia weed.

In temperate areas you can overseed spring broccoli with HUBAM annual sweetclover, let the cover grow during summer, then till it in before planting a fall crop. Alternatively, you can allow it to winterkill for a thick, lasting mulch.

In Pennsylvania, Eric and Anne Nordell seed sweetclover after early vegetables (in June or July) and allow it to grow throughout the summer. It puts down a deep taproot before winter, fixes nitrogen and may bring nutrients to the soil surface from deep in the soil profile. See *Full-Year Covers Tackle Tough Weeds*, p. 38.

Other Options

First-year forage has the same palatability and feeding value as alfalfa, although harvest can reduce second-year vigor. Second-year forage is of lower quality and becomes less palatable as plants mature, but may total 2 to 3 tons per acre (120).

Sweetclover tolerates a wide range of harsh environments, poor soils and pests.

Growers report seed yield of 200 to 400 lb./A in North Dakota. Minimize shattering of seedpods by swathing sweetclover when 30 to 60 percent of its pods are brown or black. Pollinating insects are required for good seed yield (183).

Hard seed that escapes harvest will remain in the soil seed bank, but organic farmer Rich Mazour of Deweese, Neb., sees that as a plus. A 20- to 30-percent stand in his native grass pastures comes on early each spring, giving his cattle early grazing. Once warm-season grasses start to grow, they keep the clover in check. In tilled fields, sweep cultivators and residue-management tillage implements take care of sweetclovers with other tap-rooted "resident vegetation," Mazour says.

COMPARATIVE NOTES

Sweetclover and other deep-rooted biennial and perennial legumes are not suited for the most severely drought-prone soils, as their excessive soil moisture use will depress yield of subsequent wheat crops for years to come (163).

When planting sweetclover after wheat harvest, weeds can become a problem. An organic farmer in northeastern Kansas reports that to kill cocklebur, he has to mow lower than the sweetclover can tolerate. Annual alfalfa can tolerate low mowing (205).

After 90 days' growth in a North Dakota dryland legume comparison, a June planting of yellow sweetclover produced dry matter and N comparable to alfalfa and lespedeza (*Lespedeza stipulacea Maxim*). Subclover, fava beans (*Vicia faba*) and field peas had the best overall N-fixing efficiency in the dryland setting because of quick early season growth and good water use efficiency (331).

Cultivars. Yellow cultivars include MADRID, which is noted for its good vigor and production, and its relative resistance to fall freezes. GOLDTOP has excellent seedling vigor, matures two weeks later, provides larger yields of higher quality forage and has a larger seed than MADRID (361). Yellow common and YUKON joined GOLDTOP and MADRID—all high-coumarin types—as the highest yielding cultivars in a six-year North Dakota test (269).

Leading white biennial cultivars are DENTA, POLARA and ARCTIC. POLARA and ARCTIC are adapted to very cold winters. Best for grazing are the lower-producing, low-coumarin cultivars DENTA and POLARA (white) and NORGOLD (yellow).

Seed sources. See *Seed Suppliers* (p. 195). 

WHITE CLOVER

Trifolium repens

Also called: Dutch White, New Zealand White, Ladino

Type: long-lived perennial or winter annual legume

Roles: living mulch, erosion protection, green manure, beneficial insect attraction

Mix with: annual ryegrass, red clover, hard fescue or red fescue

See charts, pp. 66 to 72, for ranking and management summary.

White clovers are a top choice for "living mulch" systems planted between rows of irrigated vegetables, fruit bushes or trees. They are persistent, widely adapted perennial nitrogen producers with tough stems and a dense shallow root mass that protects soil from erosion and suppresses weeds. Depending on the type, plants grow just 6 to 12 inches tall, but thrive when mowed or grazed. Once established, they stand up well to heavy field traffic and thrive under cool, moist conditions and shade.

Three types: Cultivars of white clover are grouped into three types by size. The lowest growing type (Wild White) best survives heavy traffic and grazing. Intermediate sizes (Dutch White, New Zealand White and Louisiana S-1) flower earlier and more profusely than the larger types, are more heat-tolerant and include most of the economically important varieties. The large (Ladino) types produce the most N per acre of any white types, and are valued for forage quality, especially on poorly drained soil. They are generally less durable, but may be two to four times taller than intermediate types.

Intermediate types of white clover include many cultivated varieties, most originally bred for forage. The best of 36 varieties tested in north-central Mississippi for cover crop use were ARAN, GRASSLAND KOPU and KITAOOHA. These ranked high for all traits tested, including plant vigor, leaf area, dry matter yield, number of seed-heads, lateness of flowering and upright stems to prevent soil contact. Ranking high were ANGEL GALLARDO, CALIFORNIA LADINO and widely used LOUISIANA S-1 (392).

White clover performs best when it has plenty of lime, potash, calcium and phosphorus, but it tolerates poor conditions better than most clovers. Its perennial nature depends on new plants continually being formed by its creeping stolons and, if it reaches maturity, by reseeding.

White clover is raised as a winter annual in the South, where drought and diseases weaken stands. It exhibits its perennial abilities north through Hardiness Zone 4. The short and intermediate types are low biomass producers, while the large ladino types popular with graziers can produce as much biomass as any clover species.

BENEFITS

Fixes N. A healthy stand of white clover can produce 80 to 130 lb. N/A when killed the year after establishment. In established stands, it also may

Tough, low-growing and shade tolerant, this perennial is often used as a living mulch in vegetable systems.

provide some N to growing crops when it is managed as a living mulch between crop rows. Because it contains more of its total N in its roots than other legumes, partial tilling is an especially effective way to trigger N release. The low C:N ratio of stems and leaves causes them to decompose rapidly to release N.

Tolerates traffic. Wherever there's intensive field traffic and adequate soil moisture, white clover makes a good soil covering that keeps alleyways green. It reduces compaction and dust while protecting wet soil against trauma from vehicle wheels. White clover converts vulnerable bare soil into biologically active soil with habitat for beneficial organisms above and below the soil surface.

Premier living mulch. Their ability to grow in shade, maintain a low profile, thrive when repeatedly mowed and withstand field traffic makes intermediate and even short-stemmed white clovers ideal candidates for living mulch systems. To be effective, the mulch crop must be managed so it doesn't compete with the cash crop for light, nutrients and moisture. White clover's persistence in the face of some herbicides and minor tillage is used to advantage in these systems (described below) for vegetables, orchards and vineyards.

Value-added forage. Grazed white clover is highly palatable and digestible with high crude protein (about 28 percent), but it poses a bloat risk in ruminants without careful grazing management practices.

Spreading soil protector. Because each white clover plant extends itself by sending out root-like stolons at ground level, the legume spreads over time to cover and protect more soil surface. Dropped leaves and clipped biomass effectively mulch stolons, encouraging new plants to take root each season. Reseeding increases the number of new plants if you allow blossoms to mature.

Fits long, cool springs. In selecting a fall-seeded N-producer, consider white clover in areas with extended cool springs. MERIT ladino clover was the most efficient of eight major legumes evaluated in a Nebraska greenhouse for N_2 fixed per unit of water at 50° F. Ladino clover, as well as hairy vetch and fava beans (*Vicia faba*) were the only legumes to grow well at the 50° F temperature (334).

Overseeded companion crop. Whether frost-seeded in early spring into standing grain, broadcast over vegetables in late spring or into sweet corn in early summer, white clover germinates and establishes well under the primary crop. It grows slowly while shaded as it develops its root system, then grows rapidly when it receives more light.

MANAGEMENT

Establishment & Fieldwork

Widely adapted. White clover can tolerate wet soil—even short flooding—and short dry spells, and survives on medium to acid soils down to pH 5.5. It volunteers on a wider range of soils than most legumes, but grows better in clay and loam soils than on sandy soils (120). Ladino prefers sandy loam or medium loam soils.

Use higher seeding rates (5 to 9 lb./A drilled, 7 to 14 lb./A broadcast) when you overseed in adverse situations caused by drought, crop residue or vegetative competition. Drill 4 to 6 lb./A when mixing white clover with other legumes or grasses to reduce competition for light, moisture and nutrients.

Frostseeding of small-seeded clovers (such as alsike and white) should be done early in the morning when frost is still in the soil. Later in the day, when soil is slippery, stand establishment will be poor. Frostseed early enough in spring to allow for several freeze-thaw cycles.

Late-summer seeding must be early enough to give white clover time to become well established, because fall freezing and thawing can readily heave the small, shallow-rooted plants. Seeding about 40 days before the first killing frost is usually enough time. Best conditions for summer establishment are humid, cool and shaded (120, 361). Legumes suffer less root damage from frost heaving when they are planted with a grass.

In warmer regions of the U.S. (Zone 8 and warmer), every seeding should be inoculated. In cooler areas, where N-fixing bacteria persist in the soil for up to three years, even volunteer wild white clover should leave enough bacteria behind to eliminate the need for inoculation (120).

Mowing no lower than 2 to 3 inches will keep white clover healthy. To safely overwinter white clover, leave 3 to 4 inches (6 to 8 inches for taller types) to prevent frost damage.

Killing

Thorough uprooting and incorporation by chisel or moldboard plowing, field cultivating, undercutting or rotary tilling, or—in spring—use of a suitable herbicide will result in good to excellent kill of white clover. Extremely close mowing and partial tillage that leaves any roots undisturbed will suppress, but not kill, white clover.

Pest Management

Prized by bees. Bees work white clover blossoms for both nectar and pollen. Select insect-management measures that minimize negative impact on bees and other pollinators. Michigan blueberry growers find that it improves pollination, as does crimson clover (see *Clovers Build Soil, Blueberry Production*, p. 182).

Insect/disease risks. White clovers are fairly tolerant of nematodes and leaf diseases, but are susceptible to root and stolon rots. Leading insect pests are the potato leafhopper (*Empoasca fabae*), meadow spittlebug (*Philaenis spumarius*), clover leaf weevil (*Hypera punctata*), alfalfa weevil (*Hypera postica*) and Lygus bug (*Lygus* spp.).

If not cut or grazed to stimulate new growth, the buildup of vegetation on aged stolons and stems creates a susceptibility to disease and insect problems. Protect against pest problems by selecting resistant cultivars, rotating crops, maintaining soil fertility and employing proper cutting schedules (361).

WHITE CLOVER (*Trifolium repens*—intermediate type)

Crop Systems

Living mulch systems. As a living mulch, white clover gives benefits above and below ground while it grows between rows of cash crops, primarily in fruits, vegetables, orchards and vineyards. Living mulch has not proved effective in agronomic crops to this point. To receive the multiple benefits, manage the covers carefully throughout early crop growth—to keep them from competing with the main crop for light, nutrients, and especially moisture—while not killing them. Several methods can do that effectively.

Hand mowing/in-row mulch. Farmer Alan Matthews finds that a self-propelled 30-inch rotary mower controls a clover mix between green pepper rows in a quarter-acre field. He uses 40-foot wide, contour strip fields and the living mulch to help prevent erosion on sloping land near Pittsburgh, Pa. In his 1996 SARE on-farm research, he logged $500 more net profit per acre on his living mulch peppers than on his conventionally produced peppers (259).

Matthews mulches the transplants with hay, 12 inches on each side of the row. He hand-seeds the cover mix at a heavy 30 lb./A between the rows. The mix is 50 percent white Dutch clover, 30 percent berseem clover, and 20 percent HUIA white clover, which is a bit taller than the white Dutch. He mows the field in fall, then broadcasts medium red clover early the next spring to establish a hay

Clovers Build Soil, Blueberry Production

In the heart of blueberry country in the leading blueberry state in the U.S., Richard James "RJ" Rant and his mother, Judy Rant, are breaking new ground and reaping great rewards. Thanks to cover crops such as white and crimson clover taking center stage on their two family farms, the blueberry crop is thriving and the farmers are reaping significant rewards.

The Rants' soil is also on the receiving end of the multiple benefits of white and crimson clover cover crops.

Judy and her husband, Richard Rant, planted their first bushes in the early 1980s while both were still working full time off the farm. They managed the farm until retirement without going into debt—something of an accomplishment during that period—but the operation never really took off. Still in high school when his father passed away, RJ Rant stepped into the operation during college.

Choosing farming over graduate school, RJ began a quest to improve the blueberry operation and its bottom line. His focus on soil-building and cover crops proved key to the success of their operation, now expanded to two farms: Double-R Blueberry Farm and Wind Dancer Farms, jointly operated by RJ and his mother, Judy.

Michigan blueberry farmers have been using cover crops for many years, and top-producing Ottawa County farmers are no exception. Growing blueberry bushes on ten-foot centers, there is a lot of space between rows that farmers try to manage as economically and efficiently as possible. Seeking something that will not compete with the cash crop, most farmers choose rye or sod. Both require significant management in terms of time and labor, not to mention seed and fuel costs to plant and kill.

RJ Rant took a different tack. Their sandy loam soils were decent but not excellent, and his research led him to focus his efforts on improving the soil by reducing tillage and planting cover crops. While rye, an annual cover crop, required tillage in fall before planting and in spring to kill and incorporate, perennial white clover could be grown for two or more years without tilling. Because it is low-growing, the clover required less labor in planting and mowing.

field and replace the berseem, which is not winterhardy (259).

New Zealand white clover provided good weed control for winter squash in the wetter of two years in a New York trial. It was used in an experimental non-chemical system relying on over-the-row compost for in-row weed control. Plants were seeded into tilled strips 16 inches wide spaced 4 feet apart. Poor seed establishment and lagging clover growth in the drier year created weed problems, especially with perennial competitors. The living mulch/compost system yielded less than a conventionally tilled and fertilized control both years, due in part to delayed crop development from the in-row compost (282).

The research showed that in dry years, mowing alone won't suppress a living mulch enough to keep it from competing for soil moisture with crops in 16-inch rows. Further, weeds can be even more competitive than the clover for water during these dry times (282).

A California study showed that frequent mowing can work with careful management. A white clover cover reduced levels of cabbage aphids in harvested broccoli heads compared with clean-cultivated broccoli. The clover-mulched plants, in strip-tilled rows 4 inches wide, had yield and size comparable to clean cultivated rows. However, only intensive irrigation and mowing prevented moisture competition. To be profitable commercially, the system would require irrigation or a less

"There are so many positive things I could say about cover crops," Rant says. "The reason I keep using them is because they save me time and money."

Although he started his research and planted his first Alsike clover crop on his own, RJ soon found research partners at Michigan State University. He cooperates with researcher Dale Mutch to fine-tune cover crop selection, planting methods and management options.

"I get really excited when I think about improving my soils," Rant says. "I see my fields as one unified system, and the biology of the soil in the inter-rows is as important as the soil and fertility up and down my blueberry rows."

Early screening of different cover crops led Rant to further test crimson clover, a winter annual cover crop that not only grows well in fall and spring, but also shows great potential to reseed itself, further reducing costs. They have also tested red clover, small white clover, mustard, rye and spring buckwheat. Rant says white clover is his favorite because it is low growing and out-competes weeds.

Soil-building remains a primary objective of Rants cover crop program. Compacted soils are a problem for blueberries, which prefer loose, friable soil. To build better soil structure, Rant is working with MSU's Mutch to improve soil organic matter with cover crop mulches and manures.

Mutch and Rant are studying crimson clover reseeding, another cost-saving measure, comparing mowing and tilling the crimson clover after it has set seed. They are also studying pH ranges for the clovers, which prefer a higher pH than the blueberry cash crop.

Mutch and Michigan State researcher Rufus Isaacs are helping to elucidate other management aspects of clover cover crops, such as whether honeybees and native bees such as mason bees and sweat bees are attracted to clovers.

"The clovers work really well for blueberry production, rather than needing to fit system to cover crop. If you really want to do this, you can make it work," says RJ.

—Andy Clark

thirsty legume, as well as field-scale equipment able to mow between several rows in a single pass (93).

Chemical suppression is unpredictable. An application rate that sets back the clover sufficiently one year may be too harsh (killing the clover) or not suppressive the next year due to moisture, temperature or soil conditions.

Partial rotary tillage. In a New York evaluation of mechanical suppression, sweet corn planting strips 20 inches wide were rotary tilled June 2 into white clover. Although mowing (even five times) didn't sufficiently suppress clover, partial rotary tilling at two weeks after emergence worked well. A strip of clover allowed to pass between the tines led to ample clover regrowth. A surge of N within a month of tilling aided the growing corn. The loss of root and nodule tissue following stress from tillage or herbicide shock seems to release N from the clover. Leaf smut caused less problem on the living-mulch corn than on the clean-cultivated check plot (170).

Healthy stands can produce 80 to 130 lb. N/A when killed the year after establishment.

Crop shading. Sweet corn shading can hold white clover in check when corn is planted in 15-inch rows and about 15 inches apart within the row. This spacing yielded higher corn growth rates, more marketable ears per plant and higher crop yields than conventional plots without clover in an Oregon study. Corn was planted into tilled strips 4 to 6 inches wide about the same time the clover was chemically suppressed. Adapted row-harvesting equipment and handpicking would be needed to make the spacing practical (139).

Unsuppressed white Dutch clover established at asparagus planting controlled weeds and provided N over time to the asparagus in a Wisconsin study, but reduced yield significantly. Establishing the clover in the second year or third year of an asparagus planting would be more effective (312).

Other Options

Seed crop should be harvested when most seed heads are light brown, about 25 to 30 days after full bloom.

Intermediate types of white clover add protein and longevity to permanent grass pastures without legumes. Taller ladino types can be grazed or harvested. Living mulch fields can be overseeded with grasses or other legumes to rotate into pasture after vegetable crops, providing IPM options and economic flexibility.

COMPARATIVE NOTES

- White clover is less tolerant of basic soils above pH 7 than are other clovers.
- In a Wisconsin comparison, ladino clover biomass was similar to mammoth red clover when spring-seeded (402).
- White clover stores up to 45 percent of its N contribution in its roots, more than any other major legume cover crop.
- Ladino and alsike are the best hay-type legumes on poorly drained soils.
- Spring growth of fall-seeded white clover begins in mid-May in the Midwest, about the same time as alfalfa.

Seed sources. See *Seed Suppliers* (p. 195). 

WOOLLYPOD VETCH

Vicia villosa ssp. dasycarpa

Also called: LANA vetch; also spelled woolypod vetch

Cycle: cool-season annual

Type: legume

Roles: N source, weed suppressor, erosion preventer, add organic matter, attract bees

Mix with: other legumes, grasses

See charts, p. 66 to 72, for ranking and management summary.

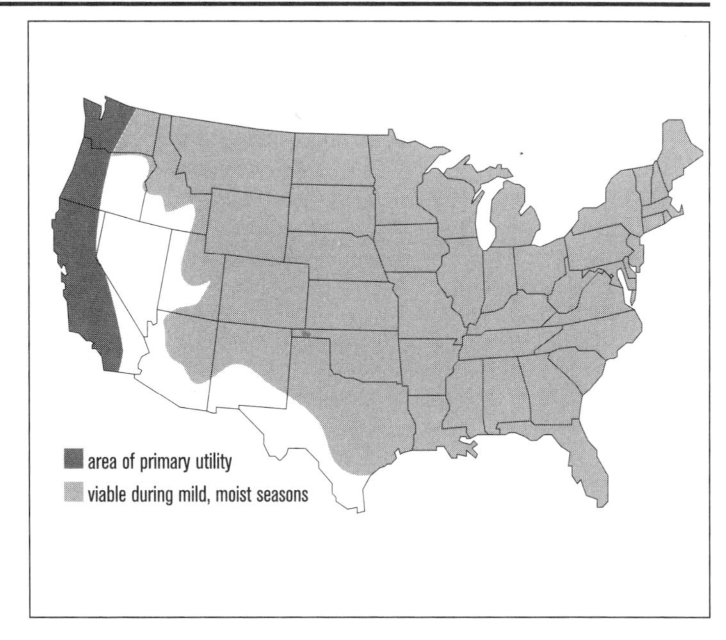

Specialty vetches such as woollypod and purple vetch (*Vicia benghalensis*) are faster-growing alternatives to hairy vetch (*Vicia villosa*) in Hardiness Zone 7 and warmer. Requiring little or no irrigation as a winter cover in these areas, they provide dependable, abundant N and organic matter, as well as excellent weed suppression.

Many growers of high-value crops in California rely on one or more vetch species as a self-reseeding cover crop, beneficial insect habitat and mulch. They can mow the vetch during winter and in late spring after it reseeds.

Some vineyard managers seed LANA woollypod vetch each year with oats or as part of a legume mix—common vetch, subterranean clover, a medic and LANA, for example. They plant the mix in alternate alleyways to save on seeding costs and reduce moisture competition, while ensuring sufficient cover that they can mow or disk. LANA's climbing tendency (even more so than purple or common vetch) and abundant biomass can become problems in vineyards and young orchards, but can be readily managed with regular monitoring and timely mowing.

In Zone 5 and colder and parts of Zone 6, woollypod vetch can serve as a winterkilled mulch—or as a quick, easy-to-mow spring cover—for weed control and N addition to vegetable transplants. It's a good choice as an overwintering cover before or after tomato crops in Zone 6 and warmer. In California, LANA provided the most N and suppressed the most weeds during two consecutive but distinctly different growing seasons, compared with purple vetch and other legume mixtures (413, 414).

BENEFITS

N Source. A first-year, overwintering stand of woollypod vetch easily will provide more than 100 pounds of N per acre in any system when allowed to put on spring growth. The popular LANA cultivar starts fixing N in as little as one week after emergence.

LANA can contribute as much as 300 pounds of N its first year or two, given adequate moisture and warm spring growing conditions (273, 396). Fall-planted LANA incorporated before a corn crop can provide a yield response equivalent to 200 lb.

N/A, a California study showed (273). Similar results have been seen in tomato research in California (396). In western Oregon, a yield response equivalent to 70 lb. N/A for sweet corn has been observed (364).

Plenty of soil-building organic matter. Woollypod typically produces more dry matter than any other vetch. LANA shows better early growth than other vetches, even during cool late fall and winter weather in Zone 7 and warmer. LANA shows explosive growth in early spring in the Pacific Northwest (364) and in late winter and early spring in California when moisture is adequate. It can provide up to 8,000 lb. DM/A, which breaks down quickly and improves soil structure (63, 273, 396).

Woollypod vetch is often used in mixes in California orchards and vineyards.

Frost protectant. Some orchard growers have found that keeping a thick floor cover before the blossom stage can help prolong a perennial crop's dormant period by up to 10 days in spring. "This reduces the risk of early frost damage (to the blossoms, by delaying blossoming) and lengthens the blossoming period of my almond trees," notes almond grower Glenn Anderson, Hilmar, Calif.

Smother crop. Woollypod's dense spring growth smothers weeds and also provides some allelopathic benefits. Of 32 cover crops in a replicated study at a California vineyard, only LANA completely suppressed biomass production of the dominant winter annual weeds such as chickweed, shepherd's purse, rattail fescue and annual ryegrass (422).

Beneficial habitat. Woollypod vetch attracts many pollinators and beneficial insects. In some orchards, these beneficials move up into the tree canopy by late spring, so you can mow the floor cover after it reseeds and not worry about loss of beneficial habitat (184).

MANAGEMENT

Establishment & Fieldwork

Woollypod does well on many soil types—even poor, sandy soil—and tolerates moderately acidic to moderately alkaline conditions. It's well-adapted to most orchard and vineyard soils in California (422).

It establishes best in recently tilled, nutrient-deficient fields. Tillage helps enhance the reseeding capability of vetches (63). LANA woollypod vetch hasn't done as well in some no-till systems as it was expected to.

Given adequate moisture, however, broadcasting LANA even at low to moderate rates—and with light incorporation—can give satisfactory results from fall seedings, especially if the stand is allowed to grow through mid-spring. If your goal is to shade out competition quickly, however, broadcast at medium to high rates and incorporate lightly.

You might not recognize the emerging plant without its characteristic multiple leaflets, says Glenn Anderson. "You should spot it within two weeks of planting, three at the latest, depending on temperature and soil conditions. Even at 6 inches, it'll still look spindly. It won't really leaf out until late winter and early spring, when more aggressive growth kicks in." That may continue until maturity in mid- to late May.

Fall planting. Most growers seed at low to medium rates, regardless of seeding method. If drilling, ½ to 1 inch deep is best, although up to 2 inches will work for early seedings. If broadcasting, follow with a cultipacker or a shallow pass of a spike-toothed harrow.

Seedbed preparation is crucial for establishing a healthy cover crop stand in vineyards. California viticulturist and consultant Ron Bartolucci recommends making two passes with a disk to kill existing vegetation and provide some soil disturbance. He cautions against using a rotary tiller, which can pulverize the soil and reduce its water-holding capacity (211).

Bartolucci prefers to drill rather than broadcast cover crops, saving on seed costs and ensuring seed-to-soil contact. He recommends the economical, alternate row planting that also ensures easy access for pruning grape vines.

Don't wait too long in fall to seed woollypod vetch in Zone 7 and warmer, however. If you wait until the soil starts getting cold, in mid-October in Oregon and early November in parts of central California, germination will be poor and the stand disappointing. Seed too early, though, and you miss the early moisture benefit of the Central Valley's fog season and will need to irrigate more before the rainy season.

Regardless of your planting method, seed woollypod vetch into moist soil or irrigate immediately after seeding to help germination (273). If irrigation is an option but you want to conserve water costs, try seeding just before a storm is forecast, then irrigate if the rain misses you.

Spring planting. Planted in early spring, woollypod vetch can provide plowdown N by Memorial Day for a summer annual cash crop in the Northeast.

Mowing & Managing

Woollypod vetch can survive freezing conditions for days, but severe cold can markedly reduce its dry matter and N production (212, 273).

In most cases, main challenges for an established woollypod vetch stand include managing its abundant growth and viny tendrils and ensuring adequate moisture for your primary crop. In wet environments such as western Oregon, however, LANA vetch can retard spring soil drying and seedbed preparation for summer crops (364).

Woollypod responds well to mowing, as long as you keep the stand at least 5 inches tall and avoid mowing during the two-month period just before it reseeds. "I can mow as late as mid-March and still see good reseeding," says Glenn Anderson, an organic almond grower in California's Central Valley. "After that, I may mow if I want to prevent some frost damage, but I know I'll lose some of the vetch through reduced reseeding."

Anderson usually mows the floor cover once or twice before mid-March and after it reseeds. He cuts in the direction of prevailing winds—which can be on a diagonal to his tree rows—to facilitate air movement throughout the orchard, especially when he anticipates moist air heading his way.

In vineyards, "high chopping" legume mixes to a 12-inch height can help keep them from trellising over vine cordons. In vineyards without sprinklers for frost protection, some growers incorporate legume mixes in spring, before the soil becomes too dry for disking. Where sprinklers are used, the covers might be allowed to grow for a longer period and provide additional N. Timing is important when disking, however, as you don't want to make equipment access difficult or compact soil during wet spring conditions (211).

Elayne Sears

WOOLLYPOD VETCH (*Vicia villosa* ssp. *dasycarpa*)

Given the high dry matter production from woollypod vetch when it's allowed to grow at least until late March, two or three diskings or mowings will encourage rapid decomposition. Power spaders can reduce soil compaction when incorporating vetches in spring conditions, compared with heavier disk harrows (421).

Moisture concerns. Many orchard and vineyard growers find it helpful to drip irrigate tree or vine rows if they are growing an aggressive cover crop such as LANA between the rows for the first time. In California vineyards where irrigation isn't used, a few growers report that vines seem to lose vigor faster when grown with cover crops. Others haven't observed this effect. After a few years of growing leguminous covers, many find that their soil is holding moisture better and they need less water to make the system work.

Reseeding concerns. Vetch mixtures often fail to reseed effectively, especially if they have been mowed at the wrong time or soil fertility is high. Some vineyard managers expect low persistence and reseed a vetch mix in alternate rows every fall, or reseed spotty patches.

Regardless of mowing regime, LANA's persistence as a self-reseeding cover diminishes over time, and other resident vegetation starts to take over. That's a sign that the cover's water-holding, fertility- and tilth-enhancing benefits have kicked in, says Glenn Anderson.

It's natural to expect a change in the resident vegetation over time, observes Anderson. After a few years of reseeding itself—and providing abundant dry matter and nitrogen—the LANA he had clear seeded at low rates between orchard rows on half his acreage eventually diminished to about 10 percent of the resident vegetation, he notes. Subclovers and other legumes he introduced have become more prominent. Those legumes may have better self-reseeding capability than LANA, other growers note.

LANA woollypod vetch shows explosive growth in early spring in the Pacific Northwest.

Pest Management

Woollypod vetch outcompetes weeds and will quickly resolve most weed problems if seeded at high rates. Woollypod also provides some allelopathic benefits. A root exudate can reduce growth in some young grasses, lettuces and peas, however.

Hard seed carryover can cause LANA to become a weed in subsequent cash crops and vineyards, however (102). Its strong climbing ability can cover grape vines or entwine sprinklers. In orchards, it's fairly easy to cut or pull LANA vines out of the canopy of young trees. Mowing or "high chopping" may be needed, especially in vineyards, even though this can reduce LANA's reseeding rate.

Insects pests aren't a major problem with woollypod vetch, in part because it attracts lady beetles, lacewings, minute pirate bugs and other beneficials insects that help keep pests in check.

LANA can be a host of *Sclerotinia minor*, a soilborne pathogen that causes lettuce drop, a fungal disease affecting lettuce, basil and cauliflower crops. In a California study involving cover crops that were deliberately infected with *S. minor*, the pathogen levels were associated with higher lettuce drop incidence the summer after LANA had been incorporated, but wasn't as problematic the following year. Woollypod vetch probably isn't a good choice if you're growing crops susceptible to this pathogen.

Other Options

Seed. Woollypod vetch is a prolific seed producer, but its pods are prone to shattering. You can increase seed harvest by raking the field (without mowing, if possible) to gather the crop into windrows for curing, before combining with a belt-type rubber pickup attachment (421).

Forage. Like most vetches, LANA is a somewhat bitter yet palatable forage when green, and the palatability increases with dryness (421). It is a nutritious forage for rangeland use (421).

For hay, it is best cut in full bloom. The leaves dry rapidly and swaths can be gathered within a day or two (421).

COMPARATIVE NOTES

- Woollypod has slightly smaller flowers than hairy vetch, and its seeds are more oval than the nearly round seeds of hairy vetch. LANA also has a higher proportion of hard seed than hairy vetch (422).
- LANA shows more early growth than common vetch, although both increase their biomass dramatically by midspring.
- LANA and purple vetch are more cold-sensitive than common vetch or hairy vetch. Once established, LANA can tolerate early frosts for a few days (especially if the temperature doesn't fluctuate widely or with some snow cover) and is hardier than purple vetch, which is more susceptible to early spring dieback (149).
- LANA flowers about three weeks earlier than purple vetch and has a better chance of setting seed in dryland conditions (273).
- LANA and LANA mixes suppress weeds better than purple vetch (149).

Seed sources. See *Seed Suppliers* (p. 195). 

Appendix A

TESTING COVER CROPS ON YOUR FARM

by Marianne Sarrantonio

To find your best cover crops, you needn't become Dr. Science or devote your life to research. It's not hard to set up valid, on-farm tests and make observations. Follow these steps:

A. Narrow your options. Aim for a limited-scale trial of just two to five species or mixtures. You can test the best one or two in a larger trial the next year. Unsure of the best place and time in your rotation? Start with small plots separated from cropped fields and plant over a range of dates, under optimal soil and weather conditions. If you're sure where and when to plant and have just two or three covers to try, put the trial right in your cropped fields, using your normal seedbed preparation. This method provides rapid feedback on how the cover crops fit into your cropping system. Keep in mind management may need some tweaking (such as seeding rate or date) to get the best results.

B. Order small seed amounts. Many companies provide 1- to 10-pound bags if you give them advance notice. If 50-pound bags are the only option, arrange to share it with other growers. Don't eliminate a species just because seed price seems high. If it works well, it could trim other costs. You could consider growing your own seed eventually, and perhaps even selling it locally. Be sure to obtain appropriate inoculants if you'll be testing legumes, which require species-specific rhizobial bacteria so the cover can capture and "fix" N efficiently. See *Nodulation: Match Inoculant to Maximize N* (p. 122).

C. Determine plot sizes. Keep them small enough to manage, yet large enough to yield adequate and reliable data. Plots two to four rows wide by 50 to 100 feet could suffice if you grow vegetables for market. If you have 10 or more acres, quarter- or half-acre plots may be feasible. If you use field-scale machinery, establish field-length plots. For row crops, use plots at least four rows wide, or strips based on your equipment width. Keep in mind the subsequent crop's management needs.

D. Design an objective trial. Plots need to be as uniform as possible, randomly selected for each option you're testing, and preferably replicated (at least two or three plots for each option). If parts of the field have major differences (such as poorer drainage or weedy spots), put blocks or groups of plots together so each treatment has equal representation in each field part, or avoid those areas for your trial entirely. Label each plot and make a map of the trial area.

E. Be timely. Regard the trial as highly as any other crop. Do as much or as little field preparation as you would for whole fields, and at an appropriate time. If possible, plant on two or more dates at least two weeks apart. In general, seed winter annuals at least six weeks before a killing frost. Wheat and rye can be planted later, although that will reduce the N-scavenging significantly.

F. Plant carefully. If seeding large plots with tractor-mounted equipment, calibrate your seeding equipment for *each* cover crop. This can prevent failures or performance differences due to incorrect seeding rates. Keep a permanent record of drill settings for future reference. A hand-crank or rotary spin seeder works well for small plots getting less than five pounds of seed. Weigh seed for each plot into a separate container. To calculate seeding rates for small plots, use this formula: 1 lb./acre = 0.35 oz (10 grams)/ 1000 ft^2 area seeded. If your cover crop seeding rate calls for 30 lb/acre, multiply 0.35 oz by 30. You will need 10.5 oz (300 grams) of seed for each 1000 ft^2 you seed. Put half the seed in the seeder and seed smoothly as you walk the length of the field and back, with a little overlap in the spread pattern.

Then seed the remainder while walking in perpendicular directions so you crisscross the plot in a gridlike pattern. If broadcasting by hand, use a similar distribution pattern. With small seed, mix in sand or fresh cat litter to avoid seeding too much at a time.

G. Collect data. Start a trial notebook or binder for data and observations. Management information could include:
- field location
- field history (crops, herbicides, amendments, unusual circumstances, etc.)
- plot dimensions
- field preparation and seeding method
- planting date and weather conditions
- rainfall after planting
- timing and method of killing the cover crop
- general comments

Growth data for each plot might include:
- germination rating (excellent, OK, poor, etc.), seven to 14 days after seeding
- early growth or vigor rating, a month after establishment
- periodic height and ground cover estimates, before killing or mowing
- periodic weed assessments
- a biomass or yield rating

Also rate residue before planting the next crop. Rate survival of winter annuals in early spring as they break dormancy and begin to grow. If you plan to mow-kill an annual, log an approximate flowering date. Regrowth could occur if most of the crop is still vegetative. Rate overall weather and record dates such as first frost. Note anything you think has a bearing on the outcome, such as weed infestations. If time allows, try killing the cover crops and continuing your expected rotation, at least on a small scale. You might need hand tools or a lawn mower. Use field markers to identify plots.

H. Choose the best species for the whole farm system. Not sure which covers did best? Whatever you found, don't be satisfied with only a single year's results. Weather and management will vary over time.

Assess performance by asking some of the questions you answered about the cover niche (see *Selecting the Best Cover Crops for your Farm*, p. 12). Also ask if a cover:
- was easy to establish and manage
- performed its primary function well
- avoided competing excessively with the primary crop
- seemed versatile
- is likely to do well under different conditions
- fits your equipment and labor constraints
- provides options that could make it even more affordable

In year two, expand the scale. Test your best-performing cover as well as a runner-up. With field crops, try one-acre plots; stick with smaller plots for high-value crops. Also try any options that might improve the cover stand or its benefits. Entries for the major cover crops in this book include some management tips that can help. Record your observations faithfully.

I. Fine-tune and be creative. Odds are, you won't be completely satisfied with one or more details of your "best" cover. You might need to sacrifice some potential benefits to make a cover work better in your farm system. For example, killing a cover earlier than you'd like will reduce the amount of biomass or N it provides, but could ensure that you plant summer crops on time. In most cases, fine-tuning your management also makes it more affordable. Lowering a seeding rate or shifting the seeding date also could reduce the tillage needed. Narrower rows in your cash crop might hinder establishment of an overseeded legume but reduce weeds and bump up the cash crop yield. Don't expect all of a cover's benefits to show up in yearly economic analyses. Some benefits are hard to assess in dollars.

Your best covers may seem well-suited to your farm, but there could be an up-and-coming species or management technique you haven't thought of testing. See *Up-and-Coming Cover Crops* (p. 191) for a few examples. Overwhelmed? You needn't be. Initiative and common sense—traits you already rely on—are fundamental to any on-farm testing program. As a grower, you already test varieties, planting dates and other management practices every year. This section offers enough tips to start testing cover crops. You also can collaborate with others in your region to pool resources and share findings. There's a good chance others in your area could benefit from your cover cropping wisdom!

[Adapted and updated in 2006 from *Northeast Cover Crop Handbook* by Marianne Sarrantonio, Rodale Institute, 1994.]

Appendix B
UP-AND-COMING COVER CROPS

Balansa clover

Identified as a promising new cover crop in screening trials throughout the Southeastern U.S., balansa clover (*Trifolium michelianum Savi*) is a small-seeded annual legume with superior reseeding potential compared with other legumes, including crimson clover. Well-adapted to a wide range of soil types, balansa performs particularly well on silty clay soil with a pH of about 6.5. Established stands tolerate waterlogging, moderate salinity, and soil pH from 4.5 to 8.0. It does not do well on highly alkaline soils (30). It is considered marginal in Zone 6B.

Balansa and other reseeding legumes were screened in Zones 6, 7, and 8 (from the Gulf Coast to northern Tennessee, and from Georgia to western Arkansas). TIBBEE crimson clover (*Trifolium incarnatum*) was used as a phenological check. Growth was terminated 2 to 3 weeks after TIBBEE bloomed at each location to identify adapted cover crops that reseed earlier than TIBBEE. Spotted burclover (*Medicago arabica*) and balansa clover were the best reseeding legumes that were hardy throughout zone 7A. Of these, only balansa clover is commercially available.

Balansa clover is open pollinated. Flowers vary from white to pink and are attractive to bees. Ungrazed, it grows up to three feet high and produces thick hollow stems that are palatable and of good feed value. It becomes more prostrate when grazed.

Balansa clover was named *Trifolium michelianum Savi* in 1798. It is sometimes called *Trifolium balansae* or *Trifolium michelianum* subsp. *balansae*. A landrace of balansa clover collected in Turkey in 1937 was released in 1952 by the Alabama office of NRCS with the name MIKE. Small amounts of seed of this accession are available from the Plant Introduction station in Athens, GA.

Balansa clover seed is quite small, so planting only 5 lb./A gives a dense stand. Seed is produced commercially only in Australia. Balansa clover requires a relatively rare inoculant, designated "Trifolium Special #2" by Liphatech, Inc., manufacturer of "Nitragin" brand inoculants. Kamprath Seed Co. imports balansa seed (See *Seed Suppliers*, p. 195). Some seed suppliers offer coated seed that is pre-inoculated. The price per pound of coated seed is about the same as bare seed, but 1/3 of the weight is coating so the seeding rate for coated seed should be increased to 8 lb./A.

PARADANA is the cultivar that has been most widely tested in the U.S. It was released in 1985 by the South Australia Department of Agriculture. It was derived from Turkish introductions crossed and tested at Kangaroo Island, NSW, Australia. Seed yields over 550 lb./A have been obtained. BOLTA is 1-2 weeks later than PARADANA and FRONTIER is 2-3 weeks earlier. FRONTIER, a selection out of PARADANA, has replaced its parent in the seed trade in recent years.

While PARADANA seed matures slightly earlier than crimson clover, it often does not produce as much biomass. Nitrogen accumulation in above ground biomass is about 60 lb./A at full bloom. Balansa can reseed for several years from a single seed crop, due to its relatively high amount of hard seed. It reseeded for four years following maturation of a seed crop in 1993 in Senatobia, Miss., and for at least two years in no-till systems at several other locations in Alabama, Georgia and Mississippi. Neither TIBBEE nor AU ROBIN crimson clover reseeded for more than one year at any location in those tests. Balansa clover does not reseed well after tillage, probably because the small seeds are buried too deeply.

Allowing balansa clover to grow for 40 days past first bloom every 3 to 4 years will allow stands to persist indefinitely in no-till systems. Reseeded stands are denser, bloom 5 to 7 days earlier, and are more productive than planted stands because growth begins as soon as conditions are favorable and seedling density is higher. However, seed cost is minor compared to opportunity cost and risk associated with delaying main crop planting. Waiting past the optimum planting date to encourage reseeding is only practical in rotations that include main crops optimally planted in May in the Southeastern U.S.

Balansa is less likely than crimson clovers to host root-knot nematodes (*Meloidogyne incognita*, race 3). Gary Windham, USDA-ARS, Starkville, Miss., found that balansa had egg mass index scores between 2.3 and 2.9. For comparison, a very resistant white clover scored 1.5, most crimson clovers score between 3 and 3.5 and very susceptible crops like REGAL white clover score 5 on a scale from 1 to 5.

—Seth Dabney
USDA-ARS National Sedimentation Lab
P.O. Box 1157 Oxford, MS 38655-2900
662-232-2975; sdabney@ars.usda.gov

Black oat

Black oat (*Avena strigosa L.*) is the No. 1 cover crop on millions of acres of conservation-tilled soybean in southern Brazil, and has potential for use in the southern USA (Zones 8-10).

Black oat produces large amounts of biomass, similar to rye. It maintains a narrower carbon to nitrogen (C:N) ratio than rye so it cycles nitrogen better than rye, important for nitrogen management in conservation tillage systems. It breaks disease cycles for wheat and soybean and is resistant to root-knot nematodes. It is very resistant to rusts and has exceptional allelopathic activity for weed control. It is easy to kill mechanically.

Black oat is adapted for use as a winter cover crop in the lower Coastal Plain of the USA, including Zones 8b-10a. It has done well in fall plantings in Zone 8b, but winterkilled one year of six at some locations within this zone, dependent on planting date.

Planting dates are similar to common oat. If planted too early, it is more susceptible to winterkill and lodging. Planting in late winter (early February) yielded good biomass and ground cover for late planted cash crops in the lower Coastal Plain.

Seed 50-70 lb./A for use as a cover crop, 40 lb./A for seed production. In the Southeast, fall plantings (November) result in seed ripening in mid May through early June. Seed yields range from 800 to 1400 lb./A. Seed is available commercially in limited amounts.

One cultivar, SOILSAVER, was selected for increased cold tolerance and released by Auburn University and IAPAR (Institute of Agronomy of Paraná, Brazil). Auburn University and USDA-ARS researchers developed it from a population of IAPAR-61-IBIPORA, a public variety from the Institute of Agronomy of Parana, Brazil (IAPAR) and the Parananese Commission for Evaluation of Forages (CPAF).

SOILSAVER black oat has several advantages as a cover crop. It tillers well, producing good soil coverage in relation to total biomass produced. It suppresses broadleaf weeds extremely well. In one study, weed control in conservation tillage cotton (*Gossypium hirsutum L.*) averaged 34% with black oat compared to 26% for rye, 19% for wheat, and 16% with no cover crop.

—D.W. Reeves
Research Leader
USDA-ARS, J. Phil Campbell Sr. Natural Resource Conservation Center
1420 Experiment Station Road
Watkinsville, GA 30677

706-769-5631 ext. 203
fax 706-769-8962
Wayne.Reeves@ars.usda.gov

Additional Information:
SoilSaver—A Black Oat Winter Cover Crop for the Lower Southeastern Coastal Plain. 2002. USDA-ARS National Soil Dynamics Lab. Conservation Systems Fact Sheet No. 1.
www.ars.usda.gov/SP2UserFiles/Place/64200500/csr/FactSheets/FS01.pdf

Lupin

Lupins are cool-season annual legumes that provide plenty of N and can be grown widely in the USA and southern Canada. Lupins have aggressive taproots, especially the narrow-leaf cultivars. You can kill lupins mechanically or with herbicides. Their hollow stems crush or break readily, making it easy to plant cash crops using conservation tillage equipment.

White lupin (*Lupinus albus L.*) and blue or narrow-leaf lupin (*Lupinus angustifolius L.*) species were originally named after their flower colors, but both species now have cultivars with white, blue or magenta/purple flowers. Blue lupin is adapted to the lower Coastal Plain and is more readily identified by its narrow leaflets (about 0.5-inch wide) rather than flower color.

As a fall and winter cover crop in the southeastern USA, white lupin is the most cold-tolerant. Some cultivars overwinter as far north as the Tennessee Valley. They typically produce 100 to 150 lb. N/A when fall-planted and killed in early spring.

Seed spring cultivars in early April in the northern U.S. and southern Canada. Kill in June when they're at peak biomass (early-bloom to early-pod stage).

For use as a cover crop, drill lupins no deeper than 1 inch at 70 lb./A for small-seeded blue varieties to 120 lb./A for larger-seeded white varieties. At $30 to $40 per acre, the seed is relatively expensive. Be sure to inoculate lupin seed with compatible rhizobia.

Three winter-hardy lupin cultivars are readily available on a commercial scale. TIFWHITE-78 white lupin and TIFBLUE-78 blue or narrow-leaf lupin were both released by USDA's Agricultural Research Service in the 1980s. These two varieties, and other modern varieties, are "sweet" types as opposed to "bitter" types that were widely grown in the South prior to 1950. Sweet varieties have a low concentration of naturally occurring alkaloids. Sweet lupin is favored by wildlife, especially deer. Sweet lupin cover crops may act as a trap crop for thrips (*Frankliniella* spp.) in cotton plantings, but this has yet to be confirmed by research.

AU HOMER bitter white lupin is a new release by Auburn University derived from Tifwhite-78. It was selected for increased alkaloid content for use as a cover crop. Alkaloids make lupin seed and forage unpalatable for livestock, but also play a major role in resistance to disease, insects and nematodes.

Lupins are susceptible to many fungal and viral diseases and should not be grown in the same field in successive years. Rather, rotate lupin cover crops with a small grain cover crop, ideally in a rotation that allows three years between lupin plantings. Lupin are intolerant to poorly drained soils.

For information about lupins and seed sources, contact: Edzard van Santen, Professor Crop Science, Agronomy & Soils Dept., 202 Funchess Hall, Auburn University, Auburn, AL 36849;
334-844-3975; fax 334-887-3945
vanedza@auburn.edu

Sunn Hemp

A tropical legume, sunn hemp (*Crotalaria juncea L.*) can produce more than 5,000 lb. dry matter/A and 120 lb. N/A in just nine to 12 weeks. It can fill a narrow niche between harvest of a summer crop and planting of a fall cash or cover crop and is especially fitted to vegetable production. Sunn hemp sown by September 1 following a corn crop in Alabama, for example, can produce an average of 115 lb. N/A by December 1.

Sunn hemp is not winter hardy and a hard freeze easily kills it. Sow sunn hemp a minimum of nine weeks before the average date of the first fall freeze. Seed at 40 to 50 lb./A, with a cowpea-type inoculant.

Sunn hemp seed is expensive, about $2.25/lb., so the cost may be prohibitive for large-scale plantings. Seed can be produced only in tropical areas, such as Hawaii, and currently is imported only by specialty seed companies.

A New Alternative for South Florida Producers

A study by the NRCS Plant Materials Center (PMC) in Brooksville, Florida, concluded that sunn hemp seed can be a viable alternative cash crop for southern Florida growers. Sunn hemp is an annual legume that suppresses some types of nematodes and can produce over 5,000 pounds of biomass and 100 pounds of nitrogen per acre within a few months. Because of its potential use in alternative pest management systems and as a sustainable biological source of nitrogen, sunn hemp is a promising cover crop for rotation with vegetables throughout the Southeastern U.S.

Unfortunately, its use has been limited by the high seed cost—most is shipped from Hawaii as seed production requires a tropical climate. Two years ago, the NRCS PMC in Brooksville initiated a study to determine which zones in Florida could most economically produce sunn hemp seed. Seed was distributed to 15 growers throughout Florida and although many locations lost their crop to frost, sunn hemp stands in coastal counties below the 27th parallel consistently produced up to 370 pounds of seed per acre. Growers in more southern areas, such as Homestead, obtained even higher yields. Your contact is Clarence Maura, Manager, NRCS Brookville Plant Materials Center, at 352-796-9600 or clarence.maura@fl.usda.gov.

A management caution: Many *Crotalaria* species contain alkaloids that are poisonous to livestock. However, the sunn hemp variety TROPIC SUN, developed jointly by the University of Hawaii and USDA-NRCS, has a very low level of alkaloid and is suitable for use as a forage.

Research suggests that sunn hemp is resistant and/or suppressive to root-knot (*Meloidogyne* spp.) and reniform (*Rotylenchulus reniformis*) nematodes.

—D.Wayne Reeves (see p. 192)

Additional Information:

Mansoer, Z., D.W. Reeves and C.W. Wood. 1997. Suitability of sunn hemp as an alternative late-summer legume cover crop. *Soil Sci. Soc. Am. J.* 61:246-253.

Balkcom, K. and D.W. Reeves. 2005. Sunn hemp utilized as a legume cover crop for corn production. *Agron. J.* 97:26-31.

Sunn Hemp: A Cover Crop for Southern and Tropical Farming Systems. USDA-NRCS Soil Quality Technical Note No. 10. May 1999. Available at: http://soils.usda.gov/sqi/files/10d3.pdf

K.-H. Wang and R. McSorley. Management of Nematodes and Soil Fertility with Sunn Hemp Cover Crop. 2004. University of Florida Cooperative Extension. Publication #ENY-717. http://edis.ifas.ufl.edu/NG043

Valenzuela, H. and J. Smith. 2002. 'Tropic Sun' Sunnhemp. University of Hawaii Cooperative Extension. Publication #SA-GM-11. www.ctahr.hawaii.edu/sustainag/GreenManures/tropicsunnhemp.asp

Appendix C
SEED SUPPLIERS

This list is for information purposes only. Inclusion does not imply endorsement, nor is criticism implied of firms not mentioned.

Adams-Briscoe Seed Co.
P.O. Box 19
325 E. Second St.
Jackson, GA 30233-0019
770-775-7826
fax 770-775-7122
abseed@juno.com
www.abseed.com
clovers, winter peas, vetches, cowpeas, wheat, rye, oats, grasses, sorghums, legume inoculants

Agassiz Seed & Supply
445 7th St. NW
West Fargo, ND 58078
701-282-8118
fax 701-282-9119
www.agassizseed.com
sorghum, sudangrass, millet, ryegrass, clovers, oats, wheat, vetch, legume inoculants

Agriliance-AFC, LLC.
P.O. Box 2207
905 Market St.
Decatur, AL 35609
256-560-2848
fax 256-308-5693
bobj@agri-afc.com
www.agri-afc.com
native grasses, wheat, rye, winter peas, vetch

Albright Seed Company
6155 Carpinteria Ave.
Carpinteria, CA 93013-3061
805-684-0436
fax 805-684-2798
Paul@ssseeds.com
www.SSSeeds.com
Specializing in native California summer dormant perennial grasses & wild flowers, handles all seeds used in western cover crops

A.L. Gilbert—Farmers Warehouse
4367 Jessup Rd.
Ceres, CA 95307
800-400-6377 or 209-632-2333
jan.bowman@farmerswarehouse.com
www.farmerswarehouse.com
forage, grass, clovers, alfalfa, pasture

Ampac Seed Co.
P.O. Box 318
Tangent, OR 97389
800-547-3230
fax 541-928-2430
info@ampacseed.com
http://ampacseed.com
forage grasses, legumes, annual ryegrass, turnip, rapeseed

Barenbrug
33477 Hwy. 99 East
Tangent, OR 97389
541-926-5801 or 800-547-4101
fax 541-926-9435
forage legumes, grasses

The Birkett Mills
Transloading Facility
North Ave.
Penn Yan, NY 14527
315-536-2594
custserv@thebirkettmills.com
www.thebirkettmills.com
buckwheat

Byron Seeds, LLC
9820 N. 750 E.
Marshall, IN 47859
765-435-7243
fax 765-435-2759
cover crops, small grains, legumes, grasses

Cache River Valley Seed, LLC
Hwy. 226 East
Cash, AR 72421
870-477-5427
crvseed@crvseed.com
www.crvseed.com
wheat, oats

Cal/West Seeds
P.O. Box 1428
Woodland, CA 95776-1428
800-824-8585
fax 530-666-5317
t.hickman@calwestseeds.com
www.calwestseeds.com
clovers, alfalfa, sudangrass, sorghum

Cedar Meadow Farm
679 Hilldale Rd.
Holtwood PA 17532
717-575-6778
fax 717-284-5967
steve@cedarmeadowfarm.com
www.cedarmeadowfarm.com
forage radish, hairy vetch

The CISCO Companies
602 N. Shortridge Rd.
Indianapolis, IN 46219
800-888-2986 ext. 319
fax 317-357-7572
www.ciscoseeds.com/nvest.php
N-Vest cover crop mixes, annual ryegrasses, clovers, winter peas, radishes, turnips, vetches, cowpeas, wheat, rye, oats, winter barley, triticale, grasses, sorghums, legume inoculants and more.

Discount Seeds
P.O. Box 84
2411 9th Ave SW
Watertown, SD 57201
605-886-5888
fax 605-886-3623
grains, forage/grain legumes, forage grasses

DLF Organic
P.O. Box 229
175 West H St.
Halsey, OR 97348
800-445-2251
info@dlforganic.com
www.dlforganic.com
red/white clover, forage peas, organic cover crop mix

Doebler's Pennsylvania Hybrids, Inc.
202 Tiadaghton Ave.
Jersey Shore, PA 17740
800-853-2676 or 570-753-3210
fax 570-753-5302
info@doeblers.com
www.doeblers.com
clovers, grasses, legume inoculants

Ernst Conservation Seeds
9006 Mercer Pike
Meadville, PA 16335
800-873-3321
fax 814-336-5191
ernstsales@ernstseed.com
www.ernstseed.com
oats, clovers, buckwheat, ryegrass, alfalfa, barley, winter peas, rye, wheat, hairy vetch, Japanese millet, foxtail millet

Fedco Seeds
P.O. Box 520
Waterville, ME 04903
207-873-7333
fax 207-872-8317
www.fedcoseeds.com
grains, forage/grain legumes, grasses, clovers, legume inoculants

Genesee Union Warehouse
P.O. Box 67
Genesee, ID 83832
208-285-1141
fax 208-285-1716
www.geneseeunion.coop
Oriental mustard, winter peas

Harmony Farm Supply & Nursery
3244 Hwy. 116 North
Sebastopol, CA 95472
707-823-9125
fax 707-823-1734
info@harmonyfarm.com
www.harmonyfarm.com
clovers, grains, grasses, legumes, legume inoculants

High Mowing Seeds
76 Quarry Rd.
Wolcott, VT 05680
802-472-6174
fax 802-472-3201
www.highmowingseeds.com
buckwheat, vetch, oats, rye, peas, clover, legume inoculants

Hobbs & Hopkins Ltd.
1712 SE Ankeny St.
Portland, OR 97214
503-239-7518
fax 503-230-0391
info@protimelawnseed.com
www.protimelawnseed.com
grasses

Horstdale Farm Supply
12286 Hollowell Church Rd.
Greencastle, PA 17225-9525
717-597-5151
fax 717-597-5185
barley, wheat, rye, oats, crimson clover, winter vetch

Hytest Seeds
2827 8th Ave. South
Fort Dodge, IA 50501
800-442-7391
jrkrenz@landolakes.com
www.hytestseeds.com
alfalfa

Johnny's Selected Seeds
955 Benton Ave.
Winslow, ME 04901
877-564-6697
fax 800-738-6314
rstore@johnnyseeds.com
www.johnnyseeds.com
buckwheat, rye, oats, wheat, vetch, clover, field peas, green manure mix, sudangrass, legume inoculants

Kamprath Seeds, Inc.
205 Stockton St.
Manteca, CA 95337
209-823-6242
fax 209-823-2582
vetches, bell beans, small grains, brassicas, subterranean clovers, balansa clover, medics, perennial clovers, grasses

Kaufman Seeds, Inc.
P.O. Box 398
Ashdown, AR 71822
870-898-3328 or 800-892-1082
fax 870-898-3302
kaufmanseeds@arkansas.net
wholesale grains, forage/grain legumes, grasses, summer annuals, sunn hemp

Keystone Group Ag. Seeds
RR 1 Box 81A
Leiser Rd.
New Columbia, PA 17856
570-538-1170
alfalfa, forages, forage radish

King's Agriseeds, LLC
96 Paradise Ln.
Ronks, PA 17572
717-687-6224 or 866-687-6224
forage radish, forages, alfalfa, clovers

Little Britain AgriSupply Inc.
398 North Little Britain Rd.
Quarryville, PA 17566
717-529-2196
lbasinc@juno.com
oats, barley, wheat, rye, triticale, intermediate ryegrass, perennial ryegrass, annual ryegrass, red clover, sorghum & sudangrass (other cover crops available by special order)

Michigan State Seed Solutions
717 N. Clinton St.
Grand Ledge, MI 48837
517-627-2164
fsiemon@seedsolutions.com
www.seedsolutions.com
forgage/grain legumes, grains, grasses

Missouri Southern Seed Corp.
P.O. Box 699
Rolla, MO 65402
573-364-1336 or 800-844-1336
fax 573-364-5963
wholesale (w/retail outlets): grains, forage legumes, grasses, summer annuals

Moore Seed Farm, LLC
8636 N. Upton Rd.
Elsie, MI 48831
989-862-4686
rye

North Country Organics
P.O. Box 372
Bradford, VT 05033
802-222-4277
fax 802-222-9661
info@norganics.com
www.norganics.com
buckwheat, ryegrass, oats, rye, wheat

Peaceful Valley Farm & Garden Supply
125 Clydesdale Ct.
P.O. Box 2209
Grass Valley, CA 95945
888-784-1722
fax 530-272-4794
helpdesk@groworganic.com
www.groworganic.com
clovers, grasses, legumes, grains, vetches, custom cover crop mixes, legume inoculants

Pennington Seeds
P.O. Box 290
Madison, GA 30650
800-285-7333
seeds@penningtonseed.com
www.penningtonseed.com
grains, grasses, forage legumes

Plantation Seed Conditioners, Inc.
P.O. Box 398
Newton, GA 39870
229-734-5466
fax 229-734-7419
lupin, rye, black oats

P.L. Rohrer & Bro. Inc.
P.O. Box 250
Smoketown, PA 17576
717-299-2571
fax 717-299-5347
info@rohrerseeds.com
www.rohrerseeds.com
grains, clovers, grasses, brassicas, vetch, rye, wildlife mix

Rupp Seeds, Inc.
17919 County Rd. B
Wauseon, OH 43567
877-591-7333
fax 419-337-5491
feedback@ruppseeds.com
www.ruppseeds.com
grains, grasses, forage legumes, vetch

Seed Solutions
2901 Packers Ave.
Madison, WI 53707
800-356-7333
fax 608-249-0695
dbastian@seedsolutions.com
www.foragefirst.com
oats, wheat, rye, hairy vetch, sweetclover, berseem clover, red clover, crimson clover, forage peas, cowpeas, ryegrass, buckwheat, millets, sorghum, sorghum-sudangrass, cover crop mixes

Seedway, LLC.
P.O. Box 250
Hall, NY 14463
800-836-3710
fax 585-526-6391
farmseed@seedway.com
www.seedway.com
ryegrass, orchardgrass, clover

Siemer/Mangelsdorf Seed Co.
515 West Main St.
P.O. Box 580
Teutopolis, IL 62401
217-857-3171
fax 217-857-3226
info@siemerent.com
www.siemerent.com
grasses, vetch, wildlife

Southern States Cooperative, Inc.
P.O. Box 26234
Richmond, VA 23260-6234
800-868-6273
wholesale (retail outlets in 6 states): grains, forage legumes, cowpeas, summer annuals, rapeseed, clover, buckwheat, rye

Sweeney Seed Company
110 South Washington St.
Mount Pleasant, MI 48858
989-773-5391
grasses, forages, legumes

Talbot Ag Supply
P.O. Box 2252
Easton, MD 21601-8944
410-820-2388
robertshaw@friendly.net
wheat, barley, rapeseed, rye

Tennessee Farmers Co-op
200 Waldron Rd.
P.O. Box 3003
LaVergne, TN 37086-1983
615-793-8400
info@ourcoop.com
www.ourcoop.com
alfalfa, grains, clover, grasses, legume inoculants

Timeless Seeds, Inc.
166 Sunrise Ln.
P.O. Box 881
Conrad, MT 59425
406-278-5722
fax 406-278-5720
info@timelessfood.com
http://timelessfood.com
dryland forage/grain legumes

The Wax Co., Inc.
212 Front St. N.
Amory, MS 38821
662-256-3511
annual ryegrass, forages

Welter Seed & Honey Co.
17724 Hwy. 136
Onslow, IA 52321-7549
800-728-8450 or 800-470-3325
fax 563-485-2764
info@welterseed.com
www.welterseed.com
oats, brassicas, rye, spring wheat, winter wheat, buckwheat, hairy vetch, sweet clovers, alfalfa, a wide variety of other seeds, legume inoculants

Wolf River Valley Seeds
N2976 County Hwy M
White Lake, WI 54491
800-359-2480
fax 715-882-4405
wrvs@newnorth.net
www.wolfrivervalleyseeds.com
clover, brassicas, grasses, annual forage, grain, legume inoculants

INOCULANT SUPPLIERS

Many of the seed suppliers listed above sell inoculated seed and/or legume inoculants.

Becker Underwood
801 Dayton Ave.
Annes, IA 50010
800-232-5907
request@beckerunderwood.com
www.beckerunderwood.com

HiStick
801 Dayton Ave.
Ames, IA 50010
800-892-2013
www.histick.com
click 'Where to Buy' for dealer locations

LiphaTech Inc. (Nitragin division)
Manufacturer of Nitragin inoculant.
www.nitragin.com
click 'Recommended Links' at bottom of page.

Appendix D

FARMING ORGANIZATIONS WITH COVER CROP EXPERTISE

This list is for information purposes only. Inclusion does not imply endorsement, nor is criticism implied of organizations not mentioned. Note: CC denotes cover crop(s) or cover cropping.

ORGANIZATIONS—NORTHEAST

The Accokeek Foundation
3400 Bryan Point Rd.
Accokeek, MD 20607
301-283-2113
fax 301-283-2049
accofound@accokeek.org
www.accokeek.org
land stewardship & ecological agriculture using CC

Chesapeake Wildlife Heritage
P.O. Box 1745
Easton, MD 21601
410-822-5100
fax 410-822-4016
info@cheswildlife.org
www.cheswildlife.org
CC in organic & sustainable farming systems; consulting & implementation in mid-shore Md area as they relate to farming & wildlife

Pennsylvania Association for Sustainable Agriculture (PASA)
P.O. Box 419
114 W. Main St.
Millheim, PA 16854
814-349-9856
fax 814-349-9840
info@pasafarming.org
www.pasafarming.org
on farm CC demonstrations

SARE Northeast Region Office
University of Vermont
10 Hills Bldg.
Burlington, VT 05405-0082
802-656-0471
fax 802-656-4656
nesare@uvm.edu
www.uvm.edu/~nesare

ORGANIZATIONS—NORTH CENTRAL

Center for Integrated Agricultural Systems (CIAS)
University of Wisconsin-Madison
1535 Observatory Dr.
Madison, WI 53706
608-262-5200
fax 608-265-3020
jhendric@facstaff.wisc.edu
www.cias.wisc.edu
CC for fresh market vegetable production; relevant publication "Grower to grower: Creating a livelihood on a fresh market vegetable farm" available on website.

Conservation Technology Information Center (CTIC)
1220 Potter Dr.
West Lafayette, IN 47906
765-494-9555
fax 765-494-5969
ctic@ctic.purdue.edu
www.conservationinformation.org
CTIC is a trusted, reliable source for information & technology about conservation in agriculture

Leopold Center for Sustainable Agriculture
Iowa State University
209 Curtiss Hall
Ames, IA 50011-1050
515-294-3711
fax 515-294-9696
leocenter@iastate.edu
www.ag.iastate.edu/centers/leopold
support research, demos, education projects in Iowa on CC systems

Northern Plains Sustainable Agriculture Society
P.O. Box 194, 100 1 Ave. SW
Lamoure, ND 58458-0194
701-883-4304
fax 701-883-4304
npsas@drtel.net
www.npsas.org
CC systems for organic/sustainable farmers

SARE North Central Region Office
University of Minnesota
120 Bio. Ag. Eng.
1390 Eckles Ave.
St. Paul, MN 55108
612-626-3113
fax 612-626-3132
ncrsare@umn.edu
www.sare.org/ncrsare

ORGANIZATIONS—SOUTH

Appropriate Technology Transfer for Rural Areas (ATTRA)
P.O. Box 3657
Fayetteville, AR 72702
800-346-9140
http://attra.ncat.org/
ATTRA is a leading information source for farmers & extension agents thinking about sustainable farming practices

Educational Concerns for Hunger Organization
ECHO
17391 Durrance Rd.
North Ft. Myers, FL 33917
239-543-3246
fax 239-543-5317
echo@echonet.org
www.echonet.org
provides technical information & seeds of tropical cover crops; maintains demonstration plots on the farm of many of the most important tropical cover crops.

The Kerr Center for Sustainable Agriculture Inc.
P.O. Box 588
Highway 271 South
Poteau, OK 74953
918-647-9123
fax 918-647-8712
mailbox@kerrcenter.com
www.kerrcenter.com
CC systems demonstrations & research

SARE Southern Region Office
University of Georgia
1109 Experiment St.
Georgia Station
Griffin, GA 30223-1797
770-412-4786
fax 770-412-4789
groland@southernsare.org
www.southernsare.org

Texas Organic Growers Association
P.O. Box 15211
Austin, TX 78761
877-326-5175
fax 512-842-1293
steve@tofga.org
www.tofga.org
TOGA is helping to make organic agriculture viable in Texas & offers a quarterly periodical.

ORGANIZATIONS—WEST

Center for Agroecology & Sustainable Food Systems
University of California
1156 High St.
Santa Cruz, CA 95064
831-459-3240
fax 831-459-2799
jonitann@ucsc.edu
www.ucsc.edu/casfs
Facilitates the training of organic farmers & gardeners, with practical hands-on & academic training

SARE Western Region Office
Utah State University
4865 Old Main Hill
Room 322
Logan, UT 84322
435-797-2257
wsare@ext.usu.edu
http://wsare.usu.edu

Small Farm Center
University of California
One Shields Ave.
Davis, CA 95616
530-752-8136
fax 530-752-7716
sfcenter@ucdavis.edu
www.sfc.ucdavis.edu
serves as a clearinghouse for questions from farmers, marketers, farm advisors, trade associations, government officials & agencies, & the academic community

University of California SAREP
One Shields Ave.
Davis, CA 95616-8716
530-752-7556
fax 530-754-8550
sarep@ucdavis.edu
www.sarep.ucdavis.edu

Appendix E
REGIONAL EXPERTS

These individuals are willing to briefly respond to specific questions in their area of expertise, or to provide referral to others in the sustainable agriculture field. Please respect their schedules and limited ability to respond. Note: CC denotes cover crop(s) or cover cropping.

NORTHEAST

Andy Clark
SARE Outreach
1122 Patapsco Building
University of Maryland
College Park, MD 20742-6715
301-405-2689
fax 301-405-7711
coordinator@sare.org
www.sare.org
technical information specialist for sustainable agriculture; legume/grass CC mixtures

Jim Crawford
New Morning Farm
22263 Anderson Hollow Rd.
Hustontown, PA 17229
814-448-3904
fax 814-448-2295
jim@newmorningfarm.net
www.newmorningfarm.net
35 years of CC for vegetable production

A. Morris Decker
5102 Paducah Rd.
College Park, MD 20740
Prof. Emeritus, University of Maryland
40 years forage mgt., CC & agronomic cropping systems research

Eric Gallandt
Weed Ecology & Management
Dept. of Plant, Soil & Environmental Sciences
University of Maine
5722 Deering Hall
Orono, ME 04469
207-581-2933
gallandt@maine.edu
www.umaine.edu/weedecology/
CC & weed management

Steve Groff
Cedar Meadow Farm
679 Hilldale Rd.
Holtwood, PA 17532
717-284-5152
fax 717-284-5967
steve@cedarmeadowfarm.com
www.cedarmeadowfarm.com
CC/no-till strategies for vegetable & agronomic crops

Vern Grubinger
University of Vermont Extension
11 University Way
Brattleboro, VT 05301-3669
802-257-7967 ext. 13
vernon.grubinger@uvm.edu
www.uvm.edu/vtvegandberry/Videos/
covercropvideo.html
vegetable & berry specialist

H.G. Haskell III
4317 S. Creek Rd.
Chadds Ford, PA 19317
tel/fax 610-388-0656
cell 610-715-7688
siwvegies@aol.com
rye/vetch mix for green manure & erosion control; uses winter rye CC on all land; 70% no-till

Zane R. Helsel
Rutgers Cooperative Extension
88 Lipman Dr.
313 Martin Hall
New Brunswick, NJ 08901-8525
732-932-5000 x585
fax 732-932-6633
helsel@aesop.rutgers.edu
www.rce.rutgers.edu
CC in field crops

Stephen Herbert
University of Massachusetts
Dept. of Plant Soil Insect Sciences
Bowditch Hall
Amherst, MA 01003
413-545-2250
fax 413-545-0260
sherbert@pssci.umass.edu
www.umass.edu/cdl
CC culture, soil fertility, crop nutrition & management & nitrate leaching

Jeff Moyer
The Rodale Institute
611 Siegfriedale Rd.
Kutztown, PA 19530
610-683-1420
fax 610-683-8548
jeff.moyer@rodaleinst.org
www.newfarm.org
CC management in biologically based no-till systems, CC in grain crop rotations, CC as a weed management tool, rolling CC

Jack Meisinger
USDA/ARS
BARC-East
Bldg. 163F Rm 6
10300 Baltimore Ave
Beltsville, MD 20705
301-504-5276
jmeising@anri.barc.usda.gov
nitrogen management in CC systems

Anne & Eric Nordell
3410 Rte. 184
Trout Run, PA 17771
570-634-3197
rotational CC for weed control

Marianne Sarrantonio
University of Maine
Dept. of Plant, Soil & Environmental Science
5722 Deering Hall
Orono, ME 04473
207-581-2913
fax 207-581-2999
mariann2@maine.edu
effects of CC on nutrient cycling & soil quality in diverse cropping systems

Eric Sideman
Maine Organic Farmers & Gardeners Assoc.
P.O. Box 1760
Unity, ME 04988
207-946-4402
fax 207-568-4141
esideman@mofga.org
www.mofga.org
CC in rotation with vegetables & small fruit on organic farms

John R. Teasdale
USDA/ARS
BARC-West
Bldg. 001 Rm 245
10300 Baltimore Ave
Beltsville, MD 20705
301-504-5504
fax 301-504-6491
john.teasdale@ars.usda.gov
CC mgmt./weed mgmt. using CC

Donald Weber
Research Entomologist
Insect Biocontrol Laboratory
10300 Baltimore Ave.
Bldg. 011A, Rm. 107
BARC-West
Beltsville, MD 20705
301-504-8369
fax 301-504-5104
Don.Weber@ars.usda.gov
CC & pest management (vegetable crops only)

David W. Wolfe
Cornell University
Dept. of Horticulture
168 Plant Science Bldg.
Ithaca, NY 14853
607-255-7888
fax 607-255-0599
dww5@cornell.edu
www.hort.cornell.edu/wolfe
CC for improved soil quality in vegetables

NORTH-CENTRAL

Rich Bennett
13 Lakeview Dr.
Napoleon, OH 43545
419-592-1100
benfarm1@excite.com
CC for soil quality & herbicide/pesticide reduction

John Cardina
Ohio State University
Dept. of Horticulture & Crop Science
1680 Madison Ave.
Wooster, OH 44691
330-263-3644
fax 330-263-3887
cardina.2@osu.edu
CC for no-till corn & soybeans

Stephan A. Ebelhar
University of Illinois, Dept. of Crop Sciences
Dixon Springs Agricultural Center
Rte. 1, Box 256
Simpson, IL 62985
618-695-2790
fax 618-695-2492
sebelhar@uiuc.edu
www.cropsci.uiuc.edu/research/rdc/dixon-springs
CC for no-till corn & soybeans

Rick (Derrick N.) Exner, Ph.D.
Iowa State University Extension
Practical Farmers of Iowa
2104 Agronomy Hall, ISU
Ames, IA 50011
515-294-5486
fax 515-294-9985
dnexner@iastate.edu
www.practicalfarmers.org
on farm CC research for the upper Midwest

Carmen M. Fernholz
2484 Hwy 40
Madison, MN 56256
320-598-3010
fax 320-598-3010
fernholz@umn.edu
minimum tillage, general cover crops, seeding soybeans into standing rye

Walter Goldstein
Michael Fields Agricultural Institute
P.O. Box 990
East Troy, WI 53120
262-642-3303 ext. 112
fax 262-642-4028
wgoldstein@michaelfieldsaginst.org
www.michaelfieldsaginst.org
CC for nutrient cycling & soil health in biodynamic, organic & conventional systems

Frederick Kirschenmann
3703 Woodland
Ames, IA 50014
515-294-3711
fax 515-294-9696
leopold1@iastate.edu
CC for cereal grains, soil quality & pest mgt.

Matt Liebman
Iowa State University
Dept. of Agronomy
3405 Agronomy Hall
Ames, IA 50011-1010
515-294-7486
fax 515-294-3163
mliebman@iastate.edu
www.agron.iastate.edu/personnel/userspage.aspx?id=646
CC for cropping system diversification, soil amendments, weed ecology & management, crop rotation, green manures, intercrops, animal manures, composts, insects & rodents that consume weed seeds

Todd Martin
MSU Kellogg Biological Station
Land & Water Program
3700 E. Gull Lake Dr.
Hickory Corners, MI 49060
269-671-2412 ext. 226
fax 269-671-4485
tmartin@kbs.msu.edu
www.covercrops.msu.edu
CC for corn, soybeans, vegetables, blueberries in Michigan

Paul Mugge
6190 470th St.
Sutherland, IA 51058-7544
712-446-2414
pmugge@midlands.net
Iowa farmer using CC for corn & soybeans

Dale R. Mutch
MSU Kellogg Biological Station
Land & Water Program
3700 E. Gull Lake Dr.
Hickory Corners, MI 49060
269-671-2412 ext. 224
800-521-2619
fax 269-671-4485
mutch@msu.edu
www.covercrops.msu.edu
general CC information provider

Rob Myers
Thomas Jefferson Agricultural Institute
601 West Nifong Blvd, Ste. 1D
Columbia, MO 65203
573-449-3518
fax 573-449-2398
rmyers@jeffersoninstitute.org
www.jeffersoninstitute.org
CC & crop diversification

Sieg Snapp
W.K. Kellogg Biological Station
Michigan State University
3700 East Gull Lake Dr.
Hickory Corners, MI 49060
fax 269-671-2104
snapp@msu.edu
www.kbs.msu.edu/faculty/snapp
using CC to enhance nitrogen fertilizer use efficiency & nutrient cycling in row crops

Richard Thompson
Thompson On-Farm Research
2035 190th St.
Boone, IA 50036-7423
515-432-1560
dickandsharon@practicalfarmers.org
CC in corn-soybean-corn-oats-hay rotation

SOUTH

Philip J. Bauer
USDA/ARS
Cotton Production Research Center
2611 West Lucas St.
Florence, SC 29501-1241
843-669-5203 x137
fax 843-662-3110
phil.bauer@ars.usda.gov
CC for cotton production

J.P. (Jim) Bostick, Ph.D.
Southern Seed Certification Assoc.
P.O. Box 357
Headland, AL 36345
334-693-3988
fax 334-693-2212
jpbostick@centurytel.net
specialty CC & seed development

Nancy Creamer
North Carolina State University
Center for Environmental Farming Systems
Box 7609
Raleigh, NC 27695
919-515-9447
fax 919-515-2505
nancy_creamer@ncsu.edu
www.cefs.ncsu.edu
CC management for no-till organic vegetable production

Seth Dabney
USDA-ARS
National Sedimentation Laboratory
P.O. Box 1157
598 McElroy Drive
Oxford, MS 38655
662-232-2975
fax 662-232-2915
sdabney@ars.usda.gov
legume reseeding & mechanical control of CC in the mid-South

Greg D. Hoyt
Dept. of Soil Science, NCSU
Mtn. Hort. Crops Res. & Ext. Center
455 Research Dr.
Fletcher, NC 28732
828-684-3562
fax 828-684-8715
greg_hoyt@ncsu.edu
CC for vegetables, tobacco & corn

D. Wayne Reeves
Research Leader
USDA/ARS
J. Phil Campbell Sr. Natural Resource Conservation Center
1420 Experiment Station Rd.
Watkinsville, GA 30677
706-769-5631 ext. 203
cell 706-296-9396
fax 706-769-8962
Wayne.Reeves@ars.usda.gov
CC management & conservation tillage for soybean, corn, cotton, peanut, & integrated grazing-row crop systems

Kenneth H. Quesenberry
University of Florida
P.O. Box 110500
Gainesville, FL 32611
352-392-1811 ex. 213
fax 352-392-1840
clover@ifas.ufl.edu
forage legume CC in S.E. USA

Jac Varco
Mississippi State University
Plant & Soil Sciences Dept.
Box 9555
Dorman Hall, Room 117
Mississippi State, MS 39762
662-325-2737
fax 662-325-8742
jvarco@pss.msstate.edu
CC & fertilizer/nutrient management for cotton production

WEST

Miguel A. Altieri
University of California at Berkeley
215 Mulford Hall
Berkeley, CA 94720-3112
510-642-9802
fax 510-643-5438
agroeco3@nature.berkeley.edu
http://nature.berkeley.edu/~agroeco3
CC to enhance biological pest control in perennial systems

Robert L. Bugg, Ph.D
U.C. Sustainable Agriculture Research & Education Program
University of California at Davis
One Shields Ave.
Davis, CA 95616-8716
530-754-8549
fax 530-754-8550
rlbugg@ucdavis.edu
www.sarep.ucdavis.edu
CC selection, growth & IPM

Jorge A. Delgado
Soil Scientist
Soil Plant Nutrient Research Unit
USDA/ARS
2150 Centre Avenue, Building D, Suite 100
Fort Collins, CO 80526
jdelgado@lamar.colostate.edu;
Jorge.Delgado@ars.usda.gov
CC in irrigated potato, vegetables and small grain systems

Richard P. Dick
Professor of Soil Microbial Ecology
Ohio State University
School of Environment & Natural Resources
Columbus, OH 43210
614-247-7605
fax 614-292-7432
Richard.Dick@snr.osu.edu
nitrogen cycling & environmental applications of CC

Shiou Kuo
Washington State University
Dept. of Crop & Soil Sciences
7612 Pioneer Way East
Puyallup, WA 98371-4998
253-445-4573
fax 253-445-4569
Skuo@wsu.edu
CC & soil nitrogen accumulation & availability to corn

John M. Luna
Oregon State University
Dept. of Horticulture
4107 Agricultural & Life Sciences Bldg.
Corvallis, OR 97331
541-737-5430
fax 541-737-3479
lunaj@oregonstate.edu
CC for integrated vegetable production & agroecology

Dwain Meyer
North Dakota State University
Loftsgard Hall, Room 470E
Extension Service
Fargo, ND 58105
701-231-8154
dmeyer@ndsuext.nodak.edu
yellow blossom sweetclover specialist

Clara I. Nicholls
University of California
Dept. of Environmental Science Policy & Management
Division of Insect Ecology
137 Hilgard Hall
Berkeley, CA 94720
510-642-9802
fax 510-643-5438
nicholls@berkeley.edu
CC for biological control in vineyards

Fred Thomas
CERUS Consulting
2119 Shoshone Ave.
Chico, CA 95926
530-891-6958
fax 530-891-5248
fred@cerusconsulting.com
CC specialist for tree & vine crops, row crops, field crops, vegetable crops & summer CC

Appendix F
CITATIONS BIBLIOGRAPHY

The publications cited in the text (in parentheses) are listed here by reference number.

1 Abdul-Baki, A.A. et al. 1997. Broccoli production in forage soybean and foxtail millet cover crop mulches. *HortSci*. 32:836-839.

2 Abdul-Baki, A.A. and J.R.Teasdale. 1993. A no-tillage tomato production system using hairy vetch and subterranean clover mulches. *HortSci*. 28:106-108.

3 Abdul-Baki, A.A. and J.R.Teasdale. 1997. Snap bean production in conventional tillage and in no-till hairy vetch mulch. *HortSci*. 32:1191-1193.

4 Abdul-Baki, A.A. and J.R.Teasdale. 1997. *Sustainable Production of Fresh-Market Tomatoes and Other Summer Vegetables with Organic Mulches*. Farmers' Bulletin No. 2279, USDA/ARS, Beltsville, Md. 23 pp. www.ars.usda.gov/is/np/tomatoes.html

5 Alabouvette, C., C. Olivain and C. Steinberg. 2006. Biological control of plant diseases: the European situation. *European J. of Plant Path*. 114:329-341.

6 Alger, J. 2006. Personal communication. Stanford, Mont.

7 Al-Sheikh, A. et al. 2005. Effects of potato-grain rotations on soil erosion, carbon dynamics and properties of rangeland sandy soils. *J. Soil Tillage Res*. 81:227-238.

8 American Forage and Grassland Council National Fact Sheet Series. Subterranean clover. http://forages.orst.edu/main.cfm?PageID=33

9 Angers, D.A. 1992. Changes in soil aggregation and organic carbon under corn and alfalfa. *Soil Sci. Soc. Am. J*. 56:1244-1249.

10 ATTRA. 2006. *Overview of Cover Crops and Green Manures*. ATTRA. Fayetteville, Ark. http://attra.ncat.org/attra-pub/PDF/covercrop.pdf

11 ATTRA. Where can I find information about the mechanical roller-crimper used in no-till production? http://attra.ncat.org/calendar/question.php/2006/05/08/p2221

12 Arshad, M.A. and K.S. Gill. 1996. Crop production, weed growth and soil properties under three fallow and tillage systems. *J. Sustain. Ag*. 8:65-81.

13 Ashford, D.L. and D.W. Reeves, 2003. Use of a mechanical roller-crimper as an alternative kill method for cover crops. *Amer. J. Alt. Ag*. 18:37-45.

14 Badaruddin, M. and D.W. Meyer. 1989. Water use by legumes and its effects on soil water status. *Crop Sci*. 29:1212-1216.

15 Badaruddin, M. and D.W. Meyer. 1990. Green-manure legume effects on soil nitrogen, grain yield, and nitrogen nutrition of wheat. *Crop Sci*. 30:819-825.

16 Bagegni, A.M. et al. 1994. Herbicides with crop competition replace endophytic tall fescue (*Festuca arundinacae*). *Weed Tech*. 8:689-695.

17 Bailey, R.G. et al. 1994. Ecoregions and subregions of the United States (map). Washington, DC: USDA Forest Service. 1:7,500,000. With supplementary table of map unit descriptions, compiled and edited by W. H. McNab and R. G. Bailey. www.fs.fed.us/land/ecosysmgmt/ecoreg1_home.html

18 Ball, D.M. and R.A. Burdett. 1977. *Alabama Planting Guide for Forage Grasses*. Alabama Cooperative Extension Service, Chart ANR 149. Auburn Univ., Auburn, Ala.

19 Ball, D.M. and R.A. Burdett. 1977. *Alabama Planting Guide for Forage Legumes*. Alabama Cooperative Extension Service, Chart ANR 150. Auburn Univ., Auburn, Ala.

20 Barker, K.R. 1996. Animal waste, winter cover crops and biological antagonists for sustained management of Columbia lance and other nematodes on cotton. SARE Project Report #LS95-060.1. Southern Region SARE. Griffin, Ga. www.sare.org/projects

21 Barnes, R.F. et al. 1995. Forages: *The Science of Grassland Agriculture*. 5th Edition. Iowa State Univ. Press, Ames, Iowa.

22 Bauer, P.J. et al. 1993. Cotton yield and fiber quality response to green manures and nitrogen. *Agron*. J. 85:1019-1023.

23 Bauer, P.J., J.J. Camberato and S.H. Roach. 1993. Cotton yield and fiber quality response to green manures and nitrogen. *Agron. J*. 85:1019-1023

24 Bauer P.J. and D.W. Reeves. 1999. A comparison of winter cereal species and planting dates as residue cover for cotton grown with conservation tillage. *Crop Sci*. 39:1824–1830.

25 Baumhardt, R.L. and R.J. Lascano. 1996. Rain infiltration as affected by wheat residue amount and distribution in ridged tillage. *Soil Sci. Soc. Am. J*. 60:1908-1913.

26 Baumhardt, R.L. 2003. The Dust Bowl Era. In B.A. Stewart and T.A. Howell (eds.) *Encyclopedia of Water Science*, pp. 187-191. Marcel-Dekker, NY.

27 Baumhardt, R.L. and R.L. Anderson. 2006. Crop choices and rotation principles. In G.A. Peterson, P.W. Unger, and W.A. Payne (eds.) *Dryland Agriculture, 2nd ed.* Agronomy Monograph No. 23. pp. 113-139. ASA, CSSA, and SSSA, Madison, WI.

28 Baumhardt, R.L. and R.J. Lascano. 1999. Water budget and yield of dryland cotton intercropped with terminated winter wheat. *Agron. J*. 91:922-927.

29 Baumhardt, R. L. and J. Salinas-Garcia. 2006. Mexico and the US Southern Great Plains. *In* G.A. Peterson, P.W. Unger, and W.A. Payne (eds.) *Dryland Agriculture, 2nd ed.* Agronomy Monograph No. 23. pp. 341-364. ASA, CSSA, and SSSA, Madison, WI.

30 Beale, P. et al. 1985. *Balansa Clover—a New Clover-Scorch-Tolerant Species*. South Australia Dept. of Ag. Fact Sheet.

31 Beste, C.E. 2007. Personal communication. Univ. of Maryland-Eastern Shore, Salisbury, Md.

32 Blackshaw, R.E. et al. 2001a. Yellow sweet-clover, green manure, and its residues effectively suppresses weeds during fallow. *Weed Sci.* 49:406-413.

33 Blackshaw, R.E. et al. 2001b. Suitability of undersown sweetclover as a fallow replacement in semiarid cropping systems. *Agron. J.* 93:863-868.

34 Blaser, B.C. et al. 2006. Optimizing seeding rates for winter cereal grains and frost-seeded red clover intercrops. *Agron J.* 98:1041-1049.

35 Bloodworth, L.H. and J.R. Johnson. 1995. Cover crops and tillage effects on cotton. *J. Prod.* Ag. 8:107-112.

36 Boquet, D.J. and S.M. Dabney. 1991. Reseeding, biomass, and nitrogen content of selected winter legumes in grain sorghum culture. *Agron. J.* 83:144-148.

37 Bordovsky, D.G., M. Choudhary and C.J. Gerard. 1998. Tillage effects on grain sorghum and wheat yields in the Texas Rolling Plains. *Agron. J.* 90:638–643.

38 Bowman, G. 1997. *Steel in the Field: A Farmer's Guide to Weed Management Tools*. USDA-Sustainable Agriculture Network (SAN). Beltsville, MD.

39 Boydston, R.A. and K. Al-Khatib. 2005. Utilizing Brassica cover crops for weed suppression in annual cropping systems. pp. 77-94. In H.P. Singh, D.R. Batish and R. K. Kohli (eds.). *Handbook of Sustainable Weed Management*. Haworth Press, Binghamton, NY.

40 Bradow, J.M. and J.C. William Jr. 1990. Volatile seed germination inhibitors from plant residues. *J. Chem. Ecol.* 16:645-666.

41 Bradow, J.M. 1993. Inhibitions of cotton seedling growth by volatile ketones emitted by cover crop residues. *J. Chem. Ecol.* 19:1085-1108.

42 Brady, N.C. 1990. *The Nature and Properties of Soils.* Macmillan Pub. Co., N.Y.

43 Brainard. D. 2005. Screening of cowpea and soybean varieties for weed suppression. Cornell Univ. www.hort.cornell.edu/organicfarm/htmls/legumecc.htm

44 Brandt, J.E., F.M. Hons and V.A. Haby. 1989. Effects of subterraneum clover interseeding on grain yield, yield components, and nitrogen content of soft red winter wheat. *J. Prod. Agric.* 2:347-351.

45 Brennan, E.B. and R.F. Smith. 2005. Winter cover crop growth dynamics and effects on weeds in the Central Coast of California. *Weed Tech.* 119: 1017-1024.

46 Brinsfield, R. and K. Staver. 1991. *Role of cover crops in reduction of cropland nonpoint source pollution*. Final Report to USDA-SCS, Cooperative Agreement #25087.

47 Brinton, W. Medics, general. Univ. of Calif. SAREP Cover Crops Resource Page. www.sarep.ucdavis.edu/ccrop

48 Brown, P.D. and M.J. Morra. 1997. Control of soil-borne plant pests using glucosinolate-containing plants. pp. 167–215. In: D.L. Sparks (ed.) *Adv. Agron.* Vol. 61. Academic Press, San Diego, CA.

49 Brown, S. et al. 2001. *Tomato Spotted Wilt of Peanut: Identifying and Avoiding High-Risk Situations*. Univ of Georgia Cooperative Extension Service Bulletin 1165. Univ of Georgia, Athens, GA http://pubs.caes.uga.edu/ caespubs/pubs/PDF/B1165.pdf

50 Bruce, R.R, P.F. Hendrix and G.W. Langdale. 1991. Role of cover crops in recovery and maintenance of soil productivity. pp.109-114. In W.L. Hargrove (ed.). *Cover Crops for Clean Water*. Soil and Water Conservation Society. Ankeny, Iowa.

51 Bruce, R.R., G.W. Langdale and A.L. Dillard. 1990. Tillage and crop rotation effect on characteristics of a sandy surface soil. *Soil Sci. Soc. Am. J.* 53:1744-1747.

52 Bruce, R.R. et al. 1992. Soil surface modification by biomass inputs affecting rainfall infiltration. *Soil Sci. Soc. Am. J.* 56:1614-1620.

53 Brunson, K.E. 1991. Winter cover crops in the integrated pest management of sustainable cantaloupe production. M.S. Thesis. Univ. of Georgia, Athens, Ga.

54 Brunson, K.E. and S.C. Phatak. 1990. Winter cover crops in low-input vegetable production. *HortSci.* 25:1158.

55 Brunson, K.E. et al. 1992. Winter cover crops influence insect populations in sustainable cantaloupe production. HortSci. 26:769.

56 Bugg, R.L. et al. 1990. Tarnished plant bug (*hemiptera: Miridae*) on selected cool-season leguminous cover crops. *J. Entomol. Sci.* 25:463-474.

57 Bugg, R.L. et al. 1991. Cool season cover crops relay intercropped with cantaloupe: Influence of a generalist predator, *Geocoris punctipes. J. Econ. Entomol.* 84:408-416.

58 Bugg, R.L. 1991. Cover crops and control of arthropod pests of agriculture. pp. 157-163. In W.L. Hargrove (ed.). *Cover Crops for Clean Water*. Soil and Water Conservation Society. Ankeny, Iowa.

59 Bugg, R.L. 1992. Using cover crops to manage arthropods on truck farms. *HortSci.* 27:741- 745.

60 Bugg, R.L. and C. Waddington. 1994. Using cover crops to manage arthropod pests of orchards: A review. *Ag, Ecosystems & Env.* 50:11-28.

61 Bugg, R.L. 1995. Cover biology: a mini-review. *SAREP Sustainable Agriculture-Technical Reviews.* 7:4. Univ. of California, Davis, Calif.

62 Bugg, R.L. et al. 1996. Comparison of 32 cover crops in an organic vineyard on the north coast of California. *Biol. Ag. and Hort.* 13:63-81.

63 Bugg, R.L., R.J. Zomer and J.S. Auburn. 1996. Cover crop profiles: One-page summaries describing 33 cover crops. In Cover Crops: Resources for Education and Extension (Chaney, D. and A.D. Mayse, eds.). SAREP. Univ. of Calif., Division of Ag. and Natural Resources, Davis, Calif.

64 Bugg, R.L. and M.V. Horn. 1997. Ecological soil management and soil fauna: Best practices in California vineyards. Australian Society for Viticulture and Oenology, Inc. Proc. Viticulture Seminar, Mildura, Victoria, Australia.

65 Burgos, N.R. and R.E. Talbert. 1996. Weed control by spring cover crops and imazethapyr in no-till southern pea (*Vigna unguiculata*). *Weed Tech.* 10:893-899.

66 Burgos, N.R., R.E. Talbert and R.D. Mattice 1999. Cultivar and age differences in the production of allelochemicals by *Secale cereale*. Weed Sci. 47:481-485.

67 Buntin, G.D. et al. 1994. Cover crop and nitrogen fertility effects on southern corn rootworm (Coleoptera: Chrysomelidae) damage in corn. *J. Econ. Entomol.* 87:1683-1688.

68 Butler, L.M. 1996. Fall-planted cover crops in western Washington: A model for sustainability assessment. SARE Project Report #SW94-008. Western Region SARE. Logan, Utah. www.sare.org/projects

69 Campbell, C.A. et al. 1993. Influence of legumes and fertilization of deep distribution of available phosphorus in a thin black chernozemic soil. *Can. J. Soil Sci.* 73:555-565.

70 Campbell, C.A. et al. 1993. Spring wheat yield trends as influenced by fertilizer and legumes. *J. Prod. Ag.* 6:564-568.

71 Canadian Organic Growers Inc. 1992. *Organic Field Crop Handbook*. Anne Macey (ed.). Canadian Organic Growers Inc., Ottawa, Ont.

72 CARR. P. Personal communication. 2007. North Dakota State Univ. Dickinson, ND.

73 Carr, P. M., W. P. Woodrow and L. J. Tisor. 2005. Natural reseeding by forage legumes following wheat in western North Dakota. *Agron. J.* 97:1270-1277.

74 Cash, D. et al. 1995. *Growing Peas in Montana*. Montguide MT 9520. Montana State Univ. Extension Service. Bozeman, Mont.

75 Cathey, H. M. 1990. *USDA Plant Hardiness Zone Map.* USDA-ARS Misc. Pub. No. 1475.

76 Chaney, D. and D. Mays. March 1997. Cover Crops: Resources for Education and Extension. UC SAREP, Division of Agriculture and Natural Resources, Davis Calif.

77 Cherr, C. M., J. M. S. Scholberg and R. McSorley. 2006. Green manure approaches to crop production: A synthesis. *Agron. J.* 98:302–319.

78 Chen, J., G.W. Bird. and R.L. Mather. 1995. Impact of multi-year cropping regimes on *Solanum tuberosum* tuber yields in the presence of *Pratylenchus penetrans* and *Verticillium dahliae*. *J. Nematol.* 27:654-660.

79 Choi, B.H. et al. 1991. Acid amide, dinitroaniline, triazine, urea herbicide treatment and survival rate of coarse grain crop seedlings. *Research Reports of the Rural Development Administration*, Upland and Industrial Crops 3:33-42.

80 Clark, A.J. 2007. Personal communication. Sustainable Agriculture Research and Education. USDA-SARE. College Park, Md.

81 Clark, A.J., A.M. Decker and J.J. Meisinger. 1994. Seeding rate and kill date effects on hairy vetch-cereal rye cover crop mixtures for corn production. *Agron. J.* 86:1065-1070.

82 Clark, A.J. et al. 1995. Hairy vetch kill date effects on soil water and corn production. *Agron. J.* 87:579-585.

83 Clark, A.J. et al. 1997a. Kill date of vetch, rye and a vetch-rye mixture: I. Cover crop and corn nitrogen. *Agron. J.* 89:427-434.

84 Clark, A.J. et al. 1997b. Kill date of vetch, rye and a vetch-rye mixture: II. Soil moisture and corn yield. *Agron. J.* 89:434-441.

85 Clark, A.J. et al. 2007a. Effects of a grass-selective herbicide in a vetch–rye cover crop system on corn grain yield and soil moisture. *Agron. J.* 99:43-48.

86 Clark, A.J. et al. 2007b. Effects of a grass-selective herbicide in a vetch–rye cover crop system on nitrogen management. *Agron. J.* 99:36-42.

87 Coale, F.J. et al. 2001. Small grain winter cover crops for conservation of residual soil nitrogen in the mid-Atlantic Coastal Plain. *Amer. J. of Alt. Ag.* 16:66-72.

88 Collins, H.P. et al. 2006. Soil microbial, fungal and nematode responses to soil fumigation and cover crops under potato production. *Biol. Fert. Soils*. 42:247-257.

89 Conservation Tillage Information Center. (2006). www.ctic.purdue.edu

90 Cooke, L. 1996. *New Red Clover Puts Pastures in the Pink*. USDA/ARS. 44:12 Washington, D.C.

91 Corak, S.J., W.W. Frye and M.S. Smith. 1991. Legume mulch and nitrogen fertilizer effects on soil water and corn production. *Soil Sci. Soc. Am.Jj*. 55(5):1395-1400.

92 Costa, J.M., G.A. Bollero and F.J. Coale. 2000. Early season nitrogen accumulation in winter wheat. *J. Plant Nutrition* 23:773-783.

93 Costello, M.J. 1994. Broccoli growth, yield and level of aphid infestation in leguminous living mulches. *Biol. Ag. And Hort.* 10:207-222.

94 Crawford, E.J. and B.G. Nankivell. Medics, general. Univ. of Calif. SAREP Cover Crops Resource Page. www.sarep.ucdavis.edu/ccrop

95 Creamer, N.G. and S.M. Dabney. 2002. Killing cover crops mechanically: Review of recent literature and assessment of new research results. *Amer. J. Alt. Ag.* 17:32-40.

96 Creamer, N.G. et al. 1995. A method for mechanically killing cover crops to optimize weed suppression. *Amer. J. Alt. Ag.* 10:157-162.

97 Creamer, N.G. et al. 1996. A comparison of four processing tomato production systems differing in cover crop and chemical inputs. *HortSci.* 121:559-568.

98 Cruse, R.M. 1995. Potential economic, environmental benefits of narrow strip intercrop ping. *Leopold Center Progress Reports* 4:14-19.

99 Cunfer, B.M. 1997. Disease and insect management using new crop rotations for sustainable production of row crops in the Southern U.S. SARE Project Report #LS94-057. Southern Region SARE. Griffin, Ga. www.sare.org/projects

100 Curran W.S. et al. 1996. *Cover Crops for Conservation Tillage Systems*. Penn State Conservation Tillage Series, Number 5.

101 Dabney, S. 1996. Cover crop integration into conservation production systems. SARE Project Report #LS96-073. Southern Region SARE. Griffin, Ga. www.sare.org/projects

102 Dabney, S.M. 1995. Cover crops in reduced tillage systems. Proc. Beltwide Cotton Conferences. pp. 126-127. 5 Jan 1995. National Cotton Council, Memphis, Tenn.

103 Dabney, S. 2007. Personal communication. USDA/ARS. Oxford, Miss.

104 Dabney, S.M. and J.L. Griffin. 1987. Efficacy of burn down herbicides on winter legume cover crops. pp. 122-125. In Power, J.F. (ed.) *The Role of Legumes in Conservation Tillage Systems*. Soil and Water Conservation Society. Ankeny, Iowa.

105 Dabney, S.M. et al. 1991. Mechanical control of legume cover crops. pp. 146-147. In W.L. Hargrove (ed.). *Cover Crops for Clean Water.* Soil and Water Conservation Society. Ankeny, Iowa.

106 Dabney, S.M., J.A. Delgado and D.W. Reeves. 2001. Using winter cover crops to improve soil and water quality. *Commun. Soil Plant Anal.* 32:1221-1250.

107 Davis, D.W. et al. 1990. Cowpea. In *Alternative Field Crops Manual*. Univ. of Wisc-Ext. and Univ. of Minnesota. Madison, Wis. and St. Paul, Minn.

108 Decker, A.M. et al. 1994. Legume cover crop contributions to no-tillage corn production. *Agron. J.* 86:126-136.

109 Decker, A.M. et al. 1992. *Winter Annual Cover Crops for Maryland Corn Production Systems.* Agronomy Mimeo 34. Univ. of Md. Cooperative Ext. Service, Md. Inst. for Ag. and Natural Resources, College Park, Md.

110 DeGregorio, R. et al. 1995. Bigflower vetch and rye vs. rye alone as a cover crop for no-till sweetcorn. *J. Sustain. Ag.* 5:7-18.

111 Delgado, J.A. 1998. Sequential NLEAP simulations to examine effect of early and late planted winter cover crops on nitrogen dynamics. *J. Soil Water Conserv.* 53:241-244.

112 Delgado, J.A. and J. Lemunyon. 2006. Nutrient Management. pp. 1157-1160. In R. Lal (ed.). *Encyclopedia Soil Sci.* Markel and Decker, New York, pp 1924. NY.

113 Delgado, J.A. et al. 2007. Cover crops-potato rotations: Part III, making the connection - green manure cover crop effects on potato yield and quality. *In Proceedings of the 25th Annual San Luis Valley Potato Grain Conference.* Jan. 30–Feb. 2, 2007. Monte Vista, CO.

114 Delgado, J.A. et al. 1999. Use of winter cover crops to conserve soil and water quality in the San Luis Valley of South Central Colorado. pp 125-142. In R. Lal (ed.). *Soil Quality and Soil Erosion*. CRC Press, Boca Raton, FL.

115 Delgado, J.A., W. Reeves and R. Follett. 2006. Winter Cover Crops. pp 1915-1917. *In* R Lal (ed.) *Encyclopedia Soil Sci.* Markel and Decker, New York, NY.

116 Doll, J. 1991. Cited in Shirley, C.D. No-till beans: rye not?! *The New Farm*. 13:12-15.

117 Doll, J. and T. Bauer. 1991. Rye: more than a mulch for weed control. pp. 146-149. *Illinois Agricultural Pesticides Conference* presentation summaries. Urbana, Ill.

118 Duiker, S.W. and W.S. Curran. 2005. Rye cover crop management for corn production in the northern Mid-Atlantic region. *Agron. J.* 97:1413-1418.

119 Duiker, S.J. and J. Myers. 2005. *Better Soils with the No-till System*. http://panutrientmgmt.cas.psu.edu/pdf/rp_better_soils_with_noTill.pdf

120 Duke, J.A. 1981. *Handbook of Legumes of World Economic Importance*. Plenum Press, N.Y.

121 Earhart, D.R. 1996. Managing soil phosphorus accumulation from poultry litter application through vegetable/legume rotations. Project Report #LS95-69. Southern Region SARE. Griffin, Ga. www.sare.org/projects

122 Eberlein, C. 1995. Development of winter wheat cover crop systems for weed control in potatoes. SARE Project Report #LW91-027. Western Region SARE. Logan, Utah. www.sare.org/projects

123 Eckert, D.J. 1991. Chemical attributes of soils subjected to no-till cropping with rye cover crops. *Soil Sci. Soc. Am. J.* 55:405-409.

124 Edwards, W.M. et al. 1993. Tillage studies with a corn-soybean rotation: Hydrology and sediment loss. *Soil Sci. Soc. Am. J.* 57:1051-1055.

125 Einhellig, F.A. and J.A. Rasmussen. 1989. Prior cropping with grain sorghum inhibits weeds. *J. Chem. Ecol.* 15:951-960.

126 Einhellig, F.A. and I.F. Souze. 1992. Allelopathic activity of sorgoleone. *J. Chem. Ecol.* 18:1-11.

127 Enache, A.J., R.D. Ilnicki and R.R. Helberg. 1992. Subterranean clover living mulch: A system approach. pp. 160-162. In *Proc. First Int'l Weed Control Congress, Vol. 2*. February 17- 21, 1992. Weed Science Society of Victoria, Melbourne, Australia.

128 Enache, A.J. 1990. Weed control by subterranean clover (*Trifolium subterraneum L.*) used as a living mulch. *Dissertation Abstracts Int., Sci. and Eng.*, 1990, 50(11):4825B.

129 Entz, M.H. et al. 2002. Potential of forages to diversify cropping systems in the Northern Great Plains. *Agron. J.* 94:240-250.

130 Evers, G.W and G.R. Smith. 2006. Crimson clover seed production and volunteer reseeding at various grazing termination dates. *Agron. J.* 98:1410-1415.

131 Evers, G.W., G.R. Smith and P.E. Beale. 1988. Subterranean clover reseeding. *Agron. J.* 80:855-859.

132 Evers, G.W., G.R. Smith and C.S. Hoveland. 1997. Ecology and production of annual ryegrass. pp. 29-43. In F.M. Rouquette, Jr. and L.R. Nelson (eds.). *Production and management of Lolium for forage in the USA*. CSSA Spec. Pub. #24. ASA, CSSA, SSA. Madison, Wisc.

133 Fairbrother, T.E. 1991. Effect of fluctuating temperatures and humidity on the softening rate of hard seed of subterranean clover (*Trifolium subterraneum L.*). *Seed Sci. Tech.* 19:93-105.

134 Fairbrother, T.E. 1997. Softening and loss of subterranean clover hard seed under sod and bare ground environments. *Crop. Sci.* 37:839-844.

135 Farahani H.J., G.A. Peterson and D.G. Westfall. 1998. Dryland cropping intensification: A fundamental solution to efficient use of precipitation. *Adv. Agron.* 64:197-223.

136 Fasching, R. 2006. Personal communication. Bozeman, Mont.

137 Feng, Y. et al. 2003. Soil microbial communities under conventional-till and no-till continuous cotton systems. *Soil Biol. Biochem.* 35(12):1693-1703.

138 Finch, C.U. Univ. of Calif. Medics, general. SAREP Cover Crops Resource Page. www.sarep.ucdavis.edu/ccrop

139 Fischer, A. and L. Burrill. 1993. Managing interference in sweet corn-white clover living mulch system. *Am. J. Alt. Ag.* 8:51-56.

140 Fisk, J.W. and O.B. Hesterman. 1996. N contribution by annual legume cover crops for no-till corn. *In 1996 Cover Crops Symposium Proceedings*. Michigan State Univ., W.K. Kellogg Biological Station, Battle Creek, Mich.

141 Fisk, J.W. et al. 2001. Weed suppression by annual legume cover crops in no-tillage corn. *Agron. J.* 93:319-325.

142 Flexner, J.L. 1990. Hairy vetch. Univ. of Calif. SAREP Cover Crops Resource Page. www.sarep.ucdavis.edu/ccrop

143 Folorunso, O. et al. 1992. Cover crops lower soil surface strength, may improve soil permeability. *Calif. Ag.* 46:26-27.

144 Forney, D.R. and L.F. Chester. 1984. Phytotoxicity of products from rhizospheres of a sorghum-sudangrass hybrid. *Weed Sci.* 33:597-604.

145 Forney, D.R. et al. 1985. Weed suppression in no-till alfalfa (*Medicago sativa*) by prior cropping of summer-annual forage grasses. *Weed Sci.* 33:490-497.

146 Fortin, M.C. and A.S. Hamill. 1994. Rye residue geometry for faster corn development. *Agron. J.* 86:238-243.

147 Foster, R.K. 1990. Effect of tillage implement and date of sweetclover incorporation on available soil N and succeeding spring wheat yields. *Can. J. Plant Sci.* 70:269-277.

148 Fox, R.H. and W.P. Piekielek. 1988. Fertilizer N equivalence of alfalfa, birdsfoot trefoil, and red clover for succeeding corn crops. *J. Prod. Agric.* 1:313-317.

149 Friedman, D. et al. 1996. Evaluation of five cover crop species or mixes for nitrogen production and weed suppression in Sacramento Valley farming systems. *Univ. of California Cover Crop Research and Education Summaries*. 1994–1996. Davis, Calif.

150 Friesen, G.H. 1979. Weed interference in transplanted tomatoes (*Lycopersicon esculentum*). *Weed Sci.* 27:11-13.

151 Frye W.W., W.G. Smith and R.J. Williams. 1985. Economics of winter cover crops as a source of nitrogen for no-till corn. *J. Soil Water Conserv.* 40:246–249.

152 Gardiner, J.B. et al. 1999. Allelochemicals released in soil following incorporation of rapeseed (*Brassica napus*) green manures. *J. Agric. Food Chem.* 47:3837-3842.

153 Gardner, J. (ed.). 1992. *Substituting legumes for fallow in U.S. Great Plains wheat production: The first five years of research and demonstration 1988-1992*. USDA/SARE and North Dakota State Univ., Michael Fields Agricultural Institute, Kansas State Univ. and Univ. of Nebraska. NDSU Carrington Research Extension Center, Carrington, N.D.

154 Geneve, R.L. and L.A. Weston. 1988. Growth reduction of Eastern redbud (*Cercis canadensis L.*) seedlings caused by interaction with a sorghum-Sudangrass hybrid (Sudax). *J. Env. Hort.* 6:24-26.

155 Ghaffarzadeh, M. 1994. Progress Report: Berseem Clover (*Trifolium alexandrinum L.*). 17 pp. Soil Management, Agronomy Department, Iowa State Univ., Ames, Iowa.

156 Ghaffarzadeh, M. 1995. *Considering annual clover?* Don't overlook berseem. Pub. #SA-7. 4 pp. Sustainable Agriculture Fact Sheet Series. The Leopold Center, Ames, Iowa.

157 Ghaffarzadeh, M. 1996. *Forage-based beef production research at the Armstrong Outlying Research farm*. Annual research report. A. S. Leaflet R1245. Iowa State Univ., Ames, Iowa.

158 Ghaffarzadeh, M. 1997. Annual legume makes comeback. pp. 24-25. *Beef Today*. January, 1997.

159 Ghaffarzadeh, M. 1997. Economic and biological benefits of intercropping berseem clover with oat in corn-soybean-oat rotations. *J. Prod. Ag.* 10:314-319.

160 Ghaffarzadeh, M. 1995. Potential uses of annual berseem clover in livestock production. Proc. *Rotational Grazing*, a conference sponsored by The Leopold Center for Sustainable Agriculture, Feb. 5-6, 1995. Published by Iowa State Univ. Extension. See also Demonstration of an annual forage crop integrated with crop and livestock enterprises. In *Progress Report*. March 1998.The Leopold Center for Sustainable Agriculture 7:6-10.

161 Gill, G.S. 1995. Development of herbicide resistance in annual ryegrass populations (*Lolium rigidum Gaud.*) in the cropping belt of Western Australia. *Austral. J. of Exp. Ag*. 35:67-72.

162 Graves, W.L. et al. 1996. *Berseem Clover: A Winter Annual Forage for California Agriculture*. Univ. of California, SAREP. Div. of Agriculture and Natural Resources. Pub. 21536. Davis, Calif.

163 Green, B.J. and V.O. Biederbeck. 1995. *Farm Facts: Soil Improvement with Legumes, Including Legumes in Crop Rotations*. Canada-Saskatchewan Agreement on Soil Conservation, Regina, SK.

164 Greene, D.K. et al.1992. Research report. *New Crop News*. Vol. 3. Purdue Univ., West Lafayette, Ind.

165 Griffin, J.L. and S.M. Dabney. 1990. Preplant postemergence herbicides for legume cover-crop control in minimum tillage systems. *Weed Tech*. 4:332-336.

166 Griffin, T. 2007. Personal communication. USDA Agricultural Research Service, Orono, Maine.

167 Griffith, K and J. Posner. 2001. Comparing Upper Midwest farming systems: Results from the first 10 years of the Wisconsin Integrated Cropping Systems Trial (WICST). Univ of Wisconsin, Center for Integrated Agricultural Systems. www.cias.wisc.edu/wicst/pubs/tenyear_report.pdf

168 Grinsted M.J. et al. 1982. Plant-induced changes in the rhizosphere of rape (Brassica napus var. Emerald) seedlings. I. pH change and the increase in P concentration in the soil solution. *New Phytol*. 91:19.

169 Groff, Steve. 1997. Cedar Grove Farm. www.cedarmeadowfarm.com

170 Grubinger, V.P. and P.L. Minotti. 1990. Managing white clover living mulch for sweet corn production with partial rototilling. *Amer. J. Alt. Ag*. 5:4-11.

171 Halvorson, A.D. et al. 2006. Nitrogen and tillage effects on irrigated continuous corn yields. *Agron. J*. 98:63–71.

172 Halvorson, A.D. and C.A. Reule. 2006. Irrigated corn and soybean response to nitrogen under no-till in northern Colorado. *Agron. J.* 98:1367–1374.

173 Hanson, J.C. et al. 1993. Profitability of no-tillage corn following a hairy vetch cover crop. *J. Prod. Ag*. 6:432-436.

174 Hanson, J.C. et al. 1997. Organic versus conventional grain production in the mid-Atlantic: An economic and farming system overview. *Amer. J. Alt. Ag*. 12:2-9.

175 Hao, J.J and K.V. Subbarao. 2006. Dynamics of lettuce drop incidence and *Sclerotinia minor* inoculum under varied crop rotations. *Plant Dis*. 90:269-278.

176 Haramoto, E.R. and E.R. Gallandt. 2004. Brassica cover cropping for weed management: A review. *Renewable Ag. and Food Sys*. 19:187-198.

177 Haramoto, E. R. and E. R. Gallandt. 2005. Brassica cover cropping: I. Effects on weed and crop establishment. *Weed Sci.* 53:695-701.

178 Haramoto, E.R. and E.R. Gallandt. 2005. Brassica cover cropping: II. Effects on growth and interference of green bean (*Phaseolus vulgaris*) and redroot pigweed (*Amaranthus retroflexus*). *Weed Sci*. 53:702-708.

179 Haramoto, E.R. and E.R. Gallandt. 2005. Brassica cover cropping: II. Effects on growth and interference of green bean (*Phaseolus vulgaris*) and redroot pigweed (*Amaranthus retroflexus*). Weed Sci. 702-708.

180 Harlow, S. 1994. Cover crops pack plenty of value. *Amer. Agriculturalist* 191:14.

181 Harper, L.A. et al. 1995. Clover management to provide optimum nitrogen and soil water conservation. *Crop Sci.* 35:176-182.

182 Hartwig, N.L. and H.U. Ammon. 2002. Cover crops and living mulches. *Weed Sci.* 50:688-699.

183 Helm, J.L. and D. Meyer. 1993. *Sweetclover production and management.* North Dakota Extension Service Publication R-862. Fargo, N.D.

184 Hendricks L.C. 1995. Almond growers reduce pesticide use in Merced County field trial. *Calif. Ag.* 49:5-10.

185 Hendrix, P.F. et al. 1986. Detritus food webs in conventional and no-tillage agroecosystems. *Bioscience* 36:374-380.

186 Hermel, R. 1997. Forage Focus: Annual legume makes comeback. pp. 24-25. *Beef.* January, 1997.

187 Herrero, E.V. et al. 2001. Use of cover crop mulches in a no-till furrow-irrigated processing tomato production system. *HortTech.* 11:43-48.

188 Hiltbold, A.E. 1991. Nitrogen-fixing bacteria essential for crimson clover. *Alabama Agricultural Experiment Station* 38:13. Auburn, Ala.

189 Hoffman, M.L. et al. 1993. Weed and corn responses to a hairy vetch cover crop. *Weed Tech.* 7:594-599.

190 Hoffman, M.L. et al. 1996. Interference mechanisms between germinating seeds and between seedlings: Bioassays using cover crop and weed species. *Seed Sci.* 44:579-584.

191 Hofstetter, B. 1988. *The New Farm*'s cover crop guide: 53 legumes, grasses and legume-grass mixes you can use to save soil and money. *The New Farm* 10:17-22, 27-31.

192 Hofstetter, B. 1992. Bank on buckwheat. *The New Farm* 14:52-53.

193 Hofstetter, B. 1992. How sweet it is: Yellow-blossom sweetclover fights weeds, adds N and feeds livestock. *The New Farm* 14:6-8.

194 Hofstetter, B. 1992. Reliable ryegrass. *The New Farm* 14:54-55, 62.

195 Hofstetter, B. 1993a. Meet the queen of cover crops. *The New Farm* 15:37-41.

196 Hofstetter, B. 1993b. Fast and furious. *The New Farm* 15:21-23, 46.

197 Hofstetter, B. 1993c. Red clover revival. *The New Farm* 15:28-30.

198 Hofstetter, B. 1993d. A quick & easy cover crop. *The New Farm* 15:27-28.

199 Hofstetter, B. 1993e. Reconsider the lupin. *The New Farm* 15:48-51.

200 Hofstetter, B. 1994a. The carefree cover. *The New Farm* 16:22-23.

201 Hofstetter, B. 1994b. Bring on the medics! *The New Farm* 16:56, 62.

202 Hofstetter, B. 1994c. Warming up to winter peas. *The New Farm* 16:11-13.

203 Hofstetter, B. 1995. Keep your covers in the pink. *The New Farm* 17:8-9.

204 Holderbaum, J.F. et al. 1990. Fall-seeded legume cover crops for no-tillage corn in the humid East. *Agron. J.* 82:117-125.

205 Holle, O. 1995. Compare the agronomic and economic benefits of 3 or 4 annual alfalfa varieties to sweetclover for forage and soil building purposes in a feed grain, soybeans, wheat/legume rotation. SARE Project Report #FNC92-004. North Central Region SARE. St. Paul, Minn. www.sare.org/projects

206 House, G.J and Alzugaray, M.D.R. 1989. Hairy vetch. Univ. of Calif. SAREP Cover Crops Resource Page. www.sarep.ucdavis.edu/ccrop

207 Howieson, J. and M.A. Ewing. 1989. Medics, general. Univ. of Calif. SAREP Cover Crops Resource Page. www.sarep.ucdavis.edu/ccrop

208 Hoyt, G.D. and W.L. Hargrove. 1986. Legume cover crops for improving crop and soil management in the southern United States. *HortSci.* 21:397-402.

209 Hutchinson, C.M. and M.E. McGiffen. 2000. Cowpea cover crop mulch for weed control in desert pepper production. *HortSci.* 35:196-198.

210 Ingels, C.A. et al. 1994. Selecting the right cover crop gives multiple benefits. *Calif. Ag.* 48:43-48.

211 Ingels, C.A. 1995. Cover cropping in vineyards. *Amer. Vineyard.* 6/95, 8/95, 9/95, 10/95. In Chaney, David and Ann D. Mayse (eds.). 1997. Cover Crops: Resources for Education and Extension. Univ. of California, Div. of Agriculture and Natural Resources, Davis, Calif.

212 Ingels, C.A. et al. 1996. *Univ. of California Cover Crop Research & Education Summaries, March 1996.* UC SAREP, Davis, Calif.

213 Ingham, R.E. et al.1994. Control of *Meloidogyne chitwoodi* with crop rotation, green manure crops and nonfumigant nematicides. *J. Nematol.* 26:553.

214 Iyer, J.G., S.A. Wilde and R.B. Corey. 1980. Green manure of sorghum-sudan: Its toxicity to pine seedlings. *Tree Planter's Notes* 31:11-13.

215 Izaurralde, R.C. et al. 1990. Plant and nitrogen yield of barley-field pea intercrop in cryoboreal-subhumid central Alberta. *Agron. J.* 82:295-301.

216 Jackson, L.E. 1995. Cover crops incorporated with reduced tillage on semi-permanent beds: Impacts on nitrate leaching, soil fertility, pests and farm profitability. SARE Project Report #AW92-006. Western Region SARE. Logan, Utah. www.sare.org/projects

217 Jacobs, E. 1995. Cover crop breathes life into old soil: Sorghum-sudangrass, used as an onion rotation crop on organic soils, cuts pesticide costs, rejuvenates soil and increases yields. *Amer. Agriculturist* 192:8.

218 Jensen, E.S. 1996. Barley uptake of N deposited in the rhizosphere of associated field pea. *Soil. Biol. Biochem.* 28:159-168.

219 Jeranyama, P., O.B. Hesterman and C.C. Sheaffer. 1998. Medic planting date effect on dry matter and nitrogen accumulation when clear-seeded or intercropped with corn. *Agron. J.* 90:616-622.

220 Jordan, J.L. et al. 1994. *An Economic Analysis of Cover Crop Use in Georgia to Protect Groundwater Quality.* Research Bulletin #419. Univ. of Georgia, College of Agric. and Environ. Sciences, Athens, Ga. 13 pp.

221 Kandel, H.J., A.A. Schneiter and B.L. Johnson. 1997. Intercropping legumes into sunflower at different growth stages. *Crop Sci.* 37:1532-1537.

222 Kandel, H.J., B.L. Johnson and A.A. Schneiter. 2000. Hard red spring wheat response following the intercropping of legumes into sunflower. *Crop Sci.* 40:731-736.

223 Kaspar, T.C., J.K. Radke and J.M. Laflen. 2001. Small grain cover crops and wheel traffic effects on infiltration, runoff, and erosion. *J. Soil Water Conserv.* 56:160-164.

224 Kelly, T.C. et al. 1995. Economics of a hairy vetch mulch system for producing fresh-market tomatoes in the Mid-Atlantic region. *HortSci.* 120:854-869.

225 Kimbrough, E., L. and W.E. Knight. *Forage 'Bigbee' Berseem Clover.* Mississippi State Univ. Extension Service. http://ext.msstate.edu/pubs/is1306.htm.

226 King, L.D. 1988. Legumes for nitrogen and soil productivity. *Stewardship News* 8:4-6.

227 Kirchmann, H. and H. Marstop. 1992. Calculation of N mineralization from six green manure legumes under field conditions from autumn to spring. *Acta Agriculturae Scand.* 41:253- 258.

228 Knight, W.E. 1985. Crimson clover. Univ. of Calif. SAREP Cover Crops Resource Page. www.sarep.ucdavis.edu/ccrop

229 Knorek, J. and M. Staton. 1996. "Red Clover." *Cover Crops: MSU/KBS* (fact sheet packet), Michigan State Univ. Extension, East Lansing, Mich. www.covercrops.msu.edu/ CoverCrops/red_clover.htm

230 Koala, S. 1982. Adaptation of Australian ley farming to Montana dryland cereal production. M.S. thesis. Montana State Univ., Bozeman

231 Koch, D.W. 1995. Brassica utilization in sugarbeet rotations for biological control of cyst nematode. SARE Project Report #LW91-022. Western Region SARE. Logan, Utah. www.sare.org/projects. See also www.uwyo.edu/Agexpstn/research.pdf

232 Koike, S.T. et al. 1996. Phacelia, Lana woollypod vetch and Austrian winter pea: three new cover crop hosts of *Sclerotina minor* in California. *Plant Dis.* 80:1409-1412.

233 Koume, C.N. et al. 1988. Screening subterranean clover (*Trifolium* spp.) germplasm for resistance to *Meloidogyne* species. *J. Nematol.* 21(3):379-383.

234 Kremen, A. and R.R. Weil. 2006. Monitoring nitrogen uptake and mineralization by Brassica cover crops in Maryland. 18th World Congress of Soil Science. http://crops.confex.com/crops/wc2006/techprogram/P17525.htm

235 Krishnan, G., D.L. Holshauser and S.J. Nissen. 1998. Weed control in soybean (Glycine max) with green manure crops. *Weed Tech.* 12:97–102.

236 Kumwenda, J.D.T. et al. 1993. Reseeding of crimson cover and corn grain yield in a living mulch system. *Soil Sci. Soc. Am. J.* 57:517-523.

237 Langdale, G.W. and W.C. Moldenhauer (eds). 1995. *Crop Residue Management to Reduce Erosion and Improve Soil Quality*. USDA-ARS, Conservation Research Report No. 39. Washington, D.C.

238 Langdale, G.W. et al. 1992. Restoration of eroded soil with conservation tillage. *Soil Tech.* 1:81-90.

239 Lanini, W.T. et al. 1989. Subclovers as living mulches for managing weeds in vegetables. *Calif. Agric.* 43:25-27. Larkin, R.P. See #458, 459, 460.

240 Leach, S.S. 1993. Effects of moldboard plowing, chisel plowing and rotation crops on the *Rhizoctonia* disease of white potato. *Amer. Potato J.* 70:329-337.

241 LeCureux, J.P. 1996. Integrated system for sustainability of high-value field crops. SARE Project Report #LNC94-64. North Central Region SARE. St. Paul, Minn. www.sare.org/projects

242 Leroux, G.D. et al. 1996. Effect of crop rotations in weed control, *Bidens cernua* and *Erigeron canadensis* populations, and carrot yields in organic soils. *Crop Protection* 15:171-178.

243 Lichtenberg, E. et al. 1994. Profitability of legume cover crops in the mid-Atlantic region. *J. Soil Water Cons.* 49:582-585.

244 Linn, D.M. and J.W. Doran. 1984. Effect of water-filled pore space on carbon dioxide and nitrous oxide production in tilled and nontilled soil. *Soil Sci. Soc. Am. J.* 48:1267-1272.

245 Litterick, A.M. et al. 2004. The role of uncomposted materials, composts, manures, and compost extracts in reducing pest and disease incidence and severity in sustainable temperate agricultural and horticultural crop production—A review. *Critical Reviews in Plant Sci.* 23: 453-479.

246 Liu, D.L. and J.V. Lovett. 1994. Biologically active secondary metabolites of barley. I. Developing techniques and assessing allelopathy in barley. *J. Chem. Ecol.* 19:2217-2230.

247 Liu, D.L. and J.V. Lovett. 1994. Biologically active secondary metabolites of barley. II. Phytotoxicity of barley allelochemicals. *J. Chem. Ecol.* 19:2231-2244.

248 Lovett, J.V. and A.H.C. Hoult. 1995. Allelopathy and self-defense in barley. pp. 170–183. In Allelopathy: *Organisms, Processes, and Applications*. American Chemical Society, Washington, D.C.

249 MacGuidwin, A.E. and T.L. Layne. 1995. Response of nematode communities to sudangrass and sorghum-sudangrass hybrids grown as green manure crops. *J. Nematol.* 27:609- 616.

250 Madden, N.M. et al. 2004. Evaluation of conservation tillage and cover crop systems for organic processing tomato production. *HortTech*. 14(2):73-80.

251 Mackay, J.H.E. 1981. Medics, general. Univ. of Calif. SAREP Cover Crops Resource Page. www.sarep.ucdavis.edu/ccrop

252 Magdoff, F. and H. van Es. 2001. Building Soils for Better Crops, 2nd Edition. Sustainable Agriculture Network. Beltsville, MD. www.sare.org/publications/soils.htm

253 Mahler, R.L. 1989. Evaluation of the green manure potential of Austrian winter pea in northern Idaho. *Agron. J.* 81:258-264.

254 Mahler, R.L. 1993. Evaluation of the nitrogen fertilizer value of plant materials to spring wheat production. *Agron J.* 85:305-309.

255 Malik, N. and J. Waddington. 1988. Polish rapeseed as a companion crop when establishing sweetclover for dry matter production. *Can. J. Plant. Sci*. 68:1009-1015.

256 Marks, C.F. and J.L. Townsend. 1973. Buckwheat. Univ. of Calif. SAREP Cover Crops Resource Page. www.sarep.ucdavis.edu/ccrop

257 Marshall, H.G. and Y. Pomeranz. 1982. Buckwheat: description, breeding, production and utilization. pp. 157-210. In *Advances in Cereal Science and Technology* (Y. Pomeranz, ed.). Amer. Assn. of Cereal Chemists. St. Paul, Minn.

258 Matheson, N. et al. 1991. Cereal-Legume Cropping Systems: Nine farm case studies in the dryland northern plains, Canadian prairies and intermountain Northwest. Alternative Energy Resources Organization (AERO), Helena, Mont. 75 pp.

259 Matthews, A. 1997. Alternative rotation system for vegetables. SARE Project Report #FNE96-146. Northeast Region SARE. Burlington, Vt. www.sare.org/projects

260 Matthiessen, J.N. and J.A. Kirkegaard. 2006. Biofumigation and enhanced biodegradation: Opportunity and challenge in soilborne pest and disease management. *Critical Reviews in Plant Sci*. 25:235–265.

261 McCraw, D. et al. 1995. Use of Legumes in Pecan Orchards. Oklahoma Cooperative Extension Service. Current Report CR-06250, 4 pp. Stillwater, OK.

262 McGuire, C.F. et al. 1989. Nitrogen contribution of annual legumes to the grain protein content of 'Clark' barley (*Hordeum disticum L.*) production. *Applied Ag. Res*. 4:118-121.

263 McLeod, E. Medics, general. Univ. of Calif. SAREP Cover Crops Resource Page. www.sarep.ucdavis.edu/ccrop

264 McSorley, R. and R.N. Gallagher. 1994. Effect of tillage and crop residue management on nematode densities on corn. *J. Nematol*. 26:669-674.

265 Meisinger, J.J. et al. 1991. Effects of cover crops on groundwater quality. pp. 57-68. In W.L. Hargrove (ed.). *Cover Crops for Clean Water*. Soil and Water Conservation Society. Ankeny, Iowa.

266 Melakeberhan, H. et al. 2006. Potential use of arugula (*Eruca sativa L.*) as a trap crop for Meloidogyne hapla. *Nematol*. 8: 793-799.

267 Merrill, S.D. et al. 2006. Soil coverage by residue as affected by ten crop species under no-till in the northern Great Plains. *J. Soil Water Conserv*. 61:7-13.

268 Merwin, I.A. and W.C. Stiles. 1998. *Integrated weed and soil management in fruit plantings*. Cornell Cooperative Extension Information Bulletin 242. http://dspace.library.cornell.edu/handle/1813/3277

269 Meyer, D.W. and W.E. Norby. 1994. Seeding rate, seeding-year harvest, and cultivar effects on sweetclover productivity. *North Dakota Farm Res*. 50:30-33.

270 Michigan State Univ. Extension. Cover Crops Program. East Lansing, Mich. www.covercrops.msu.edu

271 Michigan State Univ. 2001. Sustainable Agriculture. Futures (newsletter). Volumes 18 and 19. http://web1.msue.msu.edu/misanet/futures01.pdf

272 Miller, M.M. 7/21/97. Reduced input, diversified systems. Press release via electronic transmission from Univ. of Wisconsin Extension-Agronomy, Madison, Wis.

273 Miller, P.R. et al. 1989. *Cover Crops for California Agriculture*. Univ of California, SAREP. Div. of Agriculture and Natural Resources. Davis, Calif. 24 pp.

274 Miller, M.H. et al. 1994. Leaching of nitrogen and phosphorus from the biomass of three cover crops. *J. Environ. Qual*. 23:267-272.

275 Miller, P.R. 1989. Medics, general. Univ. of Calif. SAREP Cover Crops Resource Page. www.sarep.ucdavis.edu/ccrop

276 Millhollon, E.P. 1994. Winter cover crops improve cotton production and soil fertility in Northwest Louisiana. *La. Ag*. 37:26-27.

277 Mishanec, J. 1996. *Use of sorghum-sudangrass for improved yield and quality of vegetables produced on mineral and muck soils in New York: Part II—Sudan trials on muck soils in Orange County*. Research report. Cornell Cooperative Extension. Ithaca, N.Y.

278 Mississippi State Univ. website. http://ext.msstate.edu/pubs

279 Mitchell, J.P. et al. 1999a. Cover crops for saline soils. *J. Agron*. and *Crop Sci*. 183:167-178.

280 Mitchell, J.P., D.W. Peters and C. Shennan. 1999b. Changes in soil water storage in winter fallowed and cover cropped soils. *J. Sustainable Ag*. 15:19-31.

281 Mitchell, J.P. et al. 2005. Surface residues in conservation tillage systems in California. Abstract. In *Conservation tillage, manure and cover crop impact on agricultural systems*. ASA/CSSA/SSSA 2005. November 6-10, 2005, Salt Lake City, Utah.

282 Mohler, C.L. 1995. A living mulch (white clover) / dead mulch (compost) weed control system for winter squash. *Proc. Northeast. Weed Sci. Soc.* 49:5-10.

283 Mojtahedi, H. et al. 1991. Suppression of root-knot nematode populations with selected rapeseed cultivars as green manure. *J. of Nematol*. 23:170-174.

284 Mojtahedi, H. et al. 1993. Managing *Meloidogyne chitwoodi* on potato with rapeseed as green manure. *Plant Dis*. 77:42-46.

285 Mojtahedi, H. et al. 1993. Suppression of Meloidogyne chitwoodi with sudangrass cultivars as green manure. J. Nematol. 25:303-311.

286 Moomaw, R.S. 1995. Selected cover crops established in early soybean growth stage. *J. Soil and Water Cons*. 50:82-86.

287 Morse, R. 1998. Keys to successful production of transplanted crops in high-residue, no-till farming systems. *Proceedings of the 21st Annual Southern Conservation Tillage Conference for Sustainable Agriculture*. July, 1998.

288 Mosjidis, J.A. and C.M. Owsley. 2000. Legume cover crops development by NRCS and Auburn Univ. http://plant-materials.nrcs.usda.gov/pubs/gapmcpo3848.pdf

289 Moynihan, J.M. et al. 1996. Intercropping barley and annual medics. In *Cover Crops Symposium Proceedings*. Michigan State Univ., W.K. Kellogg Biological Station, Battle Creek, Mich.

290 Munawar, A. et al. 1990. Tillage and cover crop management for soil water conservation. *Agron. J.* 82:773-777.

291 Munoz, F.N. 1987. *Legumes for Orchard, Vegetable and Cereal Cropping Systems.* Coop. Ext., Univ. of California, San Diego, Calif.

292 Munoz, F.N. 1988. Medics, general. Univ. of Calif. SAREP Cover Crops Resource Page. www.sarep.ucdavis.edu/ccrop

293 Munoz, F.N. and W. Graves. Medics, general. Univ. of Calif. SAREP Cover Crops Resource Page. www.sarep.ucdavis.edu/ccrop

294 Murphy, A.H. et al. 1976. Bur Medic. Univ. of Calif. Cover Crops Resource Page. www.sarep.ucdavis.edu/ccrop

295 Mutch, D.R. et al. 1996a. Evaluation of low-input corn system. In *Cover Crops Symposium Proceedings.* Michigan State Univ., W.K. Kellogg Biological Station. Battle Creek, Mich.

296 Mutch, D.R. et al. 1996b. Evaluation of over-seeded cover crops at several corn growth stages on corn yield. In *Cover Crops Symposium Proceedings.* Michigan State Univ., W.K. Kellogg Biological Station, Battle Creek, Mich.

297 Mutch, D., T. Martin and K. Kosola. 2003. Red clover (*Trifolium pratense*) suppression of common ragweed (*Ambrosia artemisiifolia*) in winter wheat (*Triticum aestivum*). *Weed Tech.* 17:181-185.

298 Mwaja, V.N., J.B. Masiunas and C.E. Eastman. 1996. Rye and hairy vetch intercrop management in fresh-market vegetables. *HortSci.* 121:586-591.

299 Myers, R.L. and L.J. Meinke. 1994. *Buckwheat: A Multi-Purpose, Short-Season Alternative.* MU Guide G 4306. Univ. Extension, Univ. of Missouri-Columbia.

300 Nafziger, E.D. 1994. Corn planting date and plant population. *J. Prod. Agric.* 7:59-62.

301 National Assn. of Wheat Growers Foundation. 1995. *Best Management Practices for Wheat: A Guide to Profitable and Environmentally Sound Production.* National Assn. of Wheat Growers Foundation. Washington, D.C.

302 Nelson, L.R. Personal communication. 2007. Texas Agricultural Experiment Station. Overton, TX.

303 NewFarm. www.newfarm.org/depts/notill/index.shtml

304 Ngouajio, M. and D.R. Mutch. 2004. *Oilseed radish: A new cover crop for Michigan.* Michigan State Univ Extension bulletin E2907. www.covercrops.msu.edu/CoverCrops/O_Radish/extension_bulletin_E2907.pdf

305 Nimbal, C.I. et al. 1996. Phytotoxicity and distribution of sorgoleone in grain sorghum germplasm. *J. Agric. Food Chem.* 44:1343-1347.

306 Ocio, J.A. et al. 1991. Field incorporation of straw and its effects on soil microbial biomass and soil inorganic N. *Soil Biol. Biochem.* 23:171-176.

307 Oekle, E.A. et al. 1990. Rye. In *Alternative Field Crops Manual.* Univ. of Wisc-Ext. and Univ. of Minnesota, Madison, Wis. and St. Paul, Minn.

308 Orfanedes, M.S. et al. 1995. *Sudangrass trials—What have we learned to date?* Cornell Cooperative Extension Program, Lake Plains Vegetable Program.

309 Oregon State Univ. website, "Forage Information System" http://forages.oregonstate.edu

310 Ott, S.L. and W.L. Hargrove. 1989. Profit and risks of using crimson clover and hairy vetch cover crops in no-till corn production. *Amer. J. Alt. Ag.* 4:65-70.

311 Oyer, L.J. and J.T. Touchton. 1990. Utilizing legume cropping systems to reduce nitrogen-fertilizer requirements for conservation-tilled corn. *Agron. J.* 82:1123-1127.

312 Paine, L. et al. 1995. Establishment of asparagus with living mulch. *J. Prod. Agric.* 8:35-40.

313 Parkin, T.B., T.C. Kaspar and C.A. Cambardella. 1997. Small grain cover crops to manage nitrogen in the Midwest. *Proc. Cover Crops, Soil Quality, and Ecosystems Conference*. March 12-14, 1997. Sacramento, Calif. Soil and Water Conservation Society, Ankeny, Iowa.

314 Paxton, J.D. and J. Groth. 1994. Constraints on pathogens attacking plants. *Critical Rev. Plant Sci.* 13:77-95.

315 Payntcr, B.H. Medics, general. Univ. of Calif. SAREP Cover Crops Resource Page. www.sarep.ucdavis.edu/ccrop

316 Peet, M. 1995. *Sustainable Practices for Vegetable Production in the South*. Summer annuals and winter annuals lists. www2.ncsu.edu/ncsu/cals/sustainable/peet.

317 Pennington, B. 1997. *Seeds & Planting* (3rd Ed.) Pennington Seed, Inc., Madison, Ga.

318 Peterson, G.A. et al. 1998. Reduced tillage and increasing cropping intensity in the Great Plains conserves soil C. *Soil & Tillage Research* 47:207-218.

319 Peterson, G.A. et al. 1996. Precipitation use efficiency as impacted by cropping and tillage systems. *J. Prod. Agric.* 9:180-186.

320 Petersen, J. et al. 2001. Weed suppression by release of isothiocyanates from turnip-rape mulch. *Agron. J.* 93:37-43.

321 Phatak, S.C. 1987a. Tillage and fertility management in vegetables. *Amer. Vegetable Grower* 35:8-9.

322 Phatak, S.C. 1987b. Integrating methods for cost effective weed control. Proc. *Integ. Weed Manag. Symp.* Expert Committee on Weeds. Western Canada. 65 pp.

323 Phatak, S.C. 1992. An integrated sustainable vegetable production system. *HortSci.* 27:738-741.

324 Phatak, S.C. 1993. Legume cover crops-cotton relay cropping systems. *Proc. Organic Cotton Conf.* pp. 280-285. Compiled and edited by California Institute for Rural Studies, P.O. Box 2143, Davis, Calif. 95617.

325 Phatak, S.C. et al. 1991. Cover crops effects on weeds diseases, and insects of vegetables. pp. 153-154. In W.L. Hargrove (ed.). *Cover Crops for Clean Water*. Soil and Water Conservation Society. Ankeny, Iowa.

326 Pieters, A.J. 1927. *Green Manuring*. John Wiley & Sons, N.Y.

327 Pledger, D.J. and D.J. Pledger, Jr. 1951. *Cotton culture on Hardscramble Plantation*. Hardscramble Plantation. Shelby, Miss.

328 Posner, J. et al. 2000. Using small grain cover crop alternatives to diversify crop rotations. SARE Project Report #LNC97-116. North Central Region SARE. St. Paul, Minn. www.sare.org/projects

329 Porter, S. 1994. Increasing options for cover cropping in the Northeast. SARE Project Report #FNE93-014. Northeast Region SARE. Burlington, Vt. www.sare.org/projects

330 Porter, S. 1995. Increasing options for cover cropping in the Northeast. SARE Project Report #FNE94-066. Northeast Region SARE. Burlington, Vt. www.sare.org/projects

331 Power, J.F. 1991. Growth characteristics of legume cover crops in a semiarid environment. *Soil Sci. Soc. Am. J.* 55:1659-1663.

332 Power, J.F. 1994. Cover crop production for several planting and harvest dates in eastern Nebraska. *Agron J.* 86:1092-1097.

333 Power, J.F. et al. 1991. Hairy vetch as a winter cover crop for dryland corn production. *J. Prod. Ag.* 4:62-67.

334 Power, J.F. and J.A. Zachariassen. 1993. Relative nitrogen utilization by legume cover crop species at three soil temperatures. *Agron. J.* 85:134-140.

335 Price, A.J., D.W. Reeves and M.G. Patterson. 2006. Evaluation of weed control provided by three winter cover cereals in conservation-tillage soybean. *Renew. Agr. Food Sys.* 21(3):159-164.

336 Przepiorkowski, T. and S.F. Gorski. 1994. Influence of rye (Secale cereale) plant residues on germination and growth of three triazine-resistant and susceptible weeds. *Weed Tech.* 8:744-747.

337 Quesenberry, K.H., D.D. Baltensperger and R.A. Dunn. 1986. Screening *Trifolium* spp. for response to *Meloidogyne* spp. Crop Sci. 26:61-64.

338 Quinlivan, B.J. et al. Medics, general. Univ. of Calif. SAREP Cover Crops Resource Page. www.sarep.ucdavis.edu/ccrop

339 Ramey, B.E. et al. 2004. Biofilm formation in plant-microbe associations. *Current Opinion in Microbiol.* 7:602-609.

340 Rajalahti, R.M. and R.R. Bellinder. 1996. Potential of interseeded legume and cereal cover crops to control weeds in potatoes. pp. 349-354. *Proc. 10th Int. Conf. on Biology of Weeds*, Dijon, France.

341 Rajalahti, R. and R.R. Bellinder. 1999. Time of hilling and interseeding of cover crop influences weed control and potato yield. *Weed Sci.* 47:215-225.

342 Ranells, N.N. and M.G. Wagger. 1993. Crimson clover management to enhance reseeding and no-till corn grain production. *Agron. J.* 85:62-67.

343 Ranells, N.N. and M.G. Wagger. 1996. Nitrogen release from grass and legume cover crop monocultures and bicultures. *Agron. J.* 88:777-782.

344 Ranells, N.N. and M.G. Wagger. 1997a. Grass-legume bicultures as winter annual cover crops. *Agron. J.* 89:659-665.

345 Ranells, N.H., and M.G. Wagger. 1997b. Winter annual grass–legume bicultures for efficient nitrogen management in no-till corn. *Agric. Ecosyst. Environ.* 65:23–32.

346 Reddy, K.N., M.A. Locke and C.T. Bryson. 1994. Foliar washoff and runoff losses of lactoben, norflurazon and fluemeteron under simulated conditions. *J. Agric. Food Chem.* 42:2338-2343.

347 Reeves, D.W. 1994. Cover crops and rotations. *In* J.L. Hatfield and B.A. Stewart (eds.). *Advances in Soil Science: Crops Residue Management.* pp. 125172. Lewis Publishers, CRC Press Inc., Boca Raton, FL.

348 Reeves, D.W. and J.T. Touchton. 1994. Deep tillage ahead of cover crop planting reduces soil compaction for following crop. p. 4. *Alabama Agricultural Experimental Station Newsletter.* Auburn, Ala. www.ars.usda.gov/SP2UserFiles/Place/66120900/Reeves/Reeves_1991_Deep Tillage.pdf

349 Reeves, D.W., A.J. Price and M.G. Patterson. 2005. Evaluation of three winter cereals for weed control in conservation-tillage non-transgenic cotton. *Weed Tech.* 19:731-736.

350 Reynolds, M.O. et al. 1994. Intercropping wheat and barley with N-fixing legume species: A method for improving ground cover, N-use efficiency and productivity in low-input systems. *J. Agric. Sci.* 23:175-183.

351 Rice, E.L. 1974. *Allelopathy.* Academic Press, Inc. N.Y.

352 Rife, C.L. and H. Zeinalib. 2003. Cold tolerance in oilseed rape over varying acclimation durations. *Crop Sci.* 43:96-100.

353 Riga, E. et al. 2003. Green manure amendments and management of root knot nematodes on potato in the Pacific Northwest of USA. *Nematol. Monographs & Perspectives* 2:151-158.

354 Robertson, T. et al. 1991. Long-run impacts of cover crops on yield, farm income, and nitrogen recycling. pp. 117-120. In W.L. Hargrove (ed.). *Cover Crops for Clean Water.* Soil and Water Conservation Society. Ankeny, Iowa.

355 Robinson, R.G. 1980. *The Buckwheat Crop in Minnesota.* Station Bulletin 539. Agricultural Experiment Station, Univ. of Minnesota, St. Paul, Minn.

356 Rogiers, S.Y. et al. 2005. Effects of spray adjuvants on grape (*Vitis vinifera*) berry microflora, epicuticular wax and susceptibility to infection by *Botrytis cinerea.Australasian Plant Pathol.* 34:221-228.

357 Rothrock, C.S. 1995. Utilization of winter legume cover crops for pest and fertility management in cotton. SARE Project Report #LS94-057. Southern Region SARE. Griffin, Ga. www.sare.org/projects

358 Roylance, H.B. and K.H.W. Klages. 1959. *Winter Wheat Production*. Bulletin 314. College of Agriculture, Univ. of Idaho, Moscow, Idaho.

359 Saini, M., A.J. Price and E. van Santen. 2005. Winter weed suppression by winter cover crops in a conservation-tillage corn and cotton rotation. *Proc. 27th Southern Conservation-Tillage Conf.* pp. 124-128.

360 Sarrantonio, M. 1991. *Methodologies for screening soil-improving legumes*. Rodale Institute. Kutztown, Pa.

361 Sarrantonio, M. 1994. *Northeast Cover Crop Handbook*. Soil Health Series. Rodale Institute, Kutztown, Pa.

362 Sarrantonio, M. and E. Gallandt. 2003.The role of cover crops in North American cropping systems.*J. of Crop Prod.* 8:53-74.

363 Sarrantonio, M. and T.W. Scott. 1988.Tillage effects on availability of nitrogen to corn following a winter green manure crop. *Soil Sci. Soc. Am. J.* 52:1661-1668.

364 Sattell, R. et al. 1999. *Cover Crop Dry Matter and Nitrogen Accumulation in Western Oregon*. Extension publication #EM 8739. Oregon State Univ Extension. http://extension.oregonstate.edu/catalog/html/em/em8739

365 Sattell, R. et al. 1998. *Oregon cover crops: rapeseed*. http://extension.oregonstate.edu/catalog/html/em/em8700

366 Schmidt, W.H., D.K. Myers and R.W. Van Keuren. 2001. *Value of legumes for plowdown nitrogen*. Ohio State Univ Extension Fact Sheet AGF-111-01. http://ohioline.osu.edu/agf-fact/0111.html

367 Sholberg, P. et al. 2006. Fungicide and clay treatments for control of powdery mildew influence wine grape microflora. *HortSci.* 41:176-182.

368 Schomberg, H.H. et al. 2005. Enhancing sustainability in cotton production through reduced chemical inputs, cover crops and conservation tillage. SARE Project Report #LS01-121. Southern Region SARE. Griffin, Ga. www.sare.org/projects.

369 Schonbeck, M. and R. DeGregorio. 1990. Cover crops at a glance. *The Natural Farmer*, Fall-Winter 1990.

370 Scott, J.E. and L.A. Weston. 1991. Cole crop (*Brassica oleracea*) tolerance to Clomazone. *Weed Sci.* 40:7-11.

371 Shaffer, M.J. and J.A. Delgado. 2002. Essentials of a national nitrate leaching index assessment tool.*J. Soil and Water Conserv.* 57:327-335.

372 Shipley, P.R. et al.1992. Conserving residual corn fertilizer nitrogen with winter cover crops. *Agron. J.* 84:869-876.

373 Sheaffer, C. 1996.Annual medics: new legumes for sustainable farming systems in the Midwest. SARE Project Report #LNC93-058. North Central Region SARE. St. Paul, Minn. www.sare.org/projects

374 Sheaffer, C.C., S.R. Simmons and M.A. Schmitt. 2001.Annual medic and berseem clover dry matter and nitrogen production in rotation with corn. *Agron. J.* 93:1080-1086.

375 Shrestha, A. et al. 1996."Annual Medics." *In Cover Crops: MSU/KBS* (factsheet packet). Michigan State Univ. Extension. East Lansing, Mich.

376 Shrestha, A. et al. 1998.Annual medics and berseem clover as emergency forages. *Agron. J.* 90:197-201.

377 Sideman, E. 1991. Hairy vetch for fall cover and nitrogen: A report on trials by MOFGA in Maine. *Maine Organic Farmer & Gardener* 18:43-44.

378 Sims, J.R. 1980. "Seeding George black medic," "George black medic in rotation," "George black medic as a green manure." In *Timeless Seeds*. Conrad, Mont.

379 Sims, J.R. 1982. Progress Report. Montana Ag. Extension Service. Research Project #382. Bozeman, Mont.

380 Sims, J.R. 1988. *Research on dryland legume-cereal rotations in Montana*. Montana State Univ. Bozeman, Mont.

381 Sims, J.R. et al. 1991. Yield and bloat hazard of berseem clover and other forage legumes in Montana. *Montana AgResearch* 8:4-10.

382 Sims, J.R. 1995. *Low input legume/cereal rotations for the northern Great Plains-Intermountain Region dryland and irrigated systems*. SARE Project Report #LW89-014. Western Region SARE. Logan, Utah. www.sare.org/projects

383 Sims, J. 1996. *Beyond Summer Fallow*. Prairie Salinity Network Workshop, June 6, 1996, Conrad, Mont. Available from Montana Salinity Control Association, Conrad, Mont. 59425.

384 Singer, J.W., M.D. Casler and K.A. Kohler. 2006. Wheat effect on frost-seeded red clover cultivar establishment and yield. *Agron. J.* 98:265-269.

385 Singer, J.W. et al. 2004. Tillage and compost affect yield of corn, soybean, and wheat and soil fertility. *Agron. J.* 96:531-537.

386 Singer, J. and P. Pedersen. 2005. *Legume Living Mulches in Corn and Soybean*. Iowa State Univ Extension Publication, Iowa State Univ, Ames Iowa. http://extension.agron.iastate.edu/soybean/documents/PM_mulches_2006.pdf

387 Singogo, W., W.J. Lamont Jr. and C.W. Marr. 1996. Fall-planted cover crops support good yields of muskmelons. *HortSci*. 31:62-64.

388 Smith, R.F. et al. 2005. Mustard cover crops to optimize crop rotations for lettuce production. California lettuce research board. Annual Report. pp. 212-219. http://ucce.ucdavis.edu/files/filelibrary/1598/29483.pdf

389 Smith, S.J and A.N. Sharpley. Sorghum and sudangrass. Univ. of Calif. SAREP Cover Crops Resource Page. www.sarep.ucdavis.edu/ccrop

390 Snapp, S. et al. 2005. Evaluating cover crops for benefits, costs and performance within cropping system niches. *Agron. J.* 97:1-11.

391 Snapp, S. et al. 2006. *Mustards—A Brassica Cover Crop for Michigan*. Extension Bulletin E-2956. Michigan State Univ.

392 Snider, J. et al. 1994. Cover crop potential of white clover: Morphological characteristics and persistence of thirty-six varieties. *Mississippi Agricultural and Forestry Experiment Service Research Report* 19:1-4.

393 Soil Science Society of America. 1997. Glossary of soil science terms. Madison, WI.

394 Stark, J.C. 1995. Development of sustainable potato production systems for the Pacific NW. SARE Project report #LW91-029. Western Region SARE. Logan, Utah. www.sare.org/projects

395 Stiraker, R.J. et al. 1995. No-tillage vegetable production using cover crops and alley cropping. pp. 466-474. In *Soil Management in Sustainable Agriculture. Proc. Third Int'l Conf. on Sustainable Agriculture*. 31 August to 4 September 1993. Wye College, Univ of London, UK.

396 Stivers, L.J. and C. Shennan. 1991. Meeting the nitrogen needs for processing tomatoes through winter cover cropping. *J. Prod Ag*. 4:330-335.

397 Stoskopf, N.C. Barley. Univ. of Calif. SAREP Cover Crops Resource Page. www.sarep.ucdavis.edu/ccrop

398 Stute, J.K. 2007. Personal communication. Michael Fields Agricultural Institute. East Troy, Wis.

399 Stute, J. 1996. *Legume Cover Crops in Wisconsin*. Wisconsin Department of Agriculture, Sustainable Agriculture Program. Madison, Wis. 27 pp.

400 Stute, J.K. and J.L. Posner. 1995a. Legume cover crops as a nitrogen source for corn in an oat-corn rotation. *J. Prod. Agric.* 8:385-390.

401 Stute, J.K. and J.L. Posner. 1995b. Synchrony between legume nitrogen release and corn demand in the upper Midwest. *Agron. J.* 87: 1063-1069.

402 Stute, J.K. and J.L. Posner. 1993. Legume cover crop options for grain rotations in Wisconsin. *Agron. J.* 85:1128-1132.

403 Sumner, D.R., B. Doupik Jr. and M.G. Boosalis. 1981. Effects of reduced tillage and multiple cropping on plant diseases. *Ann. Rev. Phytopathol.* 19:167-187.

404 Sumner, D.R. et al. 1983. Root diseases of cucumber in irrigated, multiple-cropping systems with pest management. *Plant Dis.* 67:1071-1075.

405 Sumner, D.R. et al. 1986. Interactions of tillage and soil fertility with root diseases in snap bean and lima bean in irrigated multiple-cropping systems. *Plant Dis.* 70:730-735.

406 Sumner, D.R. et al. 1986. Conservation tillage and vegetable diseases. *Plant Dis.* 70:906-911.

407 Sumner, D.R., S.R. Ghate and S.C. Phatak. 1988. Seedling diseases of vegetables in conservation tillage with soil fungicides and fluid drilling. *Plant Dis.* 72:317-320.

408 Sumner, D.R. et al. 1991. Soilborne pathogens in vegetables with winter cover crops and conservation tillage. *Amer. Phytopathol.* Soc. Abstracts.

409 Sustainable Agriculture Network. 2005. *Manage Insects on Your Farm: A Guide to Ecological Strategies*. Beltsville, MD. www.sare.org/publications/insect.htm

410 Teasdale, J.R. et al. 1991. Response of weeds to tillage and cover crop residue. *Weed Sci.* 39:195-199.

411 Teasdale, J.R. and C.L. Mohler. 1993. Light transmittance, soil temperature and soil moisture under residue of hairy vetch and rye. *Agron. J.* 85:673-680.

412 Teasdale, J.R. 1996. Contribution of cover crops to weed management in sustainable agriculture systems. *J. Prod. Ag.* 9:475-479.

413 Temple, S. 1996. A comparison of conventional, low input or organic farming systems: Soil biology, soil chemistry, soil physics, energy utilization, economics and risk. SARE Project Report #SW94-017. Western Region SARE. Logan, Utah. www.sare.org/projects

414 Temple, S. 1995. A comparison of conventional, low input and organic farming systems: The transition phase and long term viability. SARE Project Report #LW89-18. Western Region SARE. Logan, Utah. www.sare.org/projects

415 Theunissen, J., C.J.H. Booij and A.P. Lotz. 1995. Effects of intercropping white cabbage with clovers on pest infestation and yield. *Entomologia Experimentalis et Applicata* 74:7-16.

416 Tillman, G. et al. 2004. Influence of cover crops on insect pests and predators in conservation tillage cotton. *J. Econ. Entom.* 97:1217-1232.

417 Townsend, W. 1994. No-tilling hairy vetch into crop stubble and CRP acres. SARE Project Report #FNC93-028. North Central Region SARE. St. Paul, Minn. www.sare.org/projects

418 Truman, C. et al. 2003. Tillage impacts on soil property, runoff, and soil loss variations from a Rhodic Paleudult under simulated rainfall. *J. Soil Water Cons.* 58:258-267.

419 Truman, C.C., J.N. Shaw and D.W. Reeves. 2005. Tillage effects on rainfall partitioning and sediment yield from an Ultisol in central Alabama. *J. Soil Water Conserv.* 60:89-98.

420 Tumlinson, J.H., W.J. Lewis and L.E.M. Vet. 1993. How parasitic wasps find their hosts. *Sci. American* 26:145-154.

421 Univ. of California Cover Crops Working Group. March 1996. *Cover Crop Research and Education Summaries*. Davis, Calif. 50 pp.

422 Univ. of Calif. SAREP Cover Crops Resource Page. www.sarep.ucdavis.edu/ccrop

423 Unger, P.W. and M.F. Vigil. 1998. Cover crops effects on soil water relationships. *J. Soil Water Cons.* 53:241-244.

424 van Bruggen, A. H. C. et al. 2006. Relation between soil health, wave-like fluctuations in microbial populations, and soil-borne plant disease management. *European J. of Plant Path.* 115:105-122

425 van Santen, E. 2007. Personal communication. Auburn Univ, Alabama.

426 Varco, J.J., J.O. Sanford and J.E. Hairston. 1991. Yield and nitrogen content of legume cover crops grown in Mississippi. *Research Report*. Mississippi Agricultural and Forestry Experiment Station 16:10.

427 Wagger, M.G. 1989. Cover crop management and N rate in relation to growth and yield of no-till corn. *Agron. J.* 81:533-538.

428 Wagger, M.G. and D.B. Mengel. 1988. The role of nonleguminous cover crops in the efficient use of water and nitrogen. p. 115-128. In W.L. Hargrove (ed.). *Cropping Strategies for Efficient Use of Water and Nitrogen*. ASA Spec. Publ. 51. ASA, CSSA, SSSA, Madison, Wisc.

429 Wander, M.M. et al. 1994. Organic and conventional management effects on biologically active soil organic matter pools. *Soil Sci. Soc. Am. J.* 58:1120-1139.

430 Washington State Univ. 2007. Mustard green manures. WSU Cooperative Extension. www.grant-adams.wsu.edu/agriculture/covercrops/green_manures

431 Weaver, D.B. et al. 1995. Comparison of crop rotation and fallow for management of *Heterodera glycines* and *Meloidogyne* spp. in soybean. *J. Nematol.* 27:585-591.

432 Weil, R. 2007. Personal communication. Univ of Maryland, College Park, MD.

433 Weil, R. and S. Williams. 2006. *Brassica cover crops to alleviate soil compaction*. www.enst.umd.edu/weilbrassicacovercrops.doc

434 Welty, L. et al. 1991. Effect of harvest management and nurse crop on production of five small-seeded legumes. *Montana Ag Research* 8:11-14.

435 Weitkamp, B. 1988, 1989. Medics, general. Univ. of Calif. SAREP Cover Crops Resource Page. www.sarep.ucdavis.edu/ccrop

436 Weitkamp, B. and W.L. Graves. Medics, general. Univ. of Calif. SAREP Cover Crops Resource Page. www.sarep.ucdavis.edu/ccrop

437 Wendt, R.C. and R.E. Burwell. 1985. Runoff and soil losses for conventional, reduced, and no-till corn. *J. Soil Water Conserv.* 40:450-454.

438 Westcott, M.P. et al. 1991. Harvest management effects on yield and quality of small-seeded legumes in western Montana. *Montana AgResearch* 8:18-21.

439 Westcott, M.P. 1995. Managing alfalfa and berseem clover for forage and plowdown nitrogen in barley rotations. *Agron. J.* 87:1176-1181.

440 Weston, L.A., C.I. Nimbal and P. Jeandet. 1998. Allelopathic potential of grain sorghum (*Sorghum* bicolor (L.) Moench) and related species. In *Principles and Practices in Chemical Ecology*. CRC Press, Boca Raton, Fla.

441 Weston, L.A. 1996. Utilization of allelopathy for weed management in agroecosystems. *Agron. J.* 88:860-866.

442 Wichman, D. et al. 1991. Berseem clover seeding rates and row spacings for Montana. *Montana AgResearch* 8:15-17.

443 Willard, C.J. 1927. *An Experimental Study of Sweetclover*. Ohio Agricultural Station Bulletin No. 405, Wooster, Ohio.

444 William, R. 1996. Influence of cover crop and non-crop vegetation on symphlan density in vegetable production systems in the Pacific NW. SARE Project Report #AW94-033. Western Region SARE. Logan, Utah. www.sare.org/projects

445 Williams, W.A. et al. 1990. Ryegrass. Univ. of Calif. SAREP Cover Crops Resource Page. www.sarep.ucdavis.edu/ccrop

446 Williams, S.M. and R.R. Weil. 2004. Crop cover root channels may alleviate soil compaction effects on soybean crop. *Soil Sci. Soc. Am. J.* 68:1403-1409.

447 Williams, W.A. et al. 1991. Water efficient clover fixes soil nitrogen, provides winter forage crop. *Calif. Ag*. 45:30-32.

448 Wingard, C. 1996. Cover Crops in Integrated Vegetable Production Systems. SARE Project Report #PG95-033. Southern Region SARE. Griffin, Ga. www.sare.org/projects

449 Wisconsin Integrated Cropping Systems Trial (WICST). www.cias.wisc.edu/wicst

450 Wolfe, D. 1994. *Management strategies for improved soil quality with emphasis on soil compaction*. SARE Project Report #LNE94-044. Northeast Region SARE. Burlington, Vt. www.sare.org/projects

451 Wolfe, D. 1997. *Soil Compaction: Crop Response and Remediation*. Report No. 63. Cornell Univ., Department of Fruit and Vegetable Science, Ithaca, N.Y.

452 Worsham, A.D. 1991. Role of cover crops in weed management and water quality. p. 141–152. *In* W.L. Hargrove (ed.). *Cover Crops for Clean Water*. Soil and Water Conservation Society. Ankeny, Iowa.

453 Wright, S.F. and A. Upadhaya. 1998. A survey of soils for aggregate stability and glomalin, a glycoprotein produced by hyphae of arbuscular mycorrhizal fungi. *Plant Soil* 198:97-107.

454 Yenish, J.P., A. D. Worsham and A.C. York. 1996. Cover crops for herbicide replacement in no-tillage corn (*Zea mays*). *Weed Tech.* 10:815-821.

455 Yoshida, H. et al. 1993. Release of gramine from the surface of barley leaves. *Phytochem*. 34:1011-1013.

456 Zhu, Y. et al. 1998. Dry matter accumulation and dinitrogen fixation of annual *Medicago* species. *Agron. J.* 90:103-108.

457 Zhu, Y. et al. 1998. Inoculation and nitrogen affect herbage and symbiotic properties of annual *Medicago* species. *Agron. J.* 90:781-786.

458 Larkin, R.P. and T.S. Griffin. 2007. Control of soilborne potato diseases with Brassica green manures. *Crop Protection.* 26:1067-1077. http://dx.doi.org/10.1016/j.cropro.2006.10.004

459 Larkin, R.P., T.S. Griffin and C.W. Honeycutt. 2006. Crop rotation and cover crop effects on soilborne diseases of potato. *Phytopath*. 96:S48.

460 Larkin, R.P. and C.W. Honeycutt. 2006. Effects of different 3-year cropping systems on soil microbial communities and *Rhizoctonia* disease of potato. *Phytopath*. 96:68-79.

Appendix G

RESOURCES FROM SARE

SARE's print and online resources cover a range of topics, from tillage tool selection to interpreting a soil test for your conditions. Most publications can be viewed in their entirety at www.sare.org/publications.

BOOKS

***Building a Sustainable Business*, 280 pp, $17**
A business planning guide for alternative and sustainable agriculture entrepreneurs that follows one farming family through the planning, implementation, and evaluation process.

***Building Soils for Better Crops*, 240 pp, $19.95**
How ecological soil management can raise fertility and yields while reducing environmental impact.

***How to Direct Market Your Beef*, 96 pp, $14.95**
How one couple used their family's ranch to launch a profitable, grass-based beef operation focused on direct market sales.

***Manage Insects on Your Farm: A Guide to Ecological Strategies*, 128 pp, $15.95**
Manage insect pests ecologically using crop diversification, biological control and sustainable soil management.

***The New American Farmer 2nd Edition*, 200 pp, $16.95**
Profiles 60 farmers and ranchers who are renewing profits, enhancing environmental stewardship and improving the lives of their families and communities by embracing new approaches to agriculture.

***The New Farmers' Market*, 272 pp, $24.95**
Covers the latest tips and trends from leading sellers, managers, and market planners to best display and sell product. (Discount rates do not apply.)

***Steel in the Field*, 128 pp; $18**
Farmer experience, commercial agricultural engineering expertise and university research combine to tackle the hard questions of how to reduce weed management costs and herbicide use.

How to Order Books

Visit www.sare.org/WebStore; call (301) 374-9696 or specify title and send check or money order to Sustainable Agriculture Publications, PO Box 753, Waldorf MD 20604.

Shipping & Handling: Add $5.95 for first book, plus $2 for each additional book shipped within the U.S.A. Call (301) 374-9696 for shipping rates on orders of 10 or more items, rush orders, or international shipments. Please allow 3-4 weeks for delivery. MD residents add 5% sales tax. (Prices are subject to change.)

Bulk Discounts: Except as indicated above, 25% discount applies to orders of 10-24 titles; 50% discount for orders of 25 or more titles.

BULLETINS

Diversifying Cropping Systems, 20 pp.
Helps farmers design rotations, choose new crops, and manage them successfully.

Exploring Sustainability in Agriculture, 16 pp.
Defines sustainable agriculture by providing snapshots of different producers who apply sustainable principles on their farms and ranches.

How to Conduct Research on Your Farm or Ranch, 12 pp.
Outlines how to conduct research at the farm level, offering practical tips for both crop and livestock producers.

Marketing Strategies for Farmers and Ranchers, 20 pp.
Offers creative alternatives to marketing farm products, such as farmers markets, direct sales, and cooperatives.

Meeting the Diverse Needs of Limited Resource Producers, 16 pp.
A guide for agricultural educators who want to better connect with and improve the lives of farmers and ranchers who often are hard to reach.

Profitable Pork, Strategies for Hog Producers, 16 pp. (También disponible en español.)
Alternative production and marketing strategies for hog producers, including pasture and dry litter systems, hoop structures, animal health and soil improvement.

Profitable Poultry, Raising Birds on Pasture, 16 pp.
Farmer experiences plus the latest marketing ideas and research on raising chickens and turkeys sustainably, using pens, moveable fencing and pastures.

Rangeland Management Strategies, 16 pp.
Methods for managing forage and other vegetation, guarding riparian areas, winter grazing and multi-species grazing to manage weeds.

SARE Highlights, 16 pp.
Features cutting-edge SARE research about profitable, environmentally sound farming systems that are good for communities.

Smart Water Use on Your Farm or Ranch, 16 pp.
Strategies for farmers, ranchers and agricultural educators who want to explore new approaches to conserve water.

Transitioning to Organic Agriculture, 20 pp.
Lays out promising transition strategies, typical organic farming practices, and innovative marketing ideas.

A Whole Farm Approach to Managing Pests, 20 pp.
Lays out ecological principles for managing pests in real farm situations.

How to Order Bulletins:

Visit www.sare.org/WebStore; call (301) 405-8020 or e-mail tech@sare.org.

Standard shipping for bulletins is free. Please allow 2-5 weeks for delivery. Rush orders that must be received within three weeks will be charged shipping fees at cost. All bulletins can be viewed in their entirety at www.sare.org/publications prior to placing your order. Bulletins are available in quantity for educators at no cost.

Visit SARE Online

SARE's web site at www.sare.org identifies grant opportunities available through the Sustainable Agriculture Research and Education (SARE) program, reports research results, posts events and hosts electronic materials such as a database-driven direct marketing resource guide.

Reader Response Form

YOUR TURN: STRAIGHT TALK ABOUT THIS BOOK

We want to know what you think about this book. Please take a few moments to complete the following evaluation. Your suggestions will help shape future SARE publications.

About the book...

1. How did you find out about *Managing Cover Crops Profitably?* (Check all that apply).

- ☐ a flyer announcing the book
- ☐ the Internet or World Wide Web
- ☐ an extension agent or other consultant
- ☐ a farmer or rancher
- ☐ a farm publication (please specify)
- ☐ other (please specify)

2. Which section(s) of the book most interested you:

- ☐ introductory chapters (please specify)
- ☐ charts
- ☐ individual cover crop sections
- ☐ appendices

3. Did you find the information you needed?

- ☐ almost all
- ☐ mostly
- ☐ somewhat
- ☐ very little

If you answered somewhat or very little, please tell us what was missing.

____

____

4. Please rate each of the following (1 = poor, 5 = great)

- ☐ readability
- ☐ amount of information
- ☐ technical level of information
- ☐ drawings and charts
- ☐ resources for further information

5. Are you doing anything differently as a result of reading this book? If so, what?

____

____

6. What topics would you like to see SARE cover in future books or bulletins?

____

____

About you...

7. I am a (check your primary occupation):

- ☐ farmer/rancher
- ☐ state/federal agency employee
- ☐ university researcher
- ☐ student
- ☐ extension employee
- ☐ seed/equipment dealer
- ☐ crop consultant
- ☐ other ____________________

8. I typically get my information about production practices from (check as many as apply):

- ☐ other farmers/ranchers
- ☐ books
- ☐ farm journals and newsletters
- ☐ extension or other agency personnel
- ☐ the Internet
- ☐ other

9. I am from ________________ (your state)

10. I primarily raise (please specify crops/livestock):

____

____

____

____

____

Please return this form to:
Sustainable Agriculture Research and Education
1122 Patapsco Building
University of Maryland
College Park, MD 20742-6715
fax 301-405-7711

INDEX

M

S

T

Y